Mechatronic & Innovative Applications

Rochdi Merzouki

Contents

Foreword
Mechatronic & Innovative Applications

I am honored by the invitation to write a foreword for the eBook entitled Mechatronic and Innovative Applications. Mechatronics is a design discipline that integrates different areas. Originally, it encompassed three areas, namely electronics, computer engineering and mechanical systems, but today it covers multiple and diverse areas in science and engineering. This eBook emphasizes the importance of mechatronics and covers theoretical and practical topics in the design and implementation of advanced mechatronic systems.

This project brought together some distinguished researchers in mechatronics. I met and worked closely with some of the contributors to this eBook. This has been through several visits to Ecole Polytechnique de Lille, Ecole Centrale de Lille, and Université d'Evry. I participated as a member of PhD thesis committees and research discussions related to several of the topics covered in this eBook. And therefore, I have had the opportunity to see first hand their research activities. Professor Rochdi Merzouki, the editor and contributor to this eBook, and his colleagues have made great progress in the areas presented in this eBook.

The interest in mechatronics has grown tremendously over the years. Many reputable professional societies have launched conferences and Journals on mechatronics. Both industrial and academic institutions have promoted many events on the subject. In addition, there are academic and industrial departments whose main focus is mechatronic technologies and related methodologies. Some universities even grant degrees in this discipline. Today, and more so in the future, progress and innovation are facilitated by disciplines, such as mechatronics, that provide a framework for multi-disciplinary research and development.

The eBook covers theoretical and practical topics in the design and implementation of advanced mechatronic systems. It focuses specifically on mechatronics for intelligent and automatic systems. The applications cover vehicles and robotic systems. The themes include vehicle braking, vehicle simulators, electric vehicles, and autonomous and intelligent vehicles. In the area of robotic systems it includes robotic prosthetic hand, robot assisted brachytherapy of prostate, hexapod robots, mobile robots, and robot for finishing operations. Throughout these different complex mechatronic applications, the eBook demonstrates the use of power/energy mathematical modeling method (the Bond Graph modeling method), control systems design and implementation, monitoring and fault detection in addition to sensing and actuation. Thus the strength of the eBook is the integrated design approach taken in the problem solving of each one of the applications considered. It presents the modeling, instrumentation, control system design, simulations and experimentations.

I believe that this eBook will be a great resource for graduate students and researchers focusing on the mechatronic areas mentioned above. I have no doubt that the theoretical and practical experiences described herein will convey perception and intuition that will enhance understanding of the subject matter. I am sure that the methods and solutions presented will initiate new research and development aspects that will help advance the general field of mechatronics.

Professor Kamal Youcef-Toumi
Director of the Mechatronics Research Laboratory
Co-Director of the center for Clean Water and Clean Energy at MIT and KFUPM
Head of the Control, Instrumentation and Robotics Area
Department of Mechanical Engineering
Massachusetts Institute of Technology
Cambridge, MA, USA

Preface

Current day, mechatronic systems are designed with synergistic integration of mechanics, electronics and computer technology to produce intelligent devices for the purposes of solving real-world problems. This requires that a mechatronic system must be robust and fault tolerant, i.e. it should have the ability to process incomplete, imprecise or uncertain information. Moreover, such systems often have to work in collaborative environments while being subjected to unknown inputs and yet adhering to strict safety norms. Examples are not restricted to production lines, but extend to extremely safety critical systems such as space and underwater robotics, autonomous transportation systems, aviation systems and medical robots. This theme discusses the fundamentals of designing such systems from the first principles and then how to embed intelligence into them.

This is an advanced academic and research theme which has been designed primarily to meet the demands of various postgraduate level courses. A reader already conversant with the basics of mechatronic systems would learn the theory and practice of how to design high end mechatronic systems, starting from modeling to final implementation, in a step-by-step manner. Moreover, this E-book theme presents the recent developments in the design of intelligent mechatronic systems, thereby providing an authoritative support for professionals having basic knowledge in mechatronics.

This E-book theme should offer detailed treatment on design, control and monitoring of advanced mechatronic systems by considering many strands of recent. The special features of this theme should include thorough:

- mathematical modeling of mechatronic systems with experimental validation,
- virtual prototyping through hardware and software in the loop simulations,
- advanced control of mechatronic systems,
- intelligent monitoring and tolerant control of mechatronic systems,
- on-line supervision of complex mechatronic systems,
- integrated design approach.

This E-book theme should regroup detailed descriptions of the mathematical models of complex mechatronic systems, which are developed from first principles, i.e., fundamental physical relationships and from a unified multi-domains approach like bond graph. This modeling is indispensable for control synthesis and monitoring system design. The virtual prototyping is a significant first validation step, considering as an interface between the mathematical models and the real mechatronic system. In this theme, a description of how to design a virtual simulator for control and monitoring is considered. Two operation aspects for mechatronic systems should be developed:

- normal mode, where adequate advanced control are developed and applied,
- faulty mode, where a robust monitoring system is designed to detect and isolate various faults from the mechatronic systems. In this case, a tolerant fault control design could be justified.

Finally, the global operating and surveillance of complex mechatronic systems should be supervised by an on-line supervision system.

Rochdi Merzouki

Contributors

Brahim Achili,
Computer Science Lab,
University of Paris 8, France.

Yacine Amirat,
Images, Signals and Intelligent Systems Lab.
LISSI University of Paris, France.

Hichem Arioui,
Informatique, Biologie Intégrative et Systèmes Complexes,
Evry University, France.

Samir Benmoussa,
Polytech-Lille, LAGIS, UMR CNRS 8219,
Avenue Paul Langevin, 59655 Villeneuve D'Ascq, France.

Tarun Kumar Bera,
Polytech-Lille, LAGIS, UMR-CNRS 8219,
Avenue Paul Langevin, 59655 Villeneuve D'Ascq, France.

Mohamed Bouaziz,
Laboratory of mechanics,
Polytechnic school, Algiers, Algeria.

Nizar Chatti,
Polytech-Lille, LAGIS, UMR CNRS 8219,
Avenue Paul Langevin, 59655 Villeneuve D'Ascq, France.

Xiaoqi Chen,
Department of Mechanical Engineering, College of Engineering,
University of Canterbury, New Zealand.

Arab Ali Chérif,
Computer Science Lab, LIASD,
University of Paris 8, France.

Abdelhakim Chibani,
Polytech-Lille, LAGIS, CNRS UMR 8219,
Avenue Paul Langevin, 59655 Villeneuve D'Ascq, France.

Vincent Coelen,
Polytech-Lille, LAGIS, UMR CNRS 8219,
Avenue Paul Langevin, 59655 Villeneuve D'Ascq, France.

Boubaker Daachi,
Images, Signals and Intelligent Systems Lab.,
LISSI University of Paris.

Karim Djouani,
Laboratory LISSI,
University Paris12, France.

iv

Anne Lise Gehin,
Polytech-Lille, LAGIS, UMR CNRS 8219,
Avenue Paul Langevin, 59655 Villeneuve D'Ascq, France.

Sunil Kumar,
Robotics and Control Laboratory at Mechanical and Industrial Engineering Department,
Indian Institute of Technology, Roorkee, India.

Eric Lartigau,
Centre Oscar Lambret de Lille,
Université Lille 2, France.

Rui Loureiro,
Polytech-Lille, LAGIS, UMR CNRS 8219,
Avenue Paul Langevin, 59655 Villeneuve D'Ascq, France.

Chawki Mahfoudi,
Faculty of Technology Ain Beida,
University Larbi ben M'Hidi, Algeria.

Rochdi Merzouki,
Polytech-Lille, LAGIS, UMR CNRS 8219,
Avenue Paul Langevin, 59655 Villeneuve D'Ascq, France.

Lamri Nehaoua,
Informatique, Biologie Intégrative et Systèmes Complexes,
Evry University, France.

Belkacem Ould Bouamama,
Polytech-Lille, LAGIS, UMR CNRS 8219,
Avenue Paul Langevin, 59655 Villeneuve D'Ascq, France.

Pushparaj Mani Pathak,
Robotics and Control Laboratory at Mechanical and Industrial Engineering Department,
Indian Institute of Technology, Roorkee, India.

Arun Kumar Samantaray,
Indian Institute of Technology,
721302 Kharagpur, India.

Nicolas Séguy,
Informatique, Biologie Intégrative et Systèmes Complexes,
Evry University, France.

Shailabh Suman,
Robotics and Control Laboratory at Mechanical and Industrial Engineering Department,
Indian Institute of Technology, Roorkee, India.

Youcef TOUATI,
Polytech-Lille, LAGIS, UMR CNRS 8219,
Avenue Paul Langevin, 59655 Villeneuve D'Ascq, France.

Danwei Wang,
EXQUISITUS, Centre for E-City, Division of Control and Instrumentation,
School of Electrical and Electronic Engineering, Nanyang Technological University, Singapore.

Abdelouaheb Zaatri,
Departement of mechanics,
University Mentouri, Constantine, Algeria.

2

CHAPTER 1

Mechatronic Vehicle Braking Systems

Tarun Kumar Bera [1],
Polytech-Lille, LAGIS, UMR-CNRS 8219
Avenue Paul Langevin, 59655 Villeneuve D'Ascq, France
tarunkumarbera@gmail.com

Arun Kumar Samantaray
Indian Institute of Technology
721302 Kharagpur, India
samantaray@lycos.com

ABSTRACT. Safety and reliability of modern automobiles can be enhanced by Antilock braking system (ABS), traction control system etc. The wheel slip is generally kept within a certain predefined range for an antilock braking system by using an on-off control strategy. In case of single wheel or bicycle model only constant normal loading on the wheels is considered, whereas, for a four wheel vehicle model dynamic normal loading on the wheels and correct lateral forces are considered for the reliable design of braking system. So the controller design needs integration with the different subsystems of the vehicle dynamics model. The vehicle braking system dynamics and its control for a four wheel vehicle is illustrated here. The evaluation of performance of the ABS system under various operating conditions is done through bond graph modeling.

Combined regenerative and antilock braking in electric/hybrid-electric vehicles provides higher safety in addition to energy storing capability. Development of control law for this type of braking system is a challenging task. A sliding mode controller (SMC) for ABS is developed to maintain the optimal slip value. The braking of the vehicle, performed by using both regenerative and antilock braking, is based on an algorithm which decides on how to distribute the braking force between the regenerative braking and the antilock braking in emergency/panic braking situations as well as in normal city driving conditions. The passenger comfort is improved when a sliding mode ABS controller is used in place of standard ABS controller for the mechanical braking part.

Keywords: Antilock braking system, Regenerative braking, Sliding mode controller, Bond graph, Pacejka' s magic formula, Longitudinal slip ratio, Sweet-spot, Bicycle model, Newton-Euler equations, Pseudo-forces, Burckhardt formulae, Hydraulic braking system, Mechanical equivalent braking system, Aerodynamic drag coefficient, Lyapunov function, Continuously variable transmission, Overturning moment, Rolling resistance moment, Self-aligning moment, State of charge.

1.1. Introduction

The road infrastructure is poor and traffic is chaotic in sub-urban, semi-urban and rural areas and this motivates the authors to consider the work on vehicle dynamics and antilock braking systems. Designing vehicles to operate in these conditions is a severely challenging task. The controllers of the mechatronic systems are tuned by trial and error through exhaustive simulations and field testing for getting the optimized performance. The antilock braking system (ABS) is suitable for different driving conditions such as frequent braking and acceleration etc.

The control over the directional stability of the vehicle during emergency braking or braking on slippery roads by preventing wheel lock-up is maintained by the antilock braking system (ABS) which is an electronically controlled braking system. The stopping distance during emergency braking or braking on slippery roads is also reduced by ABS. For this purpose, the ABS utilizes the availability of the maximum brake power to avoid locking of the wheels. There are some limitations in control and performance for the conventional ABS. In case of conventional ABS, the slip is maintained in an acceptable range rather than at the optimal value. To keep the vehicle slip in a particular range, the ABS controller is developed so that the road wheel friction coefficient is highest to achieve the optimal performance. As the controller operates in an unstable equilibrium point, it is very difficult to achieve accurate mathematical model

---

[1]Corresponding author
Part of information included in this chapter has been previously published in Simulation Modeling Practice and Theory Volume 19, Issue 10, November 2011, Pages 2131-2150.

of ABS. The road conditions cannot be used in optimizing the ABS controller performance as it is very difficult to estimate these road conditions by available sensors. The various control algorithms to optimize the ABS performance as available in the literature are sliding mode control , fuzzy logic, and neural networks.

To stop a vehicle, it is necessary to have a device which can consume and dissipate its kinetic energy. The frictional brakes or antilock brakes can perform this function for the vehicle. With mechanical friction brakes, the kinetic energy turns into heat energy and gets dissipated into the environment. Hybridization of vehicle driveline has led to the possibility of accumulating energy due to decrease of vehicle's momentum during braking and consequently make use of that accumulated energy to drive/accelerate vehicle whenever it is intended. This type of braking process is called regenerative braking. Regenerative braking improves vehicle's fuel economy, especially in city driving conditions, where frequent braking is required. To apply regenerative braking, an algorithm has to be designed to distribute the braking force into regenerative braking force and frictional braking force.

It is expected that an electric or hybrid-electric vehicle with regenerative and antilock braking would provide better fuel economy, good braking performance, and directional stability. However, combining the two braking mechanisms is not a trivial task. The main drawback of regenerative braking is that the generated electricity must be closely matched with the supply. A continuously variable transmission (CVT) , which is an automatic, step less, smooth transmission system from which infinite number of gear ratios can be obtained within the limits, is required to maintain constant speed at the generator input for various output velocities during regenerative braking process. The regenerative braking force is usually insufficient during panic braking which is why additional braking force has to be provided by mechanical brakes.

1.2. Brake System Model

The different braking system models developed for the vehicle and their control laws are described in this chapter. The brakes are used to stop or retard the motion of the wheel. In this chapter, the models of antilock braking system (ABS) and regenerative braking are developed. The ABS is used to increase the vehicle stability during maneuvering and to decrease the stopping distance. The regenerative braking may be used for city driving where frequently brakes are used and the energy which is wasted during braking may be stored through a bank of ultra-capacitors. The influential parameters of the wheels are mass, rotary inertia, radius and tire stiffness. The tire forces and moments are very important for the dynamic modeling of vehicle as the tire is key element of the wheel. The tire forces and moments generated due to contact between the road and the tire determine the vehicle handling performance. There are three models of tires i.e. physical models, analytical models, and empirical models. The tire elastic deformation and tire forces are predicted from the physical models [**24**]. In this model, the equations of motions are solved by complex numerical methods. The analytical models are useful for lower wheel slip but unsuitable for combined slip. The more accurate model is the empirical models based on experimental correlations [**19**]. The longitudinal and cornering forces and self-aligning moment can be computed by the widely used Pacejka' s magic formula [**30**].

1.2.1. Tire Slip Forces and Moments. The tire forces and moments generated at the contact of road and the tire is shown in Figure 1.2.1. The forces acting along x, y and z axes are longitudinal force F_x, lateral force F_y and normal force F_z, respectively. Similarly the moments acting along x, y and z axis are overturning moment M_x, rolling resistance moment M_y and the self-aligning moment M_z, respectively [**31**].

In the actual case, the longitudinal slip is computed from the wheel speed and linear acceleration [**28**] which are measured by the sensors. The longitudinal slip ratio (σ_x) is defined as the ratio of the difference between the circumferential velocity $\left(\dot{\theta}_{wy} r_w\right)$ and the translational velocity of the wheel ($\dot{x}_w$) to the circumferential velocity or the translational velocity depending on the traction or braking condition [**21**]. It is expressed as

$$(1.2.1) \qquad \sigma_x = \begin{cases} \frac{\dot{\theta}_{wy} r_w - \dot{x}_w}{\dot{\theta}_{wy} r_w} \left(\text{during} \quad \text{traction, assuming} \quad \dot{\theta}_{wy} \rangle 0\right) \\ \frac{\dot{x}_w - \dot{\theta}_{wy} r_w}{\dot{x}_w} \left(\text{during} \quad \text{braking, assuming} \quad \dot{x}_w \rangle 0\right) \end{cases}$$

where $\dot{\theta}_{wy}$ and r_w are the angular speed and radius of the wheel, respectively. Lateral wheel slip (σ_y) is the ratio of lateral velocity ($\dot{y}_w$) to the longitudinal velocity of wheel [**30**]. It and the side slip angle α are given as

$$(1.2.2) \qquad \sigma_y = \tan \alpha = \frac{\dot{y}_w}{\dot{x}_w}$$

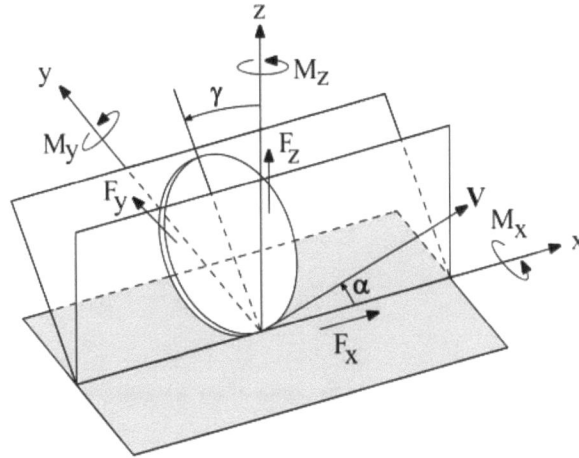

FIGURE 1.2.1. tire forces and moments.

The longitudinal force F_x and lateral force F_y can be approximated as $F_x = \sigma_x C_x$ and $F_y = \sigma_y C_y$ for smaller slip ratios, where, longitudinal tire stiffness (coefficient) and cornering coefficient are denoted by C_x and C_y, respectively, and the longitudinal slip ratio and lateral slip ratio are denoted by σ_x and σ_y, respectively [35]. However, these linear relations are not useful for large slip ratios.

As the empirical magic formula based on experimental data gives more accurate results for larger slip angles and is also applicable to wide range of operating conditions, it may be adopted for the development of a tire-road friction model. Longitudinal force F_x is developed by longitudinal slip velocity (i.e., $\dot{\theta}_{wy} r_w - \dot{x}_w$ during traction and $\dot{x}_w - \dot{\theta}_{wy} r_w$ during braking) whereas side force F_y and self-aligning moment M_z are generated by lateral or side slip velocity $\dot{y}_w$ and camber angle γ. According to Pacejka's *magic formula* the longitudinal force is function of longitudinal slip; side force and the self-aligning moment are functions of side slip:

$$(1.2.3) \qquad y_o = D \sin \left[C \tan^{-1} \left\{ B x_i - E \left(B x_i - \tan^{-1} \left(B x_i \right) \right) \right\} \right]$$

where output variable, $y_o : F_x, F_y$ or M_z and input variable, $x_i : \sigma_x$ or σ_y.

The slight variation in the function in Eq.1.2.3 due to ply-steer, rolling resistance and conicity effects may be neglected. The tire forces and moments are measured by sophisticated equipment to determine constant parameters (B, C, D, E).

Pacejka's *magic formula* is unsuitable when the vehicle's performance is significantly affected by snow and ice. This formula does not consider the variation of friction coefficient with velocity while designing brake systems. So, another friction model developed by Burckhardt [27] is often used for research on brake system design. The Burckhardt formulae are discussed later.

1.2.2. Antilock Braking System. Antilock braking system (ABS) is a vehicle autonomous system. It is used to improve stability and to reduce longitudinal stopping distance while maneuvering under braking condition. The optimum braking performance of the antilock braking system is shown in [6]. Antilock braking system is most suitable for the braking on icy or wet asphalt road or for panic braking situation. When the antilock brake is applied initially, the longitudinal slip ratio starts increasing due to the decrease of the angular velocity of the wheel faster than the vehicle's linear velocity. The grip between the tire and the road decreases at high value of slip ratio, and consequently the vehicle's deceleration rate reduces. The applied braking torque is reduced as the slip ratio reaches a maximum threshold value. Therefore, the longitudinal slip ratio decreases due to the increase of angular speed of the wheel. The grip between the tire and the road reduces at the lower value of this slip ratio. The braking torque is increased again when the slip ratio reaches a minimum threshold value. This way, the braking torque is modulated to keep the slip ratio between an upper and a lower threshold value [12]. The slip and wheel acceleration control are required for the control of ABS. The wheel angular velocities are measured and slip is controlled indirectly in case of wheel acceleration control. The main drawback for the conventional ABS is that slip cannot be kept at the exact

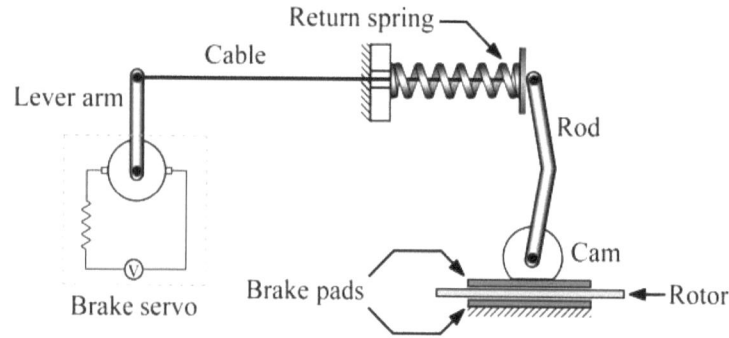

FIGURE 1.2.2. Schema of an ABS [22].

optimum value. ABS controller designed for a specific vehicle needs lots of testing and tuning. The testing of the ABS performance is done through software-in-the-loop and hardware-in-the-loop simulations [10]. The brake pressure in antilock braking systems is regulated through hydraulic valve control. The hydraulic brake system using the bond graph technique was modeled in [20]. The performance of ABS was improved using sliding mode controllers in [7]. The model dependence on velocity and wheel slip to devise a robust ABS controller were taken into account in [4]. A linear parameter varying (LPV) system for the development of wheel slip controller of a two-wheeled vehicle under straight running conditions is used to model this dependence [4].

In [34]composite ABS control technique has been used for an electrically driven vehicle with four wheels. The electric motor ABS control is combined with the hydraulic ABS control in composite ABS control. The electric motor ABS control works for the lower braking torque requirement, but both the electric and hydraulic ABS controls work for higher torque requirement [34].

The brake servo, lever arm, return spring, rod, cam, rotors, cable and brake pads are the main components of ABS [22]. The Schematic view of an ABS is shown in Figure 1.2.2. The voltage that is fed to the motor is controlled by the ABS controller. The motor is connected to the lever arm. The cable connects the arm at one end and a lever with cam is connected to the other side. The initial configuration is recovered through the return spring.

Generally all the tire friction models have a non-linear nature. The friction coefficient can be written as

$$(1.2.4) \qquad \mu = \frac{\sqrt{F_x^2 + F_y^2}}{F_z}$$

where most widely used Pacejka's [30]*magic formula* (Eq. 1.2.3)determines longitudinal force F_x and side force F_y under constant value of linear and angular velocities and F_z is the normal force. However, this model is invalid for low slip effects and wheel lockup is characterized by large forward and side slip effects. Burckhardt formula avoids these problems. The friction coefficient given by Burckhardt formula is dependent on vehicle speed. The friction model given by Burckhardt as follows [27]:

$$(1.2.5) \qquad \mu(\sigma_x, \dot{x}_w) = \left[C_1 \left(1 - e^{-C_2 \sigma_x} \right) - C_3 \sigma_x \right] e^{-C_4 \sigma_x \dot{x}_w}$$

where constants C_1, C_2, C_3 and C_4 are experimentally determined. These constant parameters for different road conditions [27] are given in Table 1. The slip-friction curve based on this model is shown in Figure 1.2.5.

The flow diagram for the ABS control is shown in Figure 1.2.3. The longitudinal slip is controlled between maximum and minimum thresholds and, as a result of this, brake torque signal (τ) varies and therefore the braking torque in the i-th step (sample) is computed as follows:

$$(1.2.6) \qquad \tau_i = \begin{cases} \tau_{\max} & \sigma_x < \sigma_{\text{low}} \\ \tau_{i-1} & \sigma_{\text{low}} \leq \sigma_x \leq \sigma_{\text{high}} \\ 0 & \sigma_x > \sigma_{\text{high}} \end{cases}$$

TABLE 1. tire-road constant parameters [27].

Surface condition	C_1	C_2	C_3	C_4
Asphalt, dry	1.029	17.16	0.523	0.03
Asphalt, wet	0.857	33.822	0.347	0.03
Cobblestones, dry	1.3713	6.4565	0.6691	0.03
Snow	0.1946	94.129	0.0646	0.03

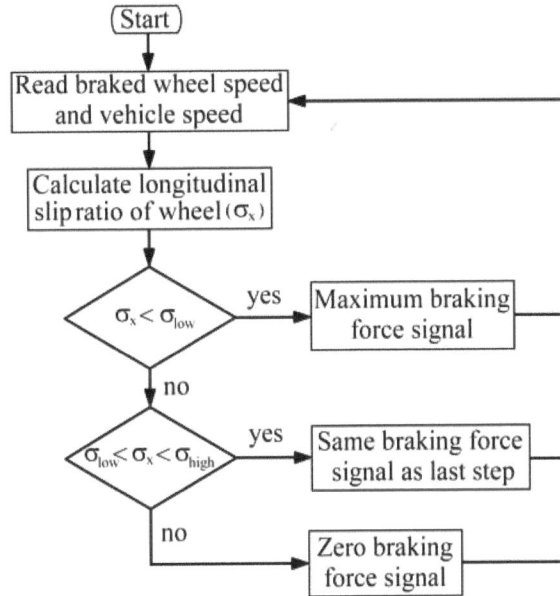

FIGURE 1.2.3. The flow chart for ABS algorithm.

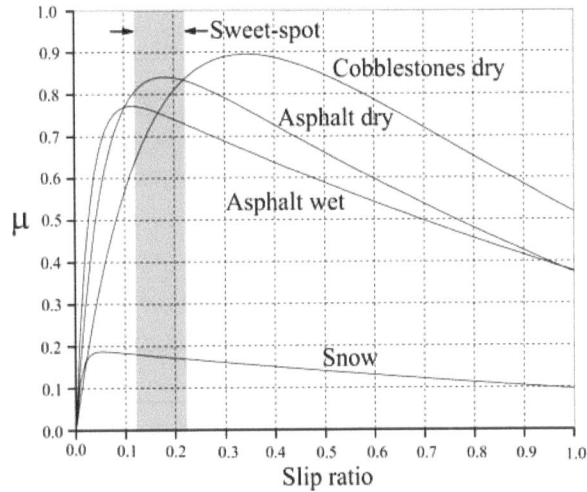

FIGURE 1.2.4. Friction coefficient versus slip ratio for different road surfaces at linear speed of 10m/s with the tire properties given in [27].

The coefficient of friction versus slip ratio curve for a wheel moving with linear speed of 10 m/s on different types of road conditions is shown in Figure 1.2.4. It is seen from this figure that the friction coefficient increases with the

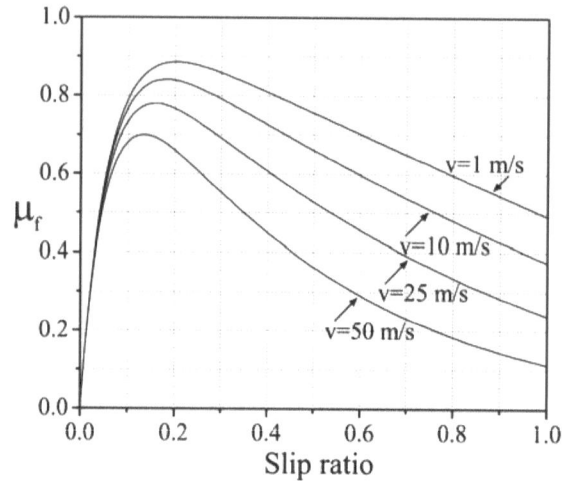

FIGURE 1.2.5. Coefficient of friction versus slip ratio curves for asphalt dry road condition for different vehicle speeds.

increase in slip ratio to a maximum value and then it decreases for further increase in the slip ratio. If the coefficient of friction is low, the wheel gets locked ($\sigma_x = 1$) and the wheel starts sliding and the steering control may be lost, which is totally undesirable. The slip value must be maintained within a range to get the higher value of friction force. Therefore, steerability and lateral stability of the vehicle increase and the stopping distance during braking decreases. The slip value is kept within a certain range which is also called sweet-spot (see the shaded area in Figure 1.2.4)due to the very fast slip dynamics and the friction curve is open loop unstable at any value after the peak. The attainable maximum friction coefficient changes with vehicle speeds (see Figure 1.2.5). Brake systems are designed in order to provide maximum braking efficiency at higher speeds. Moreover, it is difficult to accurately predict the road conditions by the available sensors. The optimal slip ratio range is considered as 0.2-0.25 in the design of ABS. Maximum friction coefficient is achieved for all kinds of road surfaces (thus, eliminating the requirement of road condition recognition) and vehicle linear speed if we consider this range of slip ratios.

1.3. Bicycle Vehicle Model

For the initial study, a bicycle model of a vehicle [33] was considered. The roll, pitch and heave motions are not taken into account by this model and the suspension dynamics is not considered. Therefore, this model does not include the load transfer during braking and maneuvering. Schematic view of the considered vehicle is shown in Figure 1.3.1 and its word bond graph is shown in Figure 1.3.2. The motion of the vehicle is planar as the road is assumed to be flat.

1.3.1. Kinematic Relations. The kinematic relations of the major part of the model are based on [26] and [33]. The calculations of wheel rotations and longitudinal and lateral slip are introduced into the model. Only front wheel steering denoted by a steering angle of δ is considered.

The velocity components in normal (v_{nfr}) and tangential (v_{tfr}) directions to the plane of rotation of the front wheel are

$$(1.3.1) \qquad \begin{aligned} v_{\mathrm{nfr}} &= (\dot{y} + \dot{\theta}_{\mathrm{cz}}a)\cos\delta - \dot{x}\sin\delta \\ v_{\mathrm{tfr}} &= (\dot{y} + \dot{\theta}_{\mathrm{cz}}a)\sin\delta + \dot{x}\cos\delta \end{aligned}$$

where $\dot{x}$, $\dot{y}$ and $\dot{\theta}_{\mathrm{cz}}$ are the velocity in x-direction, velocity in y-direction and angular speed of the vehicle, respectively; and a and δ are the distance of front axle from the vehicle cg and steering angle, respectively. Similarly, velocity components in normal (v_{nrr}) and tangential (v_{trr}) directions to the plane of rotation of the rear wheel are

$$(1.3.2) \qquad v_{\mathrm{nrr}} = (\dot{y} - \dot{\theta}_{\mathrm{cz}}b), \; v_{\mathrm{trr}} = \dot{x}$$

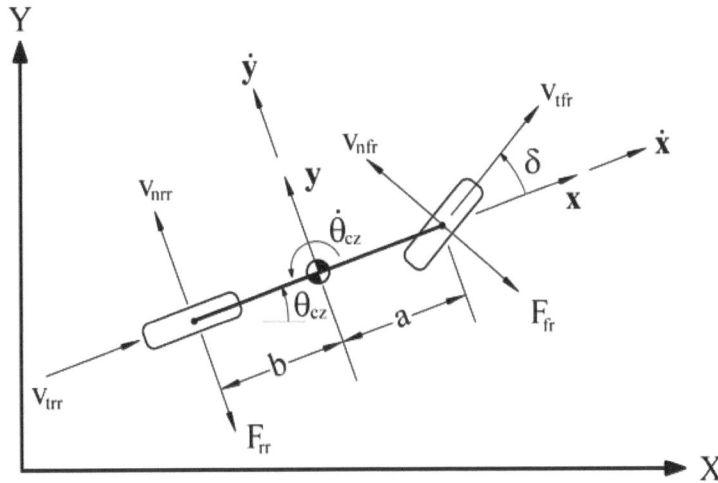

FIGURE 1.3.1. Schema of the bicycle vehicle model.

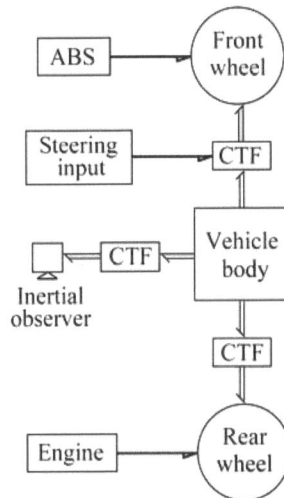

FIGURE 1.3.2. Word bond graph of the bicycle vehicle model.

where b is the distance of rear axle from the vehicle cg. From Newton-Euler equations with $z = \dot{\theta}_{cx} = \dot{\theta}_{cy} = 0$, one obtains

(1.3.3)
$$m_v \ddot{x} = m_v \dot{\theta}_{cz}\, \dot{y} + \sum F_x$$
$$m_v \ddot{y} = -m_v \dot{\theta}_{cz}\, \dot{x} + \sum F_y$$

where m_v, F_x and F_y are the mass of the vehicle, forces acting on the vehicle in x and y directions, respectively.

1.3.2. Bond Graph Model. CTF blocks represent necessary coordinate transformations [9] in the word bond graph of the bicycle vehicle model as shown in Figure 1.3.2. The complete bond graph model as shown in Figure 1.3.3 can be drawn using Eqs. 1.3.1 1.3.2 1.3.3and maintaining the model structure defined in the word bond graph. The conservative pseudo-forces in Eq.1.3.3 are implemented by a gyrator (GY) element in the bond graph model. In Figure 1.3.3), power bonds are represented by lines with half-arrow at their end and information bonds are represented by full arrow at their end. The pseudo-forces transform the vehicle inertias ($I : m_v$) modeled in the moving frame, to the inertial frame. The rotary inertia ($I : J_v$) is modeled by an I-element connected to a 1-junction. The tangential and

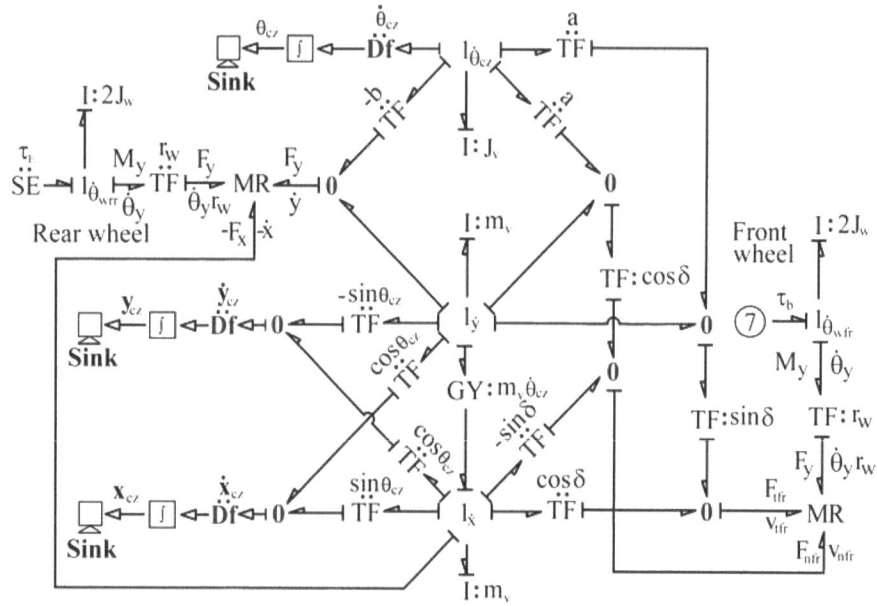

FIGURE 1.3.3. Bond graph model of bicycle vehicle model.

normal velocities at tires are computed using Eqs. 1.3.1 1.3.2 and they are modeled by transformer elements (TF). The flow detectors (Df) which are connected to velocity points in the inertial frame are not present in the actual system, i.e., the actual system is not instrumented with inertial sensors. These flow detectors are simply added to plot positions of the vehicle center of mass in the inertial frame and also to modulate the modulated transformer elements present in the junction structure.

Encircled port enumerated as 7 is interfaced with the model of the brake system which applies an external brake torque (τ_b). The engine torque (τ_E) is applied on the rear wheels whose rotary inertia is $2J_w$ where J_w is the rotary inertia per wheel/axle. The Burckhardt formulae are implemented by modulated 3-port R-fields (MR-elements). The longitudinal and lateral slip ratios are given in Eqs. 1.2.1 1.2.2 are computed with the velocity of the contact patch on the tire $\left(\dot{\theta}_y r_w\right)$, the longitudinal velocity of the tire $(\dot{x}_w)$ and the lateral velocity $(\dot{y}_w)$ as inputs. Then, the friction coefficient and the resultant friction force are calculated from the Eq. 1.2.5. The longitudinal and lateral tire forces and the reaction moment on the drive are calculated from the components of the total friction force. The self-aligning moment is not useful in the planar vehicle model. Thus, the MR-element here receives the generalized flow variable information and uses them to compute three generalized effort variables. The aerodynamic resistance (R_{aero}) can be modeled at $1_{\dot{x}}$ junction.

1.3.3. Bond Graph Model of Braking System. The bond graph model of equivalent mechanical braking system (Figure 1.2.2) and the traditional hydraulic braking system are shown in Figure 1.3.4and Figure 1.3.5, respectively. The controlled voltage from the ABS controller is fed to the motor which generates a torque (τ_m) that is represented by a Se-element as shown in the bond graph model of the mechanical equivalent braking system (Figure 1.2.2)[22]. The mechanical loss (R_{lm}) is denoted by the resistive element connected to 1-junction. Another C-element connected to 0-junction represents cable stiffness (K_{ca}). C-element represents return spring having stiffness K_{re}. The fixed end of the return spring is modeled by zero-valued flow source. The length of the arm and the brake drum radius are represented by l_a and r_{bd}, respectively.

The pedal force (F_{pedal}) is represented by the Se-element (Figure 1.3.5) in hydraulic braking system. The area of the master cylinder is represented by A_{mc}. The resistive element (R_{sv}) models the fluid flow through supply or hold valve. The brake fluid compressibility K_β (a function of bulk modulus and fluid volume)is determined by the element C connected with 0- junction, which determines pressure of brake cylinder. The resistance in the pressure relief valve is denoted by R_{rv} and the zero-valued Se-element indicates atmospheric pressure. Braking torque is dependent on the

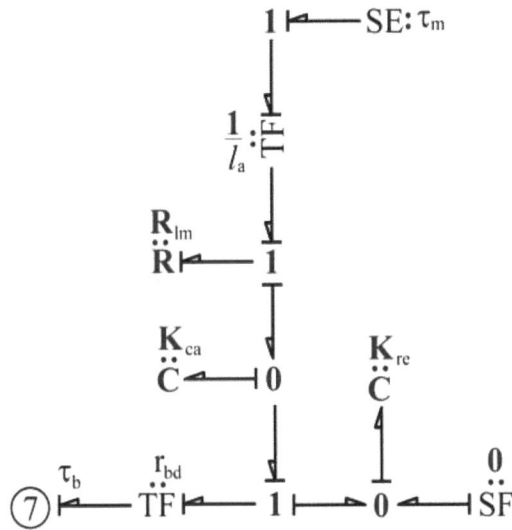

FIGURE 1.3.4. Bond graph model of mechanical equivalent braking system.

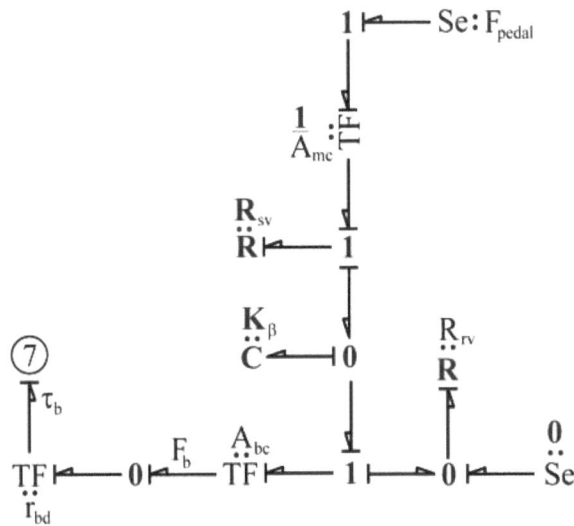

FIGURE 1.3.5. Bond graph model of hydraulic braking system.

brake cylinder area (A_{bc}) and the brake drum radius. The output braking torque is applied on the front wheel of the bicycle vehicle model (Figure 1.3.3) and wheel sub-model of a full vehicle model.

1.3.4. ABS Model Validation. The parameters which are used to simulate the bicycle vehicle model are given in Table 2. ABS on the planar bicycle model is first implemented to fine-tune and test the control laws. After satisfactory implementation and test, this ABS model can be used for the full vehicle model.

The ABS system has been validated by comparing the results with those reported in [22]. The steering angle in the bicycle model was kept zero as the steering effect was not considered in [22]. The vehicle weight and the initial linear and angular velocities of the wheels are considered to be the same as those in [22] and the rear wheel of the vehicle is taken as freely rolling (disengaged from drive and brakes). The ABS with the bicycle model developed here becomes equivalent to that given in [22]after the stated adjustments.

A comparison of the results reported in [22] with those obtained from the developed ABS bicycle model is shown in Figure 1.3.6 and Figure 1.3.7   . It is clear from both the results that the forward linear wheel speed varies almost

TABLE 2. Parameter values of bicycle vehicle model.

Subsystem	Parameter values			
Vehicle body	$m_V = 1600$ kg	$J_V = 100$ kg m^2	$a = 1.0$ m	$b = 1.0$ m
Wheel	$J_w = 15$ kg m^2	$r_w = 0.3$ m		
	$C_1 = 1.029$	$C_2 = 17.16$	$C_3 = 0.523$	$C_4 = 0.03$
Antilock brake	$\sigma_{low} = 0.04$	$\sigma_{high} = 0.5$	$K_{ca} = 10^4$ N/m	$K_{re} = 10^6$ N/m
	$r_{bd} = 0.15$ m	$R_{lm} = 0.04$ N s/m	$l_a = 1$ m	$\mu_m = 0.4$ N m/A
	$R_{aero} = 0.1$ kg/s			

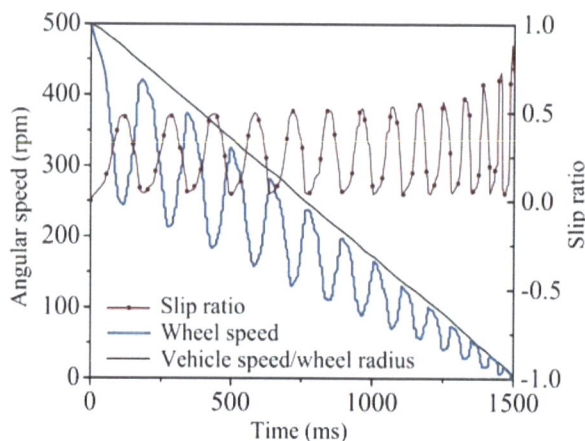

FIGURE 1.3.6. Vehicle speed, wheel speed and wheel slip ratio as reported in [**22**].

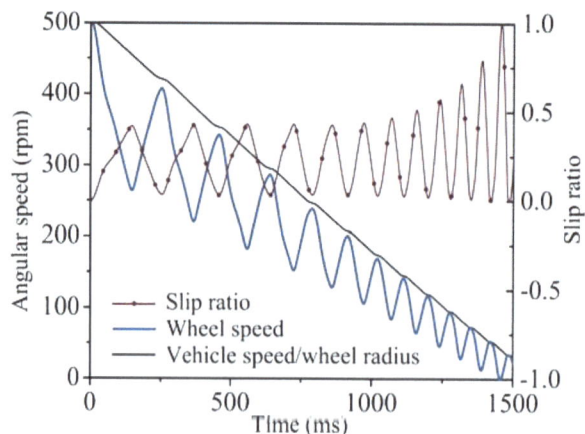

FIGURE 1.3.7. Vehicle speed, wheel speed and wheel slip ratio as obtained from the bicycle model during full braking by ABS.

linearly and the deceleration and acceleration pattern of the wheel angular speed is nearly regular. The slip ratio is initially kept within a certain range and its maximum value approaches unity as the vehicle slows down. The small variations in the results may be attributed to the differences in the models. The four wheel model of [**22**] is influenced by vehicle's pitch during braking. This causes load transfer to the front wheels (where ABS brakes are applied) and consequently it produces more braking force than the bicycle model. Therefore, Figure 1.3.6 shows a little quicker braking action in comparison to Figure 1.3.7.

1.3.5. ABS Performance during Maneuvering. The ABS provides better maneuverability while the steered wheel is braked. An example scenario has been considered here to test the vehicle's maneuverability during braking.

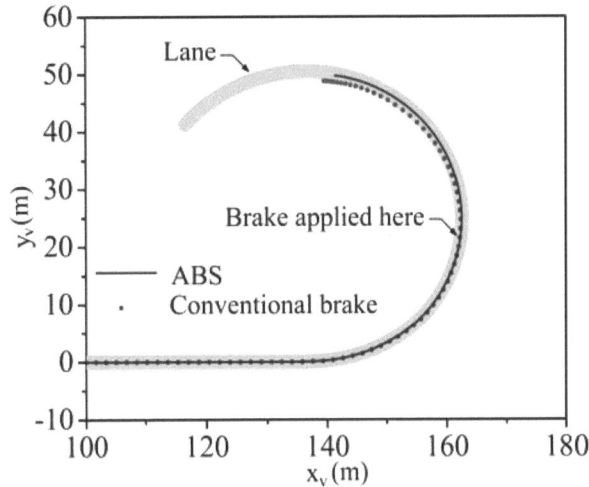

FIGURE 1.3.8. Vehicle center of mass trajectory in inertial frame for ABS and conventional braking system.

Initially, the vehicle starts from rest and attains a constant linear speed of 50 km/h along a straight path. Then the vehicle is steered with a constant steering angle of 0.1 rad to follow a circular path at $t = 12$s. The vehicle linear speed decreases due to the steering action. After the vehicle linear speed becomes constant over the circular path, the brakes are applied at the front wheel at $t = 17$s. A composite slip based formulation [**36**] developed by U.S department of transportation (DOT) has been adopted to handle large side and forward slips during wheel lockup.

The predefined path (2.5m wide lane) over which the vehicle is supposed to move is shown in Figure 1.3.8 and the actual paths taken by the vehicle under two different braking conditions (the first when ABS is used and the second when a conventional mechanical brake is used) are also shown. A slip control strategy used to restrict the slip ratios within 0.2-0.25 in the ABS. Initially, brake force is applied till the slip ratio becomes 0.2 and then the algorithm for slip control takes over. It is seen that the vehicle stays within the lane till it stops when ABS is used whereas it veers off from the lane due to application of the conventional brake.

The change in slip ratios due to braking is shown in Figure 1.3.9. The vehicle is able to steer because the slip ratios for ABS are kept within 0.2-0.25. As the slip ratio quickly approaches a value of 1.0 due to application of conventional brake, the wheels are locked up and thus the vehicle cannot steer properly.

The variations of vehicle speed due to steering and braking and the angular speed of the front wheel during that process are shown in Figure 1.3.10 and Figure 1.3.11, respectively. The wheel locks up due to application of conventional brake when the wheel speed becomes zero. On the other hand, there is an initial sudden drop in wheel speed (till the slip ratio becomes 0.2) on application of ABS and then the wheel speed changes as shown in Figure 1.3.7. The vehicle's linear speed variation is nearly linear in both cases. The deceleration (see Figure 1.3.10) is slightly faster for ABS as compared to the conventional brake. When brakes are applied both on the front and the rear wheels, this deceleration becomes almost double. ABS based on slip control mechanism aids the vehicle to stop faster and at a shorter distance. From these results, we can conclude that the control loop of an automatic lane departure control and monitoring system [**16**] should include the brake system.

1.3.6. Sliding Mode Control. Sliding Mode Control is a high-speed switching feedback control. In sliding-mode brake control, brake pressure is increased, decreased or held during the operation. The goal is to obtain a control algorithm which allows the maximal value of the tire-road friction force to be reached during emergency braking and to maintain the friction level around the sweet-spot in the friction-slip curve as shown in Figure 1.2.4. An observer can be developed to obtain the friction force values [**7**]. The brake pressure in antilock braking systems is usually regulated by hydraulic valve control. Hydraulic brake system using the bond graph technique was modeled in [**20**]. In [**7**] sliding mode controllers have been proposed to improve ABS performance.

The sliding surface (S) with incorporation of the sliding controller gain term is described as

(1.3.4)
$$S = (\sigma_x - \sigma_{des})$$

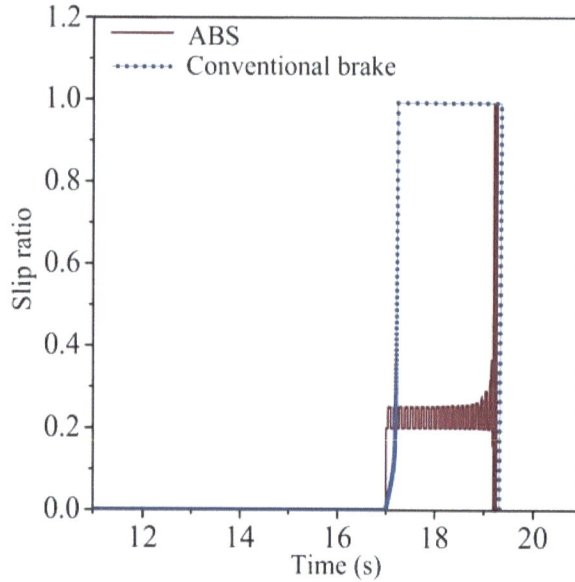

FIGURE 1.3.9. Wheel slip ratio for ABS and conventional braking system.

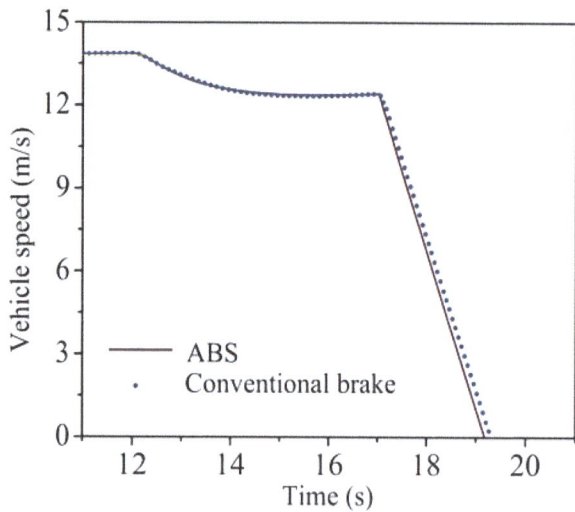

FIGURE 1.3.10. Vehicle speed for ABS and conventional braking system.

where σ_{des} is the desired slip ratio. Differentiation of Eq. 1.2.1 with respect to time and use of Eqs. 1.4.3 1.4.4, i.e. using a quarter car model, gives

(1.3.5)
$$\dot{\sigma}_{\mathrm{x}} = -\frac{1}{\dot{x}_{\mathrm{w}}}\left[\frac{\left(F_{\mathrm{x}'}+C_{\mathrm{aero}}\dot{x}_{\mathrm{w}}^{2}\right)(1-\sigma_{\mathrm{x}})}{m_{\mathrm{e}}} + \frac{F_{\mathrm{x}'}r_{\mathrm{w}}^{2}}{J_{\mathrm{wy}}}\right] + \frac{r_{\mathrm{w}}\tau_{\mathrm{b}}}{\dot{x}_{\mathrm{w}}J_{\mathrm{wy}}}$$
$$= f\left(\dot{\theta}_{\mathrm{wy}}, \dot{x}_{\mathrm{w}}\right) + g\left(\dot{x}_{\mathrm{w}}\right)\tau_{\mathrm{b}}$$

where $C_{\mathrm{aero}} = \frac{1}{2}\rho A_{\mathrm{f}}C_{\mathrm{d}}$, ρ is the density of air, A_{f} is the frontal area, C_{d} is the aerodynamic drag coefficient, m_{e} is the effective mass, J_{wy} is the rotary inertia of the wheel and axle, $\dot{x}_{\mathrm{w}}$ is the linear wheel speed, $\dot{\theta}_{\mathrm{wy}}$ is the angular velocity of wheel, and f (.) and g(.) are two functions. To evaluate stability, the Lyapunov function (V_{L}) can be written as

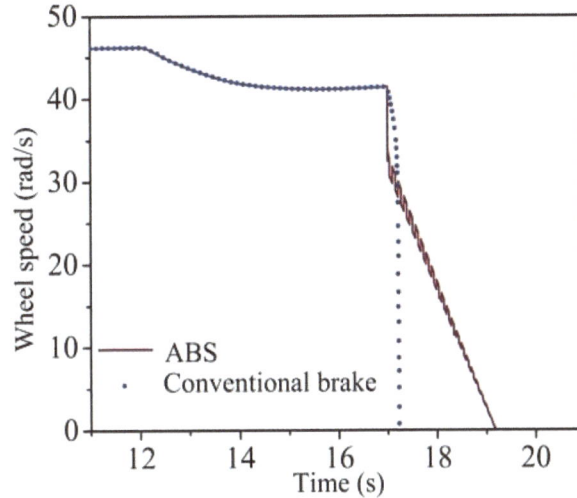

FIGURE 1.3.11. Wheel speed for ABS and conventional braking system.

(1.3.6)
$$V_{\mathrm{L}} = \frac{S^2}{2}.$$

Then $\dot{V}_{\mathrm{L}}$ will be negative definite iff

(1.3.7)
$$\left[f\left(\dot{\theta}_{\mathrm{wy}}, \dot{x}_{\mathrm{w}} \right) + g\left(\dot{x}_{\mathrm{w}} \right) \tau_{\mathrm{b}} \right] \begin{cases} < 0 & \forall\, S > 0 \\ = 0 & \forall\, S = 0 \\ > 0 & \forall\, S < 0 \end{cases}$$

which is satisfied if a controller gain $\eta > 0$ is defined such that

(1.3.8)
$$\dot{S} = -\eta.\mathrm{sgn}\left(S \right),$$

where sgn() is the signum function.

Using Eqs. 1.3.71.3.8, and noting that $F_{\mathrm{x}'} = \mu_{\mathrm{f}}(\sigma_{\mathrm{x}}, \dot{x}_{\mathrm{cb}})F_{\mathrm{z}}$, where $\dot{x}_{\mathrm{cb}}$ is car body linear speed and F_{z} is the axle load, $m_{\mathrm{e}} = m_{\mathrm{cb}}/4 + m_{\mathrm{w}}$ where m_{cb} is the car body mass and m_{w} is the wheel/axle mass, and $\dot{x}_{\mathrm{cb}} \cong \dot{x}_{\mathrm{w}}$ (as the suspension stiffness in the longitudinal direction is very high), the brake torque is given as

(1.3.9)
$$\begin{aligned}
\tau_{\mathrm{b}} &= \frac{J_{\mathrm{wy}}}{r_{\mathrm{w}}} \left[\frac{\left(\mu_{\mathrm{f}}(\sigma_{\mathrm{x}}, \dot{x}_{\mathrm{cb}})F_{\mathrm{z}} + C_{\mathrm{aero}}\dot{x}_{\mathrm{w}}^2 \right)(1-\sigma_{\mathrm{x}})}{m_{\mathrm{e}}} + \frac{\mu(\sigma_{\mathrm{x}}, \dot{x}_{\mathrm{cb}})F_{\mathrm{z}}r_{\mathrm{w}}^2}{J_{\mathrm{wy}}} \right] - \frac{J_{\mathrm{wy}}}{r_{\mathrm{w}}}\eta\dot{x}_{\mathrm{cb}}\mathrm{sgn}\left(S \right) \\
&= q\left(\dot{\theta}_{\mathrm{wy}}, \dot{x}_{\mathrm{cb}} \right) - k_{\mathrm{g}}\dot{x}_{\mathrm{cb}}\mathrm{sgn}\left(S \right),
\end{aligned}$$

where $k_g = \eta J_{\mathrm{wy}}/r_{\mathrm{w}}$ and q(.) is a function. Sliding motion occurs when states $\left(\dot{\theta}_{\mathrm{wy}}, \dot{x}_{\mathrm{cb}} \right)$ reach the sliding surface. Normal brake is applied initially to bring the system states to the neighborhood of the sliding surface and then the sliding mode controller is applied.

1.3.7. Validation of the Sliding Mode ABS Controller. A sliding mode ABS controller with integral feedback has been proposed in [25]. Although that controller has a different structure, it can be used to validate the controller developed here. A quarter-car model is used in [25] to test the controller. Moreover, Dugoff's tire model is used therein with dry asphalt road condition and the wheel slip is maintained at 0.15 on the sliding surface. By considering zero steering angle, half vehicle weight and freely rolling rear wheel, the developed bicycle vehicle model and the quarter-car model becomes equivalent. The parameter values for vehicle mass, initial forward velocity and angular speed etc. are taken from [25] and the parameters reported in [25] were used to simulate the bicycle model of the vehicle. In [25], the initial linear velocity is taken to be 20 m/s (72 km/h) and the results are plotted for 1.7 s duration.

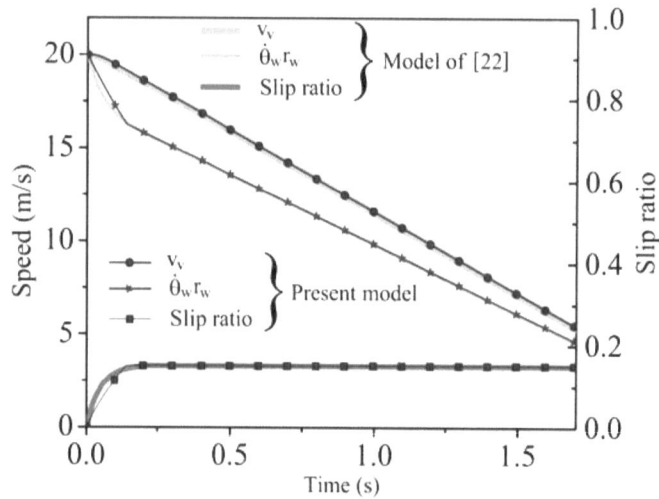

FIGURE 1.3.12. Comparison of sliding mode ABS controller results in [**25**] with the present model [**3**].

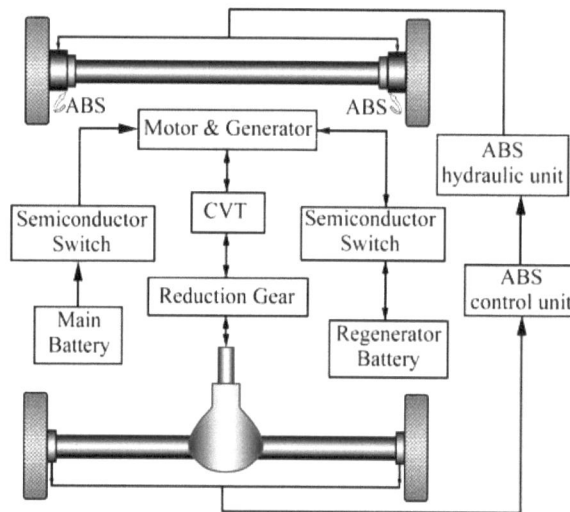

FIGURE 1.4.1. Schema of regenerative braking.

A comparison of the simulation results with settings is given in Figure 1.3.12 [**3**]. It is found that the deceleration in the vehicle linear speed is almost the same. However, the variations in wheel speed and slip ratios are slightly different at the initial stage [**3**]. This discrepancy may be attributed to the initial application of the brake force to reduce the slip ratio to a value marginally below 0.15 and the difference in the two tire force models and the two sliding mode controller algorithms, especially due to the integral feedback scheme considered in [**25**].

1.4. Regenerative Braking

The schema of regenerative braking system is shown in Figure 1.4.1. An electric vehicle setup is considered for this analysis on regenerative braking.

The vehicle is considered to have rear wheel drives. The drive torque is applied to the differential through motor, continuously variable transmission (CVT) and final reduction gear. The semiconductor switch in the transmission line behaves like a clutch. CVT is used to maintain the generator input speed constant during regeneration. During regenerative braking, the vehicle uses the motor as generator and the output of it is used to charge the regenerative battery. The braking of the vehicle is performed by using both regenerative and ABS for the present study. Pedal input,

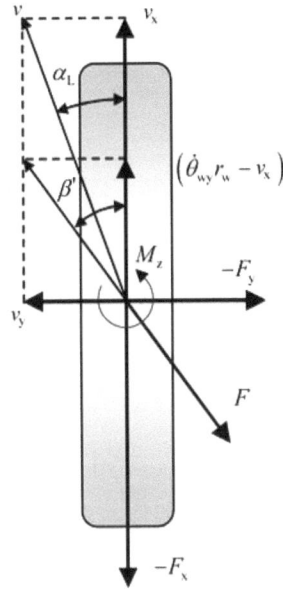

FIGURE 1.4.2. Top-view of tire forces and moments.

vehicle's longitudinal velocity and CVT ratio are used as input for calculation of regenerative braking and maximum braking torque. When the brake is applied, the controller unit calculates the required torque and its distribution.

Usually, empirical formulae developed from experiments are used to model the tire-road interaction forces [30]. The friction model given by Burckhardt [18] is used here. The forces acting on a tire are shown in Figure 1.4.2.

The resultant slip is $\sigma = \sqrt{\sigma_x^2 + \sigma_y^2}$ whose orientation with respect to x-axis is $\beta' = \tan^{-1}(\sigma_y/\sigma_x)$. The slip angle $\alpha_L = \tan^{-1}(v_y/v_x)$ and the resultant tire velocity $v = \sqrt{v_x^2 + v_y^2}$. The velocity dependent friction force at tire-road interface is given by Burckhardt in Eq. 1.2.5.

The resultant tire force is

$$(1.4.1) \qquad\qquad F = \mu_f F_z = \sqrt{F_x^2 + F_y^2},$$

where $F_x = -\mu_f F_z \cos\beta'$ and $F_y = -\mu_f F_z \sin\beta'$. The tire forces acting at the contact patch between the tire and the ground also produce the following moments about the x, y and z axes:

$$(1.4.2) \qquad\qquad M_x = F_y r_w, \quad M_y = -F_x r_w, \text{ and } M_z = -F_y \delta_t = \mu_f F_z \delta_t \sin\beta',$$

where M_x is the overturning moment, M_y is the rolling resistance moment, M_z is the self-aligning moment, r_w is the wheel radius and δ_t is the pneumatic trail of the contact patch.

The schema of a quarter car model and its bond graph model are shown in Figure 1.4.3 and Figure 1.4.4, respectively. The equations of motion for linear and rotational dynamics of the quarter car model during braking are given by

$$(1.4.3) \qquad\qquad \left(m_w + \frac{m_c}{4}\right)\ddot{x} = -F_{x'} - C_{aero}\dot{x}^2$$

$$(1.4.4) \qquad\qquad J_{wy}\ddot{\theta}_{wy} = F_{x'} r_w - \tau_b$$

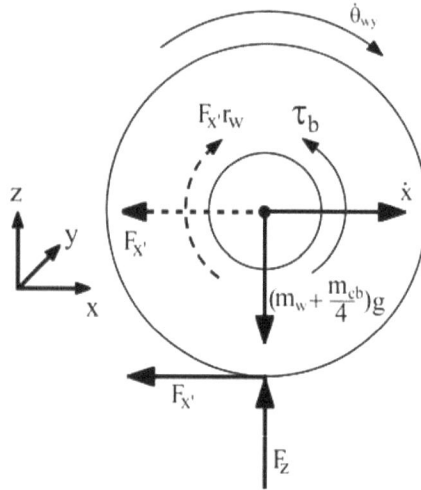

FIGURE 1.4.3. Schema of wheel under braking condition.

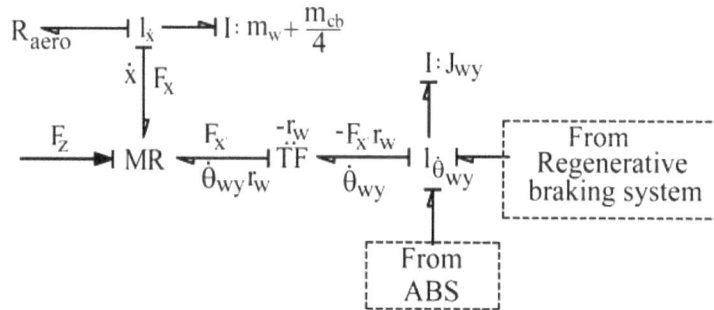

FIGURE 1.4.4. Bond graph model of wheel under braking condition.

where m_w, m_c and $F_{x'}$ are the mass of the wheel, mass of the vehicle body and reactive longitudinal force, respectively. J_{wy} and $\ddot{\theta}_{wy}$ are the polar moment of inertia and angular acceleration of the wheel in the y-direction, respectively. $C_{aero} = \frac{1}{2}\rho A_f C_d$, ρ is the density of air, A_f is the frontal area and C_d is the aerodynamic drag coefficient. τ_b is the braking torque.

The inertia (m_w) of the wheel and the vehicle body (m_c) is modeled by I-element at a 1-junction. Similarly the rotary inertia (J_{wy}) is modeled at another 1-junction. The three port modulated R-field implements Burckhardt's formulae given in Eq. 1.2.5, i.e. $F_{x'}$ $(\sigma_x, \dot{x}_c) = \mu(\sigma_x, \dot{x}_c)F_z$. The normal reaction force on the wheel ($F_z = (m_w + m_c/4)\,g$) modulates the R-field as a constant external signal input.

The controller model and associated mechatronic systems may be modeled in bond graph domain (see [5][11] [13] [15][23][32] for various applications) and integrated with the wheel's or full vehicle's bond graph model.

1.4.1. Regenerative Braking Algorithm. An algorithm is required to decide on how to distribute the braking force between regenerative braking and friction or antilock braking in an emergency braking situation. The emergency braking is differentiated from normal or slow braking in terms of the brake pedal position or force. The flow diagram is used here to decide how to distribute the braking forces depending on various input parameters as shown in Figure 1.4.5. If the regenerative braking force F_{reg} is less than the required maximum braking force F_b then both regenerative and anti-lock braking will work in union. If regenerative braking force F_{reg} is more than the required maximum braking force F_b then regenerative braking alone will carry out the job.

In the present study, the regenerative braking torque (τ_w) applied at the wheels is calculated as $\tau_w = i_{CVT}G\tau_{reg}W = F_{reg}r_w$ where W is the weight factor for battery state of charge (SOC) whose value is constant up to certain per cent

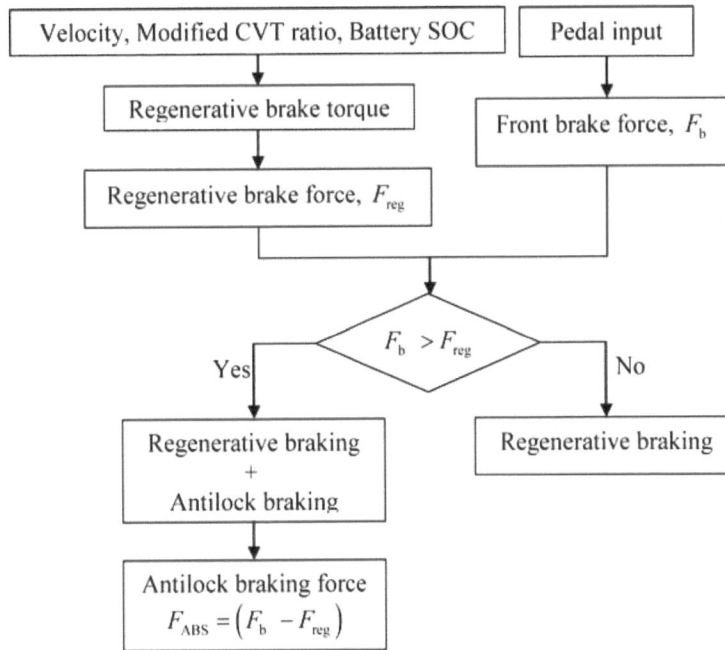

FIGURE 1.4.5. Flow diagram of regenerative braking.

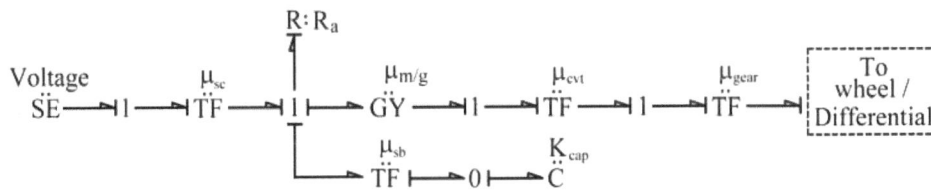

FIGURE 1.4.6. Bond graph model of regenerative braking.

of SOC and after that value of SOC, it decreases linearly [**37**]. G is the final reduction gear ratio. The value of the regenerative torque (τ_{reg}) is a constant which is determined from the motor characteristics curve. The CVT ratio (i_{CVT}) is modulated as $i_{CVT} = \omega_m r_w/(\dot{x}_v G)$, where ω_m is the desired generator speed. The desired generator speed is a fixed value such that the voltage developed is the required charging voltage, e.g. about 13.8-14.4 V charging voltage for a 12 V (12.6 V open-circuit voltage at full charge) battery. The CVT ratio is accordingly modulated to maintain constant input speed at the generator end. When the vehicle is driven at constant speed, the CVT ratio is maintained constant at 0.58 (as per data in [**37**]) so that it works like a simple reduction gear and during regeneration process, the CVT ratio is varied within the range of 0.58-2.47.

The bond graph developed for the regenerative module is shown in Figure 1.4.6. The voltage of the main battery, represented by SE-element, is used to drive the vehicle. The semiconductor switch denoted by TF-element (μ_{sc}) is modulated and it acts like a clutch. This is also used as means to model the pulse width modulation based control of input power. The power supply is disconnected ($\mu_{sc} = 0$) when brakes are applied. The R-element at the 1-junction represents armature resistance, which decides the current required for driving the vehicle and for charging the battery in the regeneration process. The regenerator-battery is modeled as a bank of high-capacity capacitors represented as a single equivalent capacitor. The motor or generator is modeled as a modulated gyrator with modulus $\mu_{m/g}$. The TF-element used to modulate the CVT ratio (μ_{cvt}) has a constant value in the forward path (i.e. motor mode) and its value is modulated continuously during regeneration process (generator mode) to maintain constant speed at the generator input. During charging and boosting (i.e. when additional power is required, the capacitor is connected to the main energy source), the charging capacitor or battery is connected to the armature by a semiconductor switch represented

TABLE 3. Parameter values for quarter-car model.

Sub-system	Parameter values			
Vehicle body	$m_{cb} = 1320\,\text{kg}$	$C_d = 0.346$	$\rho = 1.225\text{kg} \cdot \text{m}^{-3}$	$A_f = 1.964\text{m}^2$
Wheel	$m_w = 15\,\text{kg}$	$J_{wy} = 0.291\text{kgm}^2$	$r_w = 0.279\text{m}$	
tire	$C_1 = 1.029$	$C_2 = 17.16$	$C_3 = 0.523$	$C_4 = 0.03$
Regenerative braking	$G = 5.763$	$(i_{CVT})_{max} = 2.47$	$k_g = 80Ns$	$C_{RB} = 1666.67\,\text{F}$
	$R_a = 0.3\,\Omega$	$\mu_{m/g} = 0.04\,\text{V} \cdot \text{s/rad}$		

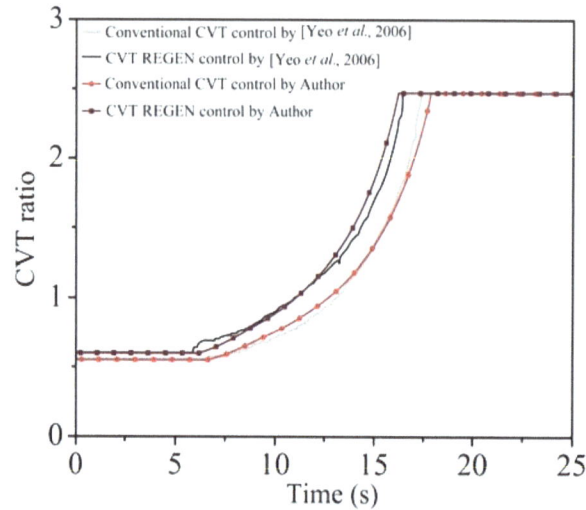

FIGURE 1.4.7. Comparison of CVT Ratio for conventional CVT control and CVT REGEN control with the experimental results of [37].

by a modulated TF-element (μ_{sb}). This semiconductor switch also protects the battery against loss of charge (when charging voltage is below a certain limit) and over charging (when charging voltage is above a certain limit).

1.4.2. Validation of Regenerative Braking. The parameter values used to simulate the quarter car model are listed in Table 3 and they are mostly taken from [37]. Because the tire model is not considered in [37], asphalt-dry road condition data listed in Table 1 are used. The regenerative braking algorithm is validated through comparison of our results with the results given in [37]. For this, the same initial angular and linear speed of wheels as taken in [37], i.e., 41.4 rad/s 11.5 m/s, respectively, are considered. This causes the quarter-vehicle model with regenerative braking system developed here and the model presented in [37].

For plotting purpose, the experimental results given in [37]] were digitized through data scanning software. Other results were obtained through numerical simulation of the model generated by combining sub-models given in Figure 1.4.4 and Figure 1.4.6. In conventional CVT control scheme, the braking torque is provided by the generator and conventional friction brakes according to the algorithm given in Figure 1.4.5 so that constant speed is maintained at the generator end. Another control scheme called CVT REGEN control is proposed in [37] wherein the input generator speed is varied dynamically (the range of variation is small) with the generator torque such that maximum generator efficiency is achieved. Those results are also validated with the present model. Although CVT REGEN control has not been used later in this chapter, it has been adopted here for model validation. Readers may consult [37] for details of optimum efficiency generator operation and the determination of the desired generator input speed.

Figure 1.4.7, 1.4.8 and 1.4.9 show the comparison between the results given in [37] and the results from the regenerative quarter-car model developed here. The CVT ratio and battery SOC with conventional CVT control and CVT REGEN control are compared with results from [37] in Figure 1.4.7 and Figure 1.4.8, respectively. The CVT ratio with REGEN control increases much faster than the conventional CVT control process. The battery state of charge increases during regeneration process. The battery SOC increases marginally more with CVT REGEN control than that with conventional CVT control. The difference of performance between the two control laws is marginal.

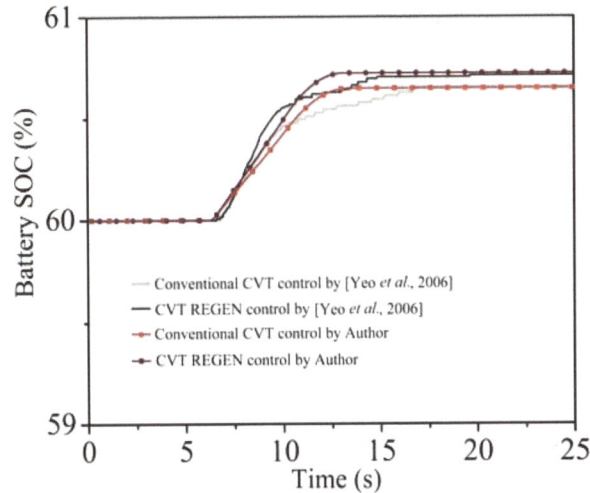

FIGURE 1.4.8. Comparison of battery SOC for conventional CVT control and CVT REGEN control with the experimental results of [**37**].

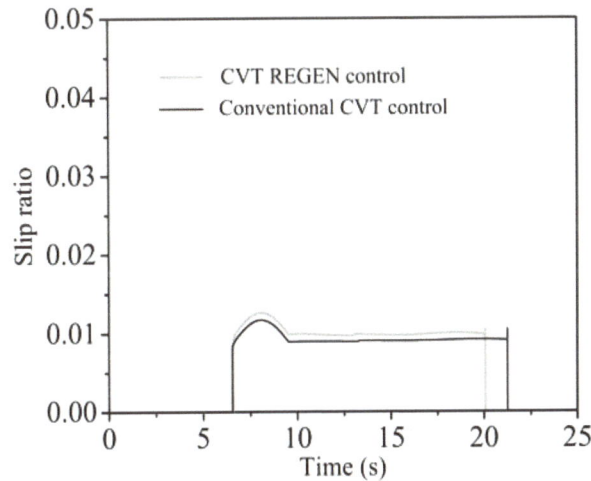

FIGURE 1.4.9. Comparison of slip ratio for conventional CVT control and CVT REGEN control with the experimental results of [**37**].

[**37**] showed that the time taken to stop the vehicle is nearly 13.6 s. As a result of this, the distance covered after applying brake is nearly 74.3 m. In this work, more importance is placed on the reduction in the stopping distance through ABS than marginal gain in energy efficiency. Therefore, CVT REGEN control is not considered further in this work.

The stopping time and distance are high because the slip ratio (see Figure 1.4.9) is small and the coefficient of friction value between the road and the wheel is very less. In order to decrease the stopping distance, slip ratio control is required. Therefore, regenerative anti-lock braking may be considered [**2**].

1.4.3. Performance of ABS with Regeneration. ABS is used to reduce the longitudinal stopping distance during emergency braking situation and to improve the maneuvering capacity during braking. In order to take the advantage of ABS, conventional braking part has been substituted by ABS in regeneration process. The validated model was simulated in this new configuration. The model parameters values are listed in Table 1 and Tables 3-4.

The vehicle forward speed was brought to a constant value of 25 m/s (90 km/h) in a straight path and thereafter ABS application on the wheel was performed. The control algorithm (see the flow chart given in Figure 1.2.3) is

TABLE 4. Additional parameter values for ABS.

Subsystem	Parameter values			
Antilock brake	$\sigma_{low} = 0.2$	$\sigma_{high} = 0.25$	$K_{ca} = 10^4$ N/m	$K_{re} = 10^6$ N/m
	$r_{bd} = 0.15$ m	$R_{lm} = 0.04$ Ns/m	$l_a = 1$ m	

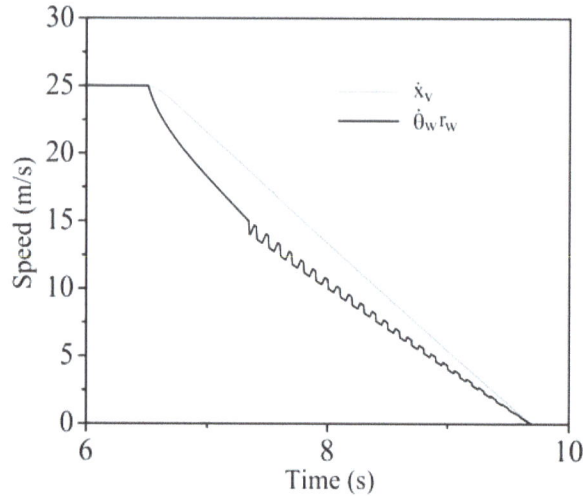

FIGURE 1.4.10. Vehicle speed and wheel speed for CVT REGEN control during braking by ABS.

implemented to restrict the slip ratio within the specified limits, i.e. 0.2 to 0.25. The CVT ratio modulation described as $i = \omega_m r_w / (\dot{x}_v G)$ in [37] can be applied when the slip ratio is small, i.e., based on the assumption that $\dot{x}_v \simeq \dot{\theta}_w r_w$. However, the generator input speed depends on the wheel's angular speed but not directly on the vehicle's linear speed. Therefore, the correct CVT ratio is described as $i = \omega_m / (\dot{\theta}_w G)$. Just after application of brake, the the CVT ratio is calculated from the expression $i = \omega_m / (\dot{\theta}_w G)$ as long as the slip ratio remains below 0.2. However, when the slip ratio reaches 0.2, ABS action produces fluctuating wheel speed. Thus, the CVT ratio would fluctuate and damage the CVT. To keep CVT ratio variation as smooth as possible, it is modulated as $i = \omega_m r_w / (\dot{x}_v G)$ for all values of slip ratio above 0.2 because the vehicle speed variation remains smooth. Note that this CVT ratio variation during the ABS phase does not keep the generator input speed constant. However, that is a price one has to pay as a compromise. During regeneration process, the CVT ratio is varied within the range 0.58-2.47 [37].

    Figure 1.4.10 shows the variation of the wheel linear and angular velocities during an emergency (sudden) braking. The CVT ratio and battery SOC with CVT REGEN control are shown in Figure 1.4.11 and Figure 1.4.12, respectively. The variation of slip ratio and generator input speeds are shown in Figure 1.4.13.

    The approximate braking time for complete stoppage of the car is 3.2s (Figure 1.4.10). At the same time, the about one percent increase in battery SOC is also a good performance. Note that if original battery SOC is small then braking from the same initial speed to complete rest produces more change in battery SOC as compared to the case when battery is initially at higher SOC. For very large battery SOC, no further charging may be possible and regenerative braking fails. The performance of ABS with regeneration is comparable with conventional braking system. But it is seen that the generator input speed (Figure 1.4.13) fluctuates during regeneration process. If we want to keep generator input speed smooth then the CVT ratio has to fluctuate. The latter is a more unwanted phenomenon. This is why regenerative braking with SMC based ABS is considered next.

1.4.4. Performance of SMC based ABS with Regeneration. The sliding mode ABS control produces smoother variations in wheel rotational speed as compared to the conventional ABS system, thereby improving passenger comfort. This also means it is possible to obtain almost constant input speed at the generator input while smoothly varying CVT ratio. This concept was verified through simulation of the quarter car model. Because the wheel speed varies smoothly with SMC based ABS, the CVT ratio is calculated from the expression $i = \omega_m / (\dot{\theta}_w G)$ for all values of slip

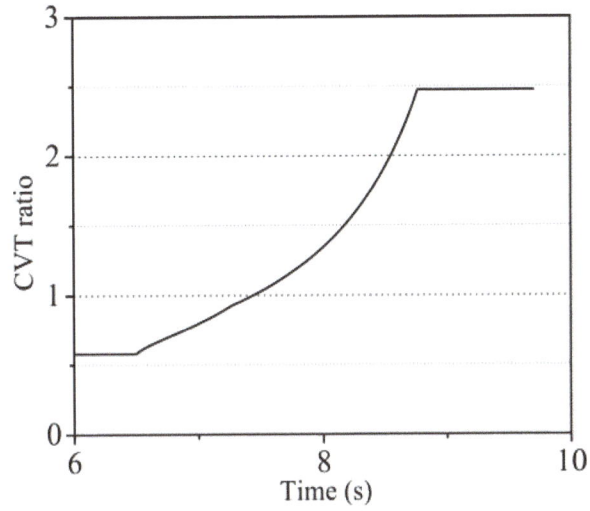

FIGURE 1.4.11. CVT ratio for CVT REGEN control during braking by ABS.

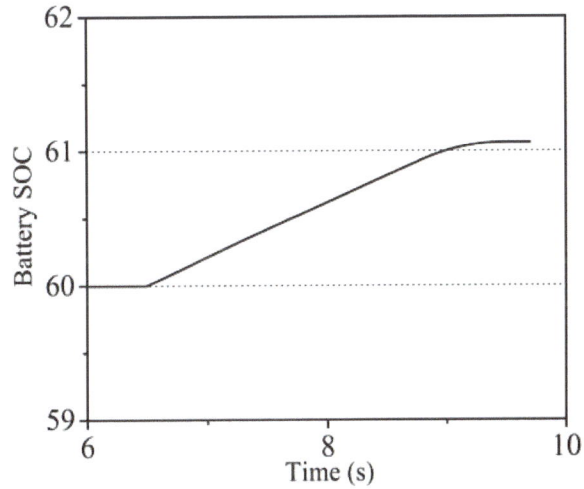

FIGURE 1.4.12. battery SOC for CVT REGEN control during braking by ABS.

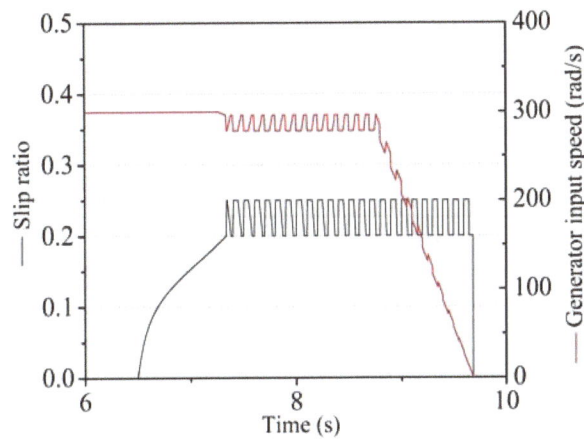

FIGURE 1.4.13. slip ratio and generator input speed for CVT REGEN control during braking by ABS.

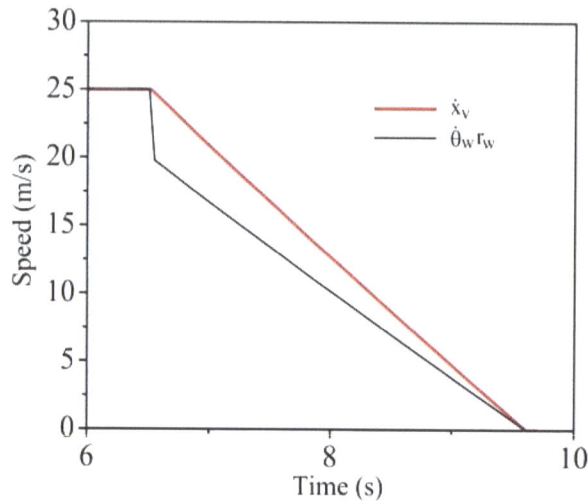

FIGURE 1.4.14. Vehicle speed and wheel speed for CVT REGEN control during braking by SMC based ABS.

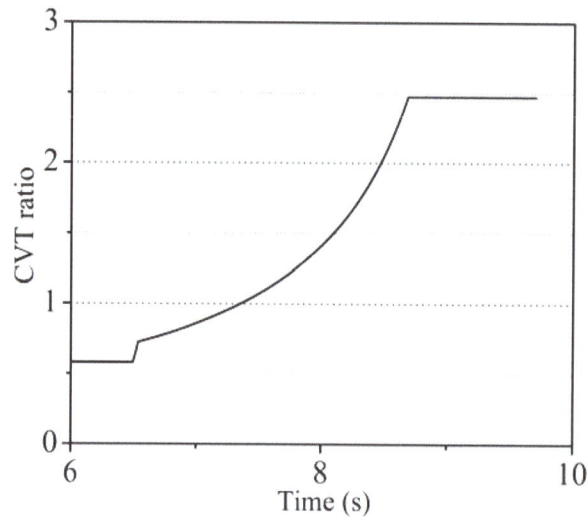

FIGURE 1.4.15. CVT ratio for CVT REGEN control during braking by SMC based ABS.

ratio. The vehicle speed was brought to a constant value of 25 m/s (90 km/h) and then front wheel SMC ABS brake with regeneration was used. The SMC tries to keep the slip ratio at the optimal value 0.2 while keeping the braking force bounded within an upper and a lower force limits till the sliding surface is reached around 0.04s after application of brakes. The initial action of SMC ABS is similar to conventional braking till the vehicle slip ratio reaches 0.2 after which the sliding mode controller truly takes over.

Simulation result reveals that the approximate braking time for complete stoppage of the car is 3.1 s (see Figure 1.4.14). Initially, there is a sharp fall in the rotational speed which is due to the initial conventional braking. The CVT ratio and battery SOC with CVT REGEN control are shown in Figure 1.4.15 and Figure 1.4.16, respectively. The slip ratio variation is shown in Figure 1.4.17.

The change CVT ratio reaches limiting value slightly faster (0.08 s less) than that for regenerative braking with conventional ABS. The battery state of charge also increases slightly more than that for regenerative braking with conventional ABS. The input speed of the generator is maintained almost constant between 296.36-300.02 rad/s as shown in Figure 1.4.17.

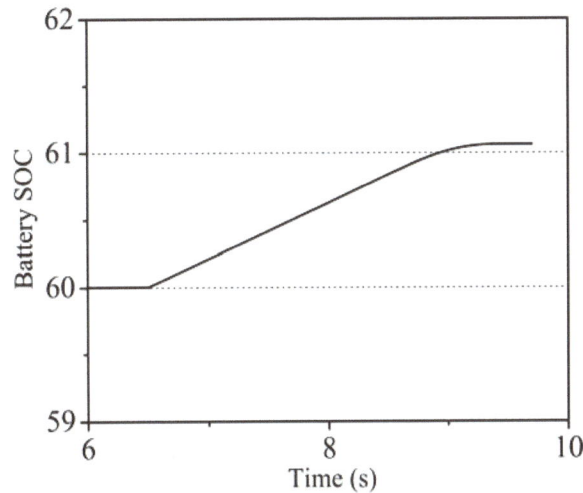

FIGURE 1.4.16. battery SOC for CVT REGEN control during braking by SMC based ABS.

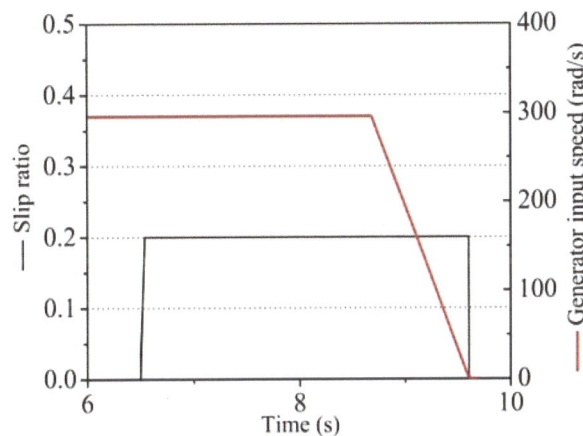

FIGURE 1.4.17. slip ratio and generator input speed for CVT REGEN control during braking by SMC based ABS.

The time taken for stopping the car and the stopping distance are lesser and increase in battery SOC is slightly more when regenerative braking with SMC ABS is used as compared to regenerative braking with conventional ABS (Figure 1.4.10, 1.4.11,1.4.12,1.4.13). At the same time, SMC ABS maintains almost constant generator input speed (Figure 1.4.17) during regeneration process. Therefore, the performance of SMC based ABS with regeneration is far better than using regeneration with conventional ABS.

The effect of load transfer on the performance of the braking system cannot be studied with the quarter car model. Moreover, after satisfactory performance of the controller in the simple quarter car model, it may be ported to the full vehicle model. Therefore, the performance of combined SMC based ABS with regenerative braking is discussed next by using a four-wheel vehicle model.

1.5.   Four Wheel Vehicle Model

A road vehicle is a multibody system. The vehicle body, wheel, axle, gearbox, differential, etc can be each considered as a 6-DOF rigid body. These rigid bodies are constrained by means of various joints. In the present case, the degrees-of-freedom of the system have been reduced by neglecting trivial dynamics. It is assumed that the engine, differential, etc are rigidly mounted on the vehicle body. Wheels are rigidly fixed to the axles which are attached with the body through suspension systems. The schema of the four-wheel vehicle is given in Figure 1.5.1.

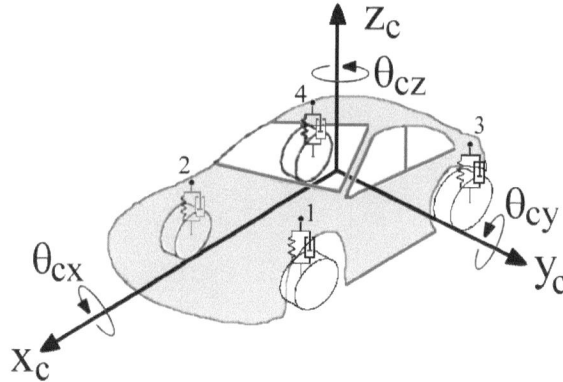

FIGURE 1.5.1. Four-wheel vehicle model.

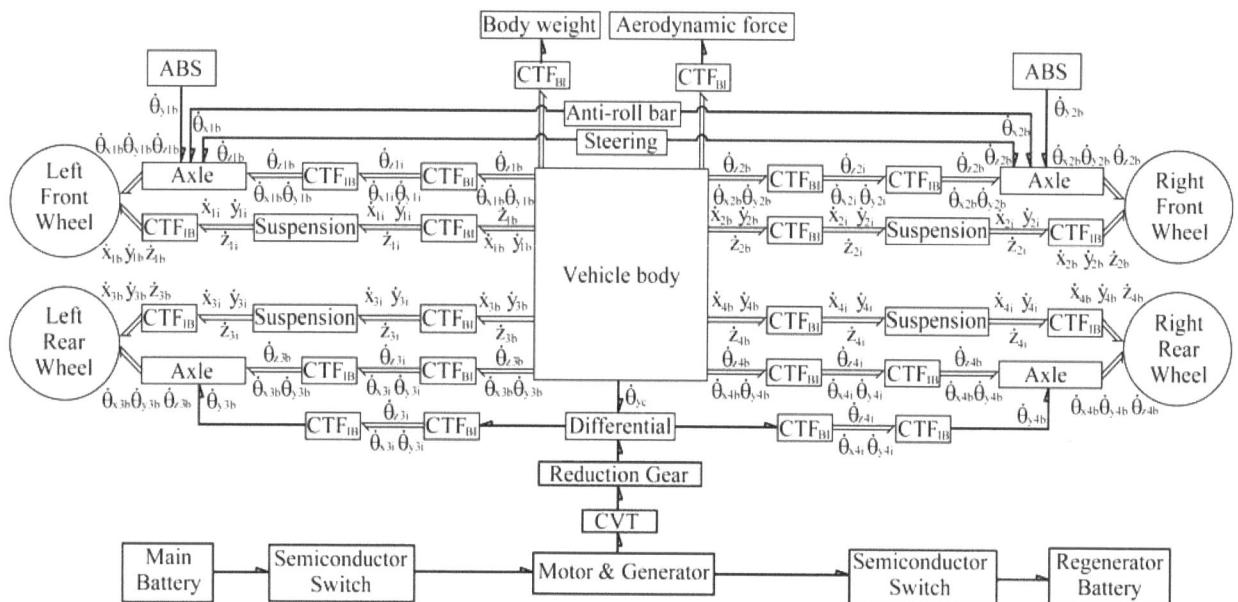

FIGURE 1.5.2. Word bond graph of full vehicle model.

1.5.1. Word Bond Graph Representation. The word bond graph of the four-wheel vehicle is given in Figure 1.5.2 where multi-bonds are represented by two parallel lines. At this point, an electric vehicle is considered; an internal combustion engine model will be developed later. The word bond graph and the detailed bond graph model are based on the models developed in [**8**][**14**][**29**]. In the word bond graph, the global system is decomposed into seven sub-systems. These are: wheel, suspension, steering, differential drive, antilock braking system, vehicle body and regenerative braking system. The flow variables between different sub-systems are shown in Figure 1.5.2. The complimentary power variables or generalized effort variables (force for linear velocity and torque for angular velocity) are not shown in Figure 1.5.2 for maintaining clarity of the figure. The vehicle body is connected with the four wheels through suspensions. The ABS and steering are connected with the axle by scalar bonds. Likewise, the rear wheels and the vehicle body are connected to the differential through scalar bonds. The CTF-blocks indicate coordinate transformations from one frame to the other. The engine and the gearbox are not shown here as they are disengaged during braking although they are included in the full model. The reduction of the integrated model is further possible for specific operational conditions [**9**].

1.5.2. Vehicle Body. The vehicle body is considered as a 6-DOF rigid body. Wheel-axles and wheels are attached with the body. Wheel-axles are attached with the body through suspension systems. Three linear displacements along body-fixed x, y and z axes (see Figure 1.5.1) and rotations about same axes describe the motion of the vehicle body. The three Cardan angles define the orientation of the vehicle. The Newton-Euler equations of the vehicle body [8] with attached body fixed axes x,y,z aligned with the principal axes of inertia are given by

$$(1.5.1) \qquad \sum F_{\mathrm{x}} = m_{\mathrm{cb}}\ddot{x}_{\mathrm{cb}} + m_{\mathrm{cb}}(\dot{z}_{\mathrm{cb}}\dot{\theta}_{\mathrm{cby}} - \dot{y}_{\mathrm{cb}}\dot{\theta}_{\mathrm{cbz}})$$

$$(1.5.2) \qquad \sum F_{\mathrm{y}} = m_{\mathrm{cb}}\ddot{y}_{\mathrm{cb}} + m_{\mathrm{cb}}(\dot{x}_{\mathrm{cb}}\dot{\theta}_{\mathrm{cbz}} - \dot{z}_{\mathrm{cb}}\dot{\theta}_{\mathrm{cbx}})$$

$$(1.5.3) \qquad \sum F_{\mathrm{z}} = m_{\mathrm{cb}}\ddot{z}_{\mathrm{cb}} + m_{\mathrm{cb}}(\dot{y}_{\mathrm{cb}}\dot{\theta}_{\mathrm{cbx}} - \dot{x}_{\mathrm{cb}}\dot{\theta}_{\mathrm{cby}})$$

$$(1.5.4) \qquad \sum M_{\mathrm{x}} = J_{\mathrm{cbx}}\ddot{\theta}_{\mathrm{cbx}} - \dot{\theta}_{\mathrm{cby}}\dot{\theta}_{\mathrm{cbz}}(J_{\mathrm{cby}} - J_{\mathrm{cbz}})$$

$$(1.5.5) \qquad \sum M_{\mathrm{y}} = J_{\mathrm{cby}}\ddot{\theta}_{\mathrm{cby}} - \dot{\theta}_{\mathrm{cbz}}\dot{\theta}_{\mathrm{cbx}}(J_{\mathrm{cbz}} - J_{\mathrm{cbx}})$$

$$(1.5.6) \qquad \sum M_{\mathrm{z}} = J_{\mathrm{cbz}}\ddot{\theta}_{\mathrm{cbz}} - \dot{\theta}_{\mathrm{cbx}}\dot{\theta}_{\mathrm{cby}}(J_{\mathrm{cbx}} - J_{\mathrm{cby}})$$

where F_{x}, F_{y} and F_{z} are the forces acting on the vehicle body in x, y and z directions, respectively. M_{x}, M_{y} and M_{z} are the moments acting on the vehicle body in x, y and z directions, respectively. m_{cb} is the mass of the vehicle body. $\dot{x}_{\mathrm{cb}}$, $\dot{y}_{\mathrm{cb}}$ and $\dot{z}_{\mathrm{cb}}$ are the velocities of the vehicle body in x, y and z directions, respectively. $\ddot{x}_{\mathrm{cb}}$, $\ddot{y}_{\mathrm{cb}}$ and $\ddot{z}_{\mathrm{cb}}$ are the accelerations of the vehicle body in x, y and z directions, respectively. J_{cbx}, J_{cby} and J_{cbz} are the polar moment of inertia of the vehicle body in x, y and z directions, respectively. $\dot{\theta}_{\mathrm{cbx}}$, $\dot{\theta}_{\mathrm{cby}}$ and $\dot{\theta}_{\mathrm{cbz}}$ are the angular speed of the vehicle body in x, y and z directions, respectively. The equations for three linear velocities ($\dot{x}_1, \dot{y}_1, \dot{z}_1$) of left-front suspension reference point (x_1, y_1, z_1) (point 1 in Figure 1.5.3) in the moving system of axes are given below:

$$(1.5.7) \qquad \dot{x}_1 = \dot{x}_{\mathrm{cb}} + z_1\dot{\theta}_{\mathrm{cby}} - y_1\dot{\theta}_{\mathrm{cbz}}$$

$$(1.5.8) \qquad \dot{y}_1 = \dot{y}_{\mathrm{cb}} + x_1\dot{\theta}_{\mathrm{cbz}} - z_1\dot{\theta}_{\mathrm{cbx}}$$

$$(1.5.9) \qquad \dot{z}_1 = \dot{z}_{\mathrm{cb}} + y_1\dot{\theta}_{\mathrm{cbx}} - x_1\dot{\theta}_{\mathrm{cby}}$$

Similarly, the equations for three linear velocities of other suspension reference points can be calculated.

The velocity transformation from the non-inertial frame to the inertial frame can be written using successive multiplication of rotation matrices as follows:

$$(1.5.10) \qquad \left\{ \begin{array}{c} \dot{X} \\ \dot{Y} \\ \dot{Z} \end{array} \right\} = \mathbf{T}_{\psi,\theta,\phi} \left\{ \begin{array}{c} \dot{x} \\ \dot{y} \\ \dot{z} \end{array} \right\} \text{ and } \left\{ \begin{array}{c} \omega_{\mathrm{X}} \\ \omega_{\mathrm{Y}} \\ \omega_{\mathrm{Z}} \end{array} \right\} = \mathbf{T}_{\psi,\theta,\phi} \left\{ \begin{array}{c} \omega_{\mathrm{x}} \\ \omega_{\mathrm{y}} \\ \omega_{\mathrm{z}} \end{array} \right\}$$

$$(1.5.11) \qquad \text{where } \mathbf{T}_{\psi,\theta,\phi} = \underbrace{\begin{bmatrix} \cos\psi & -\sin\psi & 0 \\ \sin\psi & \cos\psi & 0 \\ 0 & 0 & 1 \end{bmatrix}}_{\mathbf{T}_{\psi}} \underbrace{\begin{bmatrix} \cos\theta & 0 & \sin\theta \\ 0 & 1 & 0 \\ -\sin\theta & 0 & \cos\theta \end{bmatrix}}_{\mathbf{T}_{\theta}} \underbrace{\begin{bmatrix} 1 & 0 & 0 \\ 0 & \cos\phi & -\sin\phi \\ 0 & \sin\phi & \cos\phi \end{bmatrix}}_{\mathbf{T}_{\phi}}$$

and ψ, θ and φ are the Z-Y-X Cardan angles (Euler angles). $\dot{x}, \dot{y}$ and $\dot{z}$ are the velocities in the moving frame in respective directions. $\dot{X}, \dot{Y}$ and $\dot{Z}$ are the velocities in the inertial frame in respective directions. ω_{x}, ω_{y} and ω_{z} are the angular velocities in the moving frame in respective directions. ω_{X}, ω_{Y} and ω_{Z} are the angular velocities in the inertial frame in respective directions. A bond graph junction structure (CTF or coordinate transformation block) [9] represents this power conserving Coordinate Transformation. Similarly, the transformation of velocities in inertial frame to body-fixed frame can be represented by a transformer junction structure [9]. The CTF block simultaneously transforms forces in inertial frame such as suspension forces to the body-fixed frame.

The Cardan angles (Euler angles)are used in Eq. 1.5.11. The body-fixed angular velocities can be transformed to Euler angle rates as follows:

$$
(1.5.12) \quad
\left\{ \begin{array}{c} \dot{\phi} \\ \dot{\theta} \\ \dot{\psi} \end{array} \right\}
=
\left[\begin{array}{ccc} 1 & 0 & -\sin\theta \\ 0 & \cos\phi & \cos\theta\sin\phi \\ 0 & -\sin\phi & \cos\theta\cos\phi \end{array} \right]^{-1}
\left\{ \begin{array}{c} \omega_x \\ \omega_y \\ \omega_z \end{array} \right\}
=
\left[\begin{array}{ccc} 1 & \tan\theta\sin\phi & \tan\theta\cos\phi \\ 0 & \cos\phi & -\sin\phi \\ 0 & \sin\phi/\cos\theta & \cos\phi/\cos\theta \end{array} \right]
\left\{ \begin{array}{c} \omega_x \\ \omega_y \\ \omega_z \end{array} \right\}
$$

A transformer junction structure (EATF or Euler angle transformation block) represents the transformations given in Eq. 1.5.12 in bond graph form. The Euler angles are used in all coordinate transformations, i.e. in CTF and EATF bond graph sub-models and they are obtained by integrating the Euler angle rates defined in Eq.1.5.12. Note that , $\omega_x = \dot{\theta}_{cx}$, $\omega_y = \dot{\theta}_{cy}$, $\omega_z = \dot{\theta}_{cz}$, $I_{xx} = J_{cx}$, $I_{yy} = J_{cy}$, and $I_{zz} = J_{cz}$ for the vehicle body. The suspension reference point 1 coordinates are $x_1 = a$, $y_1 = c$ and $z_1 = -h$ where a, c and h are the distance of front axle from the vehicle cg, half of track width and height of vehicle cg from suspension reference point, respectively. The other three reference points are similarly expressed and used in kinematic relations.

It is assumed that the aerodynamic effects, although considered in the model, is not very significant and that the vehicle is symmetrical with respect to its longitudinal axis. The vehicle body inertia and transformations of the three linear and three angular velocities into velocities at relevant suspension reference points are modeled in Figure 1.5.3. The vehicle is considered as a six-degrees-of-freedom rigid body. The motion of the rigid body is expressed in a coordinate frame rotating and translating with the rigid body. The inertial principal axes are assumed to be aligned with this local coordinate frame attached at the center of mass of the body. A pair of gyrator rings (Euler junction structure ([17][26]) couple the inertias. One gyrator ring belongs to the translational and the other belongs to the rotational velocities, according to Eqs. 1.5.1-1.5.6. The Euler angle calculations are not shown explicitly in the figure. However, they are present in the actual model and are used for various coordinate transformations.

The vehicle body is subjected to three sets of forces and moments. The body weight and aerodynamic forces (modeled by R_{aero}-element) which are in the inertial frame act on the vehicle body in non-inertial frame through coordinate transformation (CTF). Moreover, the engine torque produced in the vehicle-body frame is twice transformed to act on the wheel through the differential and the reactive torque is applied on the vehicle body. Finally, the moments and suspension forces (constraint forces) which are expressed in the inertial frame are transformed to obtain their components along the vehicle body fixed coordinates. The forces at the suspension reference points are multiplied with the respective moment arms to generate the moments that are acting on the vehicle body. The suspension forces acting at the center of gravity of the body are obtained by adding the forces at all the four corners (suspension reference points) of the body.

1.5.3. Wheel. The most important parts of wheels are their tire stiffness, radius, rotary inertia and mass. As the tire forces and moments significantly influence the vehicle dynamics, tire is the most significant part in a wheel model. The vehicle is controlled by tire forces. There are three tire models i.e. analytical, empirical and physical models. The tire elastic deformation and tire forces [24]can be predicted from the physical models. In physical models, the equations of motion are solved through complex numerical methods. Analytical models are unsuitable for combined slip and large slip conditions. The more accurate model is the empirical model based on experimental correlations [19]. The longitudinal self-aligning moment and cornering and longitudinal forces can be computed from the generally accepted Pacejka's magic formula [30].

Some distinctions can be made between the modeling of vehicle body and the tire. The suspension reference points in the vehicle body are fixed points, whereas, the tire-road contact point is not a fixed point on the wheel. Moreover, the orientation of the wheel body-fixed frame with respect to the inertial frame is unaffected by the rotation of the wheel about its axle. The wheel and its axle are considered as axisymmetric bodies.

The wheel's bond graph model is given in Figure 1.5.4. The wheel is considered as a six degrees-of-freedom rigid body. Similar to the modeling of vehicle body, the inertias are coupled by a pair of gyrator rings for the translational $(\dot{x}, \dot{y}, \dot{z})$ and rotational velocities $\left(\dot{\theta}_x, \dot{\theta}_y, \dot{\theta}_z \right)$. The wheel is represented by m_w. The polar moment of inertias of the wheel in respective directions are J_{wx}, J_{wy} and J_{wz}. The self-weight and ground contact force act along inertial

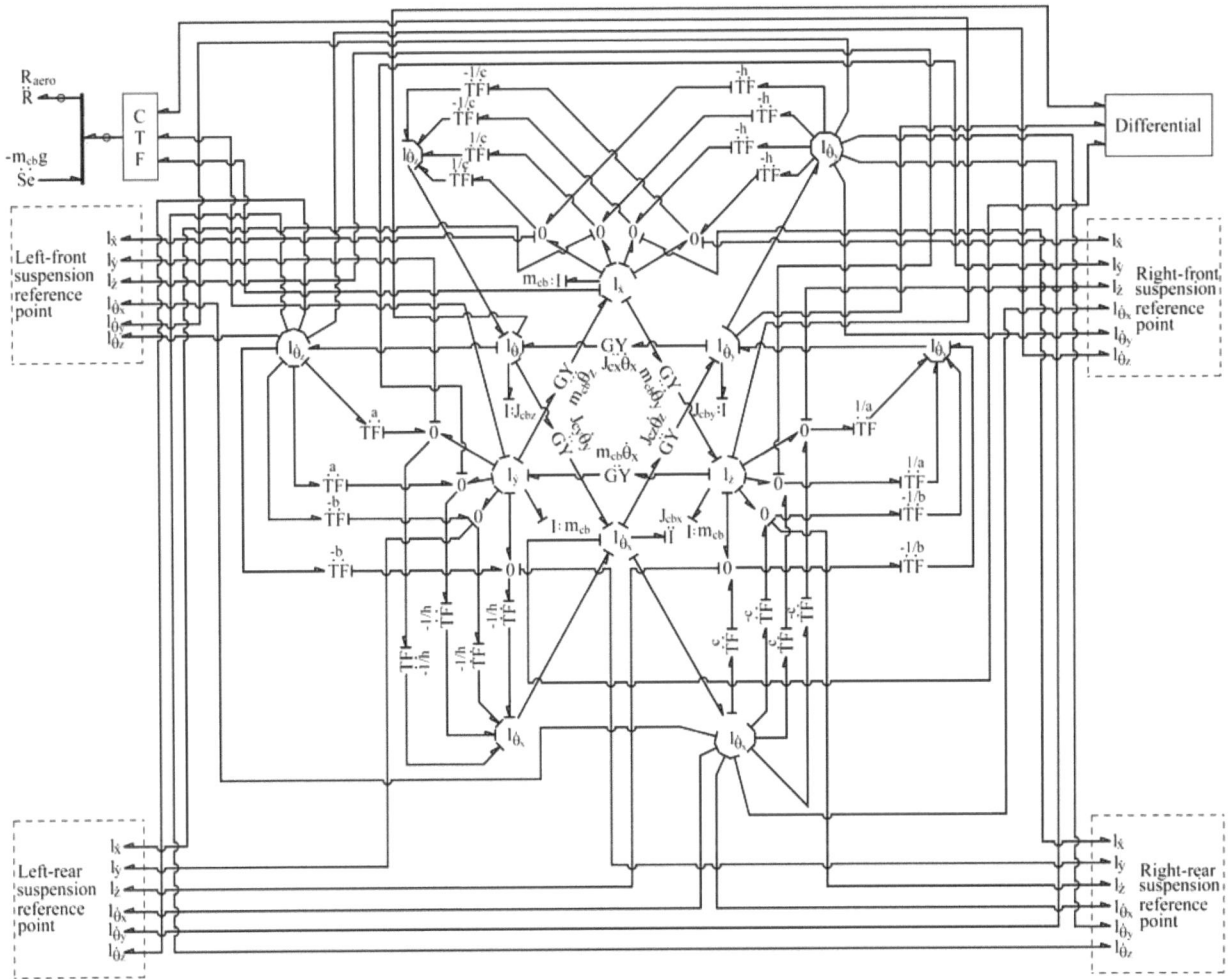

FIGURE 1.5.3. Bond graph model of vehicle body.

Z-axis. In this way, the wheel longitudinal and cornering dynamics are decoupled from vertical dynamics. The longitudinal and lateral forces acting on the wheel are represented by F_x, F_y, respectively. The longitudinal and cornering dynamics are modulated by the ground reaction force (F_z) and wheel radius (r_w). The vertical velocity of the wheel is obtained through a set of transformations at a 0-junction. This junction is connected to a 1-junction where the vertical stiffness (K_t) and damping (R_t) of tire are modeled. The overturning moment, rolling resistance moment and self-aligning moment are M_x, M_y and M_z, respectively. The angle ϕ which is the wheel tilt with respect to the ground is used in the transformations to obtain wheel vertical motion at contact patch and to model self-weight of the wheel and axle. This angle is the angle of rotation of the wheel about X-axis of a momentarily conceived inertial frame whose Z-axis is along inertial vertical direction and y-axis is aligned along the axle. Additional camber may be included by the tilt angle. The transformer structures perform the coordinate transformations as shown in the model.

The normal reaction in the inertial frame modulates the MR-element which is used for modeling road-tire interactions. The vertical dynamics of un-sprung mass consider kinetic phenomenon and weight of the wheel. The wheel longitudinal and cornering dynamics consider tire-road interactions and braking action. The vertical load and lateral slip angle are required for the cornering force and self-aligning moment. The longitudinal force is dependent on vertical load and longitudinal slip rate. Burckhardt formulae [27] or the composite slip based formulation [36] describe the characteristic relations for the MR-element.

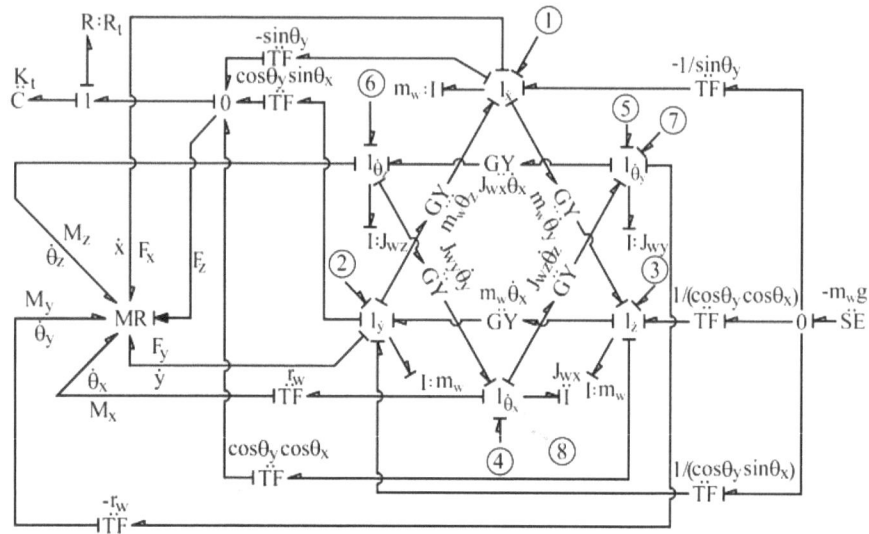

FIGURE 1.5.4. Bond graph model of a wheel.

The corresponding velocities of the suspension reference points are connected to ports 1-6 shown in Figure 1.5.4. The port 7 of front wheel is connected to the brake and similarly, the port 7 of rear wheel is connected to the engine. Port 8 (shown as by dotted bond)is used to interface the front wheel model to the anti-roll bar model. The anti-roll bar constrains the relative roll between two front wheel axles. The steering torque from the steering column model is applied at port 4. The modeling of other components are not discussed here which can be consulted in [**3**].

1.5.4. Antilock Braking System in Four Wheel Model. The model parameters used in simulations for full vehicle model are given in Table 5. J_{sw}, δ and τ are the polar moment of inertia of the steering wheel, steering angle and torque applied to the steering wheel, respectively. A_{sp} and R_{sp} are the area and frictional damping of the engine cylinder, respectively. r_{ck} and l_c are radius of crank rod and length of the connecting rod, respectively. C_{d_i} and C_{d_o} are the coefficient of discharge of inlet and outlet valve, respectively. A_{V_i} and A_{V_o} are the area of the inlet and outlet valve, respectively. m_{bo} and $C_{P_{bo}}$ are the mass and specific heat capacity of the cylinder body, respectively. λ_{gb} and λ_{ba} are the overall heat transfer coefficient between gas and the body and between the body and the environment, respectively. J_{fw} and ω_{st} are the polar moment of inertia of the flywheel and fixed angular velocity of the self-motor, respectively. P_{atm} and T_{atm} are the atmospheric pressure and temperature, respectively. C_v, γ and R' are the specific heat at constant volume, ratio of specific heats and specific gas constant, respectively. $\dot{m}_{fu}$ and q_{hc} are fuel mass injection rate and heat of combustion, respectively. The valve timing diagram and engine model are not shown here. They may be consulted in [**1**]. Note that the engine model is as such not necessary for brake system simulation because the vehicle and wheel inertias can be given the initial velocities as initial conditions and it can be assumed that the drives are disengaged.

In all following simulations, dry asphalt road condition has been considered. From Figure 1.2.4, it is seen that in case of dry asphalt road condition, the coefficient of friction reaches maximum when the slip ratio is nearly 0.2. A constant vehicle forward velocity of 33.3 m/s (120 km/h) was achieved along a straight path after which the ABS was activated on the front wheels. Figure 1.5.5 shows the changes in vehicle forward speed and wheel angular speed under emergency braking using the ABS. The result in Figure 1.5.6 shows that the slip ratio lies between the desired limits, i.e. 0.2 to 0.25 (the sweet-spot of the slip-friction curve). The quick switching of braking force is shown in Figure 1.5.7. The time taken for stopping the car is approximately 4.16s.

1.5.5. Sliding Mode Control in Four Wheel Model. Pulsating effect, which causes passenger discomfort and also leads to quick wear and fatigue failure of brake system components, is one of the major issues in antilock braking system while in operation (see Figure 1.5.10). To reduce pulsating effect but to maintain the same performance parameters, sliding mode controller was studied and the brake torque requirement was calculated. Dynamic responses

TABLE 5. Additional parameter values for ABS.

Subsystem	Parameter values			
Vehicle body	m_{cb}= 1600 kg	J_cbx = 260 kg m^2	J_cby = 1110 kg m^2	J_cbz = 1370 kg m^2
	$a = 0.9$ m	$b = 1.5$ m	$c = 0.7$ m	$h = 0.1$ m
Suspension	$K_{sx} = 10^7$ N/m	R_{sx}=2000 N s/m	$K_{sy} = 10^{74}$ N/m	R_{sy}=2000 N s/m
	$K_{sz} = 80^3$ N/m	$K_{stx} = 10^7$ Nm/rad	$K_{sty} = 0$ Nm/rad	R_{stx}= 2000 Nm s/rad
	$R_{sty} = 0$ Nm s/rad	$K_{stz} = 10^6$ Nm/rad	R_{sz}= 500 N s/m	R_{stz}=360 Nm s/rad
Wheel	m_w= 15 kg	$J_wx = 0.15$ kg m^2	$J_wy = 0.2$ kg m^2	$J_w = 0.1$ kg m^2
	r_w= 0.3 m			
tire	$K_t = 3.05^5$ N/m	R_t= 200 N s/m		
	$C_1 = 1.029$	C_2= 17.16	C_3= 0.523	C_4= 0.03
Brake	σ_{low}= 0.2	σ_{high}= 0.25	s_g= 0.01	k_g= 250 Nm
	r_{bd}=0.15 m	R_{ml1}= 0.04 Ns/m	$K_{ca} = 10^4$ N/m	$K_R = 10^6$ N/m
	l_a= 1 m	τ_m=user control	F_{pedal}=user control	
Steering wheel	$J_sw = 1$ kg m^2	δ=user control	τ=user control	
IC Engine	$A_{sp} = 5 \times 10^{-3}$ m^2	R_{sp}=0	r_{ck1}= 0.08 m	l_c= 0.2 m
	$C_{d_i} = 0.1115$	$C_{d_o} = 0.01$	$A_{V_i} = 0.008$ m^2	$A_{V_o} = 0.008$ m^2
	m_{bo}= 10 kg	$C_{P_{bo}} = 460 Jkg^{-1}K^{-1}$	$\lambda_{gb} = 1 Wm^{-2}/K$	$\lambda_{ba} = 25 Wm^{-2}/K$
	$J_fw = 10$ kg m^2	$\omega_{st} = 10 rad/s$	$P_{atm} = 10^5 N/m^2$	T_{atm}= 300 K
	C_v= $720 Jkg^{-1}K^{-1}$	$R' = 287.1 Jkg^{-1}K^{-1}$	$\gamma = 1.41$	$\dot{m}_{fu}$ = usercontrol
	$q_{hc} = 4.73 \times 10^7 Jkg^{-1}$			

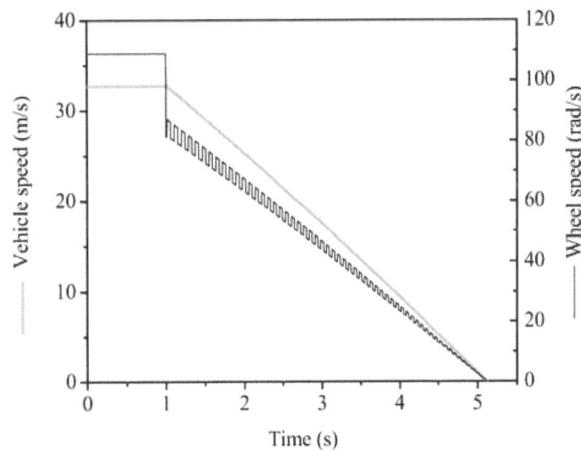

FIGURE 1.5.5. Vehicle and wheel Speed for antilock braking system in four wheel model.

are obtained to investigate the performance of the proposed sliding mode controller. It has been assumed that the optimal slip (0.2) is a known parameter.

Initially, a constant vehicle forward speed of 13.88 m/s was attained and then brakes are applied to the front wheels at 12 s and stopping time is about 1.544 s (see Figure 1.5.8). Initially, there is a sharp fall in the rotational speed (at $t = 12s$) which is due to conventional braking. The controller tries to keep the slip ratio at 0.2 while keeping the braking force bounded within an upper and a lower force limits till the sliding surface is reached around 0.06 s after application of brakes (see the inset in Figure 1.5.9). After the sliding surface is reached, the braking force increases smoothly (see Figure 1.5.9) and there is no pulsating force. It is seen from the results given in Figure 1.5.10 that the slip ratio is maintained at the desired value of 0.2 after the application of the brake till the vehicle comes to complete halt.

It is found that ABS with sliding mode control takes a little longer distance and time to bring the vehicle to complete halt when compared with conventional ABS as shown in Figure 1.5.11. The duration of pulsating effects (of wheel's angular acceleration and deceleration) due to braking is smaller in sliding mode control. This has a profound effect other than simply improving the passenger comfort. Brakes are too often used in urban driving conditions

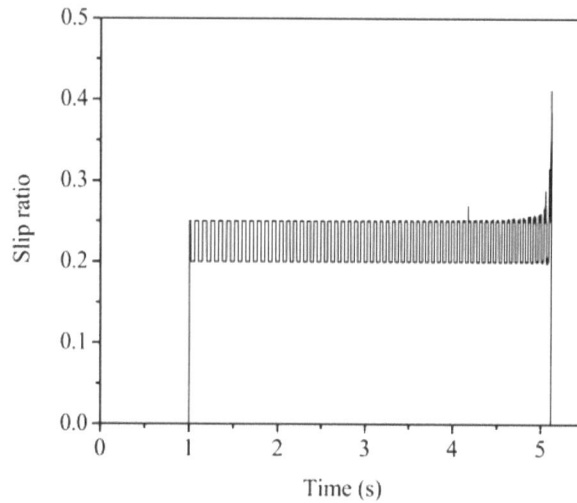

FIGURE 1.5.6. Wheel slip ratio for antilock braking system in four wheel model.

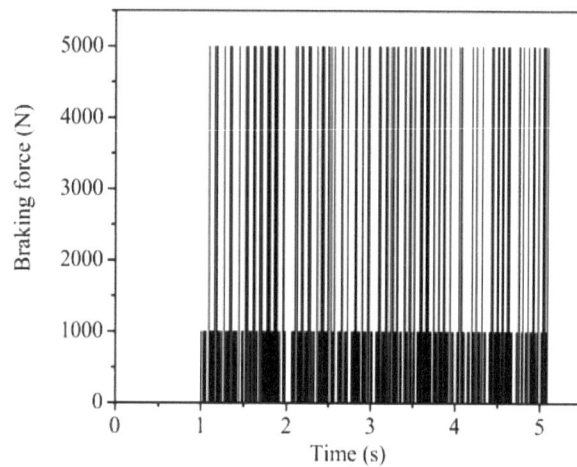

FIGURE 1.5.7. Braking force for antilock braking system in four wheel model.

with generally chaotic and congested traffic. Reduction in physiological and psychological fatigue of driver and improvement in brake component life/reliability lead to enhancement of the total road safety.

As an extension of this work, various other brake systems like ABS regenerative braking and sliding mode ABS regenerative braking were considered. The results showed that sliding mode ABS regenerative braking gives improved fuel efficiency, passenger comfort and increases the life of the CVT. Those results are not included in this chapter. Those may be consulted in [2].

1.6. Conclusions

The detailed bond graph models of different components of a vehicle, especially the braking system are developed in this chapter. The controllers for ABS are designed and tested. The procedure for analysis, modeling and simulation of ABS controller which can restrain the wheel slip between desired thresholds are developed. An equivalent mechanical braking system of the actual hydraulic system of ABS was used for modeling purpose. The control laws and vehicle models have been thoroughly validated through comparison of the simulation results with experimental and numerical results reported in various literatures.

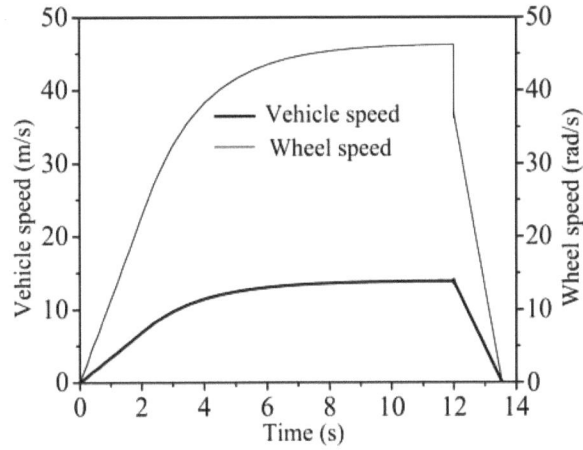

FIGURE 1.5.8. Vehicle and wheel Speed for sliding mode control in four wheel model.

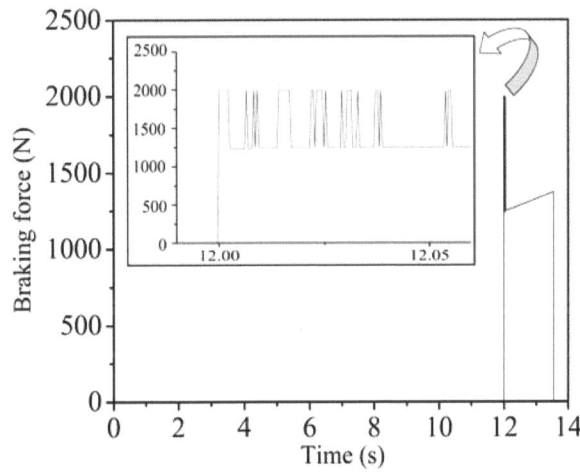

FIGURE 1.5.9. Braking force for sliding mode control in four wheel model.

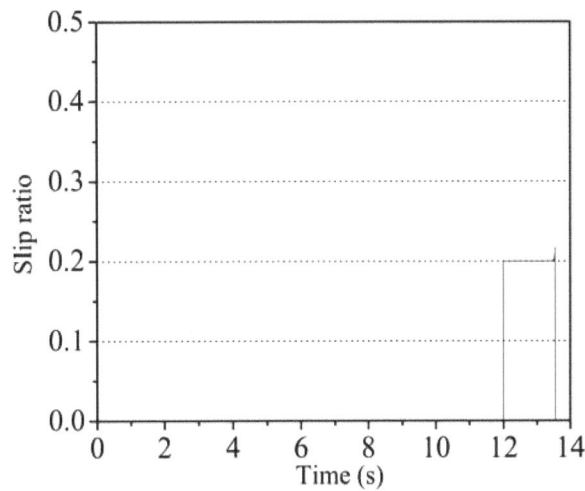

FIGURE 1.5.10. Wheel slip ratio for sliding mode control in four wheel model.

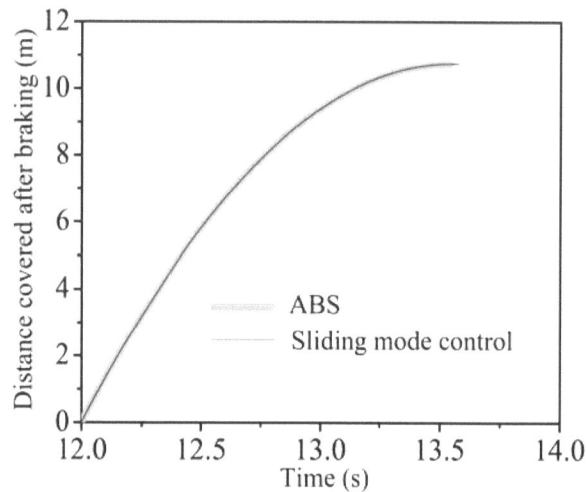

FIGURE 1.5.11. Distance covered after braking for ABS and sliding mode control.

The algorithm and model developed to design the controller are extremely good to account for different road and operating conditions. Furthermore, as the bond graph approach is a multi-disciplinary modeling tool, it is convenient to integrate models in other energy domains such as fuel cells and electrical drives into the presented model. Simulation results for adverse operating conditions indicate that the actuation rate has to be very quick. The maximum brake force and rate of actuation (actuator slew rate) are important parameters during mechanical design in order to achieve required response time and peak force. In addition, these design input parameters are critical for designing the brake system for reduced wear and increased fatigue strength. The parameters of the developed simulation model can be changed to obtain different vehicle configurations and thus the developed simulation model is useful as a controller prototype and design platform. The developed integrated control and physical system numerical prototyping approach minimizes costly field trials at the initial stages of tuning of the control algorithms and sizing of the actuators.

A bond graph model of the vehicle with ABS and regenerative braking module has also been developed here. Initially, the performance of the combined regenerative and ABS braking was evaluated by using a quarter car model. The CVT being the key component of regenerative braking needs to be protected against damage by ensuring that the CVT ratio variation is as smooth as possible. It was found that the standard ABS controller requires quick CVT ratio fluctuations to maintain constant speed at the generator. If smooth CVT ratio variation is imposed with a standard ABS controller then the speed at the generator end fluctuates. Therefore, a sliding mode controller for ABS was developed so that optimum slip value can be maintained and smooth braking operation can be performed. In regenerative braking, it was possible to recover kinetic energy of the vehicle during braking operation while maintaining the same performance parameters of ABS.

Bibliography

[1] T. B. Bera, A.K. Samantaray, and R. Karmakar. Bond graph modeling of planar prismatic joints. *Mechanism and Machine Theory*, DOI: 10.1016/j.mechmachtheory.2011.07.016.:DOI: 10.1016/j.mechmachtheory.2011.07.016., 2011.

[2] T. K. Bera, K. Bhattacharya, and A.K. Samantaray. Bond graph model-based evaluation of a sliding mode controller for a combined regenerative and antilock braking system. *Journal of Systems and Control Engineering*, 225:918–934, 2011.

[3] T. K. Bera, K. Bhattacharya, and A.K. Samantaray. Evaluation of antilock braking system with an integrated model of full vehicle system dynamics. *Simulation Modelling Practice and Theory*, 19(10):2131–2150, 2011.

[4] M. Corno, S.M. Savaresi, and G.J. Balas. Linear, parameter-varying wheel slip control for two-wheeled vehicles. In *In: proceedings of the 47th IEEE Conference on Decision and Control Cancun, Mexico*, 2008.

[5] G. Dauphin-Tanguy, A. Rahmani, and C. Sueur. Bond graph aided design of controlled systems. *Simulation Practice and Theory*, 7(5-6):493–513, 1999.

[6] M. Denny. The dynamics of antilock brake systems. *European Journal of Physics*, 26:1007–1016, 2005.

[7] S. Drakunov, U. Ozguner, P. Dix, and B. Ashrafi. Abs control using optimum search via sliding modes. *IEEE Transactions on Control Systems Technology*, 3(1):79–85, 1995.

[8] W. Drozde and H.B. Pacejka. Development and validation of a bond graph handling model of an automobile. *Journal of the Franklin Institute*, 328(5/6):941–957, 1991.

[9] T. Ersal, H.k. Fathy, and J.L. Stein. Structural simplification of modular bond-graph models based on junction inactivity. *Simulation Modelling Practice and Theory*, 17:175–196, 2009.

[10] H.K. Fathy, Z.S. Filipi, J. Hagena, and J.L. Stein. Review of hardware-in-the-loop simulation and its prospects in the automotive area. In *Proceedings of SPIE-The international Society for Optical Engineering*, 2006.

[11] P.J. Gawthrop, D.J. Wagg, and S.A. Neild. Bond graph based control and substructuring. *Simulation Modelling Practice and Theory*, 17(1):211–227, 2009.

[12] T.D. Gillespie. *Fundamentals of Vehicle Dynamics*. SAE International, 1992.

[13] J.J. Granda. The role of bond graph modeling and simulation in mechatronics systems: An integrated software tool: Camp-g, matlab-simulink. *Mechatronics*, 12(9-10):1271–1295, 2002.

[14] D. Hrovat, J. Asgari, and M. Fodor. *Automotive Mechatronic Systems*, chapter First, pages 1–98. In: Mechatronic Systems Techniques and Applications. Gordon and Breach Science Publishers, Amsterdam, 2000.

[15] G.A. Hubbard and K. Youcef-Toumi. Modeling and simulation of a hybrid-electric vehicle drivetrain. In *Proceedings of the American Control Conference 1*, 1997.

[16] K. Huh, D. Hong, and J.L. Stein. Development of a lane departure monitoring and control system. *Journal of Mechanical Science and Technology*, 19(11):1998–2006, 2005.

[17] D.C. Karnopp, D.L. Margolis, and R.C. Rosenberg. *System Dynamics, Modeling and Simulation of Mechatronic Systems*. John Wiley & Sons, NY, 2000.

[18] U. Kiencke and L. Nielsen. *Automotive Control Systems: For Engine, Driveline, and Vehicle*. Springer-Verlag, 2000.

[19] S.L. Koo, H.S. Tan, and M. Tomizuka. An improved tire model for vehicle lateral dynamics and control. In *Proceedings of the American Control Conference, Minneapolis, Minnesota, USA*, 2006.

[20] M.L. Kuang, M. Fodor, D. Hrovat, and M Tran. Hydraulic brake system modeling and control for active control of vehicle dynamics. In *In: proceedings of the American Control Conference, San Diego, California*, 1999.

[21] K. Li, J.A. Misener, and K. Hedrick. On-board road condition monitoring system using slip-based tyre-road friction estimation and wheel speed signal analysis. *J. Multi-body Dynamics, IMechE*, 221:129–146, 2007.

[22] R.G. Longoria, A. Al-Sharif, and C.B. Patil. Scaled vehicle system dynamics and control: a case study in anti-lock braking. *Int. J. Vehicle Autonomous Systems*, 2(1/2):18–39, 2004.

[23] D. Margolis and T. Shim. A bond graph model incorporating sensors, actuators, and vehicle dynamics for developing controllers for vehicle safety. *Journal of the Franklin Institute*, 338:21–34, 2001.

[24] R. Merzouki, B. Ould-Bouamama, M.A Djeziri, and M. Bouteldja. Modelling and estimation of tire-road longitudinal impact efforts using bond graph approach. *Mechatronics*, 17(2-3):93–108, 2007.

[25] H. Mirzaeinejad and M. Mirzaei. A novel method for non-linear control of wheel slip in anti-lock braking systems. *Control Engineering Practice*, 18(8):918–926, 2010.

[26] A. Mukherjee, R. Karmakar, and A.K. Samantaray. *Bond Graph in Modeling, Simulation and fault Identification*. CRC Press, FL, 2006.

[27] M. Oudghiri, M. Chadli, and A.E. Hajjaji. Robust fuzzy sliding mode control for antilock braking system. *Int. Journal on Sciences and Techniques of Automatic control*, 1(1):13–28, 2007.

[28] B. Ozdalyan. Development of a slip control anti-lock braking system model. *International Journal of Automotive Technology*, 9(1):71–80, 2008.

[29] H.B. Pacejka. Modelling complex vehicle systems using bond graphs. *Journal of the Franklin Institute*, 319(1-2):67–81, 1985.

[30] H.B. Pacejka. *Tyre and Vehicle Dynamics*. Butterworth-Heinemann. Elsevier, UK, 2006.

[31] R. Rajamani. *Vehicle dynamics and control*. Springer, US, 2006.

[32] A.K. Samantaray and B. Ould-Bouamama. *Model-based Process Supervision*. Springer Verlag, London, 2008.

[33] A. Sanyal and R. Karmakar. Directional stability of truck-dolly-trailer system. *Vehicle System Dynamics*, 24(8):617–637, 1995.

[34] C. Song, J. Wang, and L. Jin. Study on the composite abs control of vehicles with four electric wheels. *Journal of computers*, 6(3):618–626, 2011.

[35] J. Svendenius. *Tire Modeling and Friction Estimation*. Lund University, Lund, Sweden, 2007.

[36] H.T. Szostak and W.R. Allenand T.J. Rosenthal. Analytical modeling of driver response in crash avoidance maneuvering. volume ii: An interactive model for driver/vehicle simulation. Technical report, U.S Department of Transportation Report NHTSA DOT HS-807-271, 1988.

[37] H. Yeo, S. Hwang, and H. Kim. Regenerative braking algorithm for a hybrid electric vehicle with cvt ratio control. *Journal of Automobile Engineering, IMechE*, 220(11):1589–1600, 2006.

CHAPTER 2

Development of Low Cost Electromyography (EMG) Controlled Prosthetic Hand

Shailabh Suman[1],
Robotics and Control Laboratory at Mechanical and Industrial Engineering Department,
Indian Institute of Technology,
Roorkee, India
shailabhsuman@gmail.com

Sunil Kumar
Robotics and Control Laboratory at Mechanical and Industrial Engineering Department,
Indian Institute of Technology,
Roorkee, India
fynsnman@gmail.com

Pushparaj Mani Pathak
Robotics and Control Laboratory at Mechanical and Industrial Engineering Department,
Indian Institute of Technology,
Roorkee, India,
pushp_pathak@yahoo.com

ABSTRACT. Myoelectric signals are weak signals (order of micro-volts) generated by the muscular activity. The contraction of muscles can generate microvolt level electrical signals within the muscles. These signals can be realized on the surface of skin above the muscles or within the muscles using a needle. Based on this phenomenon, it was thought of implementing a myoelectric control to prosthetic hands which would use a surface electrode to receive myoelectric signals from disabled persons' residual limb. A prothetic hand is a customized limb that can be fitted on a disabled persons' residual limb to assist them in manipulation, which can be only performed by hands. This chapter demonstrates the principle of controlling the prosthetic hand by Surface EMG signals on a wooden prototype. The developed prosthetic hand also uses force sensor and proximity sensor to aid the EMG sensor for hybrid control. The chapter describes the various steps for development of low cost surface EMG controlled prosthetic hand.

Keywords: Electromyography, force sensor, proximity sensor, robotic manipulator, prosthetic hand.

2.1. Introduction

The aim of this chapter is to develop a low cost surface EMG controlled prosthetic hand. When muscles contract microvolt level electrical signals are generated within muscles. Prosthetic hand can be controlled by these surface EMG signals. The prosthetic hand discussed in the chapter also uses force sensor and proximity sensor to aid the surface EMG sensor for hybrid control. The prosthetic hand is modeled as an under-actuated robotic manipulator. The integration of the EMG signal data with force and proximity sensor data has been demonstrated to implement hybrid control that could enable the basic manipulation (like grasping, holding) by the prosthetic hand. The design of bio-amplifier for weak myoelectric signals proposed in this chapter explains a low cost versatile surface EMG signal amplifier design, which is compact and convenient to operate. The control system addresses the multiple issues of torque required at joint and angles of link. The surface EMG signal acts as interrupt to the controller with proximity

---

[1]Corresponding author

and force sensor assisting in further manipulation.

The physical hardware that are built is as follows

(1) Prosthetic hand prototype: Modeled as kinematic chain Links are under actuated but the design uses dimension of finger links to obtain basic manipulation orientation. Actuation of motor attached to the link would give a curl shape to hold a cylindrical or spherical shaped object.

(2) Single channel Bio-Sensor: EMG electrode-amplifier with comparator.

(3) Force Sensor on each finger: FlexiForceš were used. Very thin sensor and amplifier is easy to design

(4) Proximity Sensor: Developed using Infrared Light Emitting Diodes (LED) and TSOP 1738, easily available commercially.

(5) Microcontroller: ATMEL Atmega32 is used. This has 32 input/output lines, Analog to Digital Converter (ADC), enough Timer channels to control all servomotors and interrupt channels.

(6) Actuators: All the actuators used in the prosthetic hand are servomotors. These motors are light and have built in control circuit. Servomotors are good in maintaining certain angular position.

(7) Power Supply: Comprising of voltage regulators.

The Figure 2.1.1 gives an overview of system. The mechatronic components are shown in the broader scheme of things.

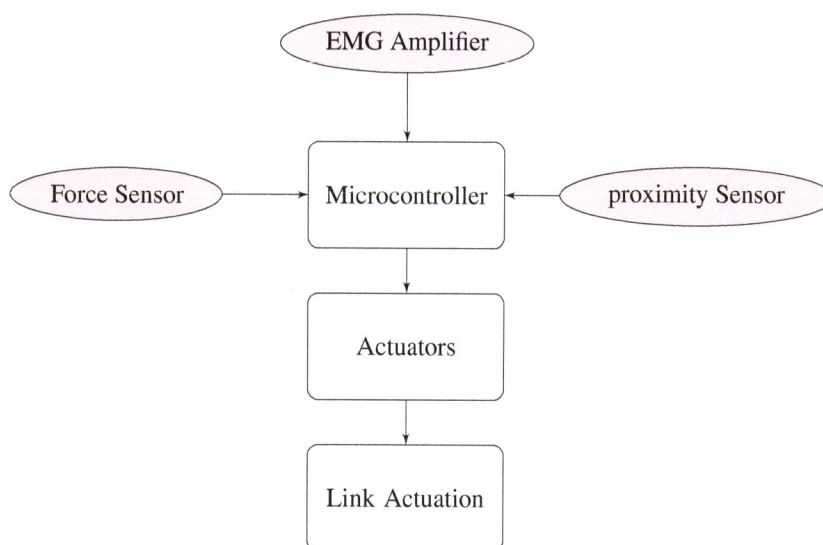

FIGURE 2.1.1. Overview of System.

2.2. Background of electromyography

2.2.1. Electromyography. Electromyography (EMG) is a technique for evaluating and recording the activation signal of muscles. Myoelectric signals are formed by physiological variations in the state of muscle fiber membranes. EMG is performed using an instrument called an electromyograph, to produce a record called an electromyogram. An electromyograph detects the electric potential generated by the muscle cells when they are active and also when they are at rest.

EMG can also be classified into two categories: Needle EMG and Surface EMG

(1) Needle EMG: A needle electrode is inserted through the skin into the muscle tissue to perform intramuscular EMG. The insertion activity provides valuable information about the state of the muscle and its innervating nerve. There is some electrical activity when muscle contracts compared to muscle at rest. Needle EMG helps in monitoring the activity of only a few fibers and is invasive.

(2) Surface EMG: A surface electrode is used to monitor the potential of muscle activation in surface EMG. This technique is used in a number of settings, for example, in the physiotherapy clinic, muscle activation

is monitored using surface EMG and patients have an auditory or visual stimulus to help them know when they are activating the muscle (biofeedback).

Here the non-invasive surface EMG signals are used to design biosensor. As muscles contract, microvolt level electrical signals are created within the muscle that may be measured from the surface of the body. A procedure that measures muscle activity from the skin (non-invasive) is referred to as surface electromyography (SEMG) . Here the attention is given to detect the peak, which arises from contraction of muscle to interrupt controller. In case of prosthetic hand, the remaining limb after amputation of patient's hand/elbow is used for detection of surface EMG. However discussion here will be limited to flexor carpi radialis , a muscle in human forearm.

2.2.2. Characteristics of the EMG Signal. It is well established that the amplitude of the EMG signal is stochastic (random) in nature and can be reasonably represented by a Gaussian distribution function [1]. The amplitude of the signal can range from 0 to 10 mV (peak-to-peak) or 0 to 1.5 mV (rms). The usable energy of the signal is limited to the 0 to 500 Hz frequency range [1], with the dominant energy being in the 50-150 Hz range. Usable signals are those with energy above the electrical noise level. Figure 2.2.1 shows the frequency spectrum of the EMG signal detected from the Tibialis Anterior muscle during a constant force isometric contraction at 50 % of voluntary maximum [1].

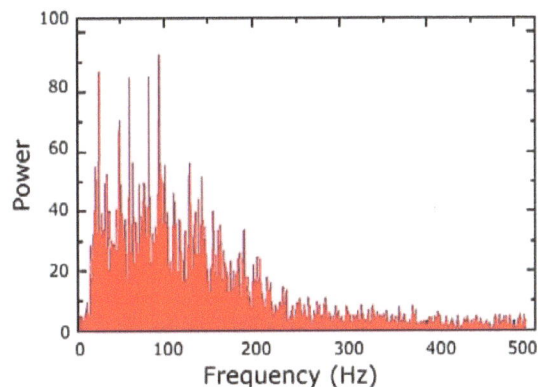

FIGURE 2.2.1. Frequency spectrum of the EMG signal detected from the Tibialis Anterior muscle during a constant force isometric contraction at 50 % of voluntary maximum [1]

2.2.3. Raw Surface EMG Signal. The smallest functional unit to describe the neural control of the muscular contraction process is called a Motor Unit. It is defined as the cell body and dendrites of a motor neuron, the multiple branches of its axon, and the muscle fibers that innervates it [2]. The term units outlines the behavior, that all muscle fibers of a given motor unit act Òas oneÓ within the innervations process. An unfiltered and unprocessed signal detecting the superposed Motor Unit Action Potential (MUAP) is called a raw EMG Signal. Figure 2.2.2 shows a raw surface EMG recording (sEMG) for three static contractions of biceps brachii muscle, a two-headed muscle located on the upper arm [2].

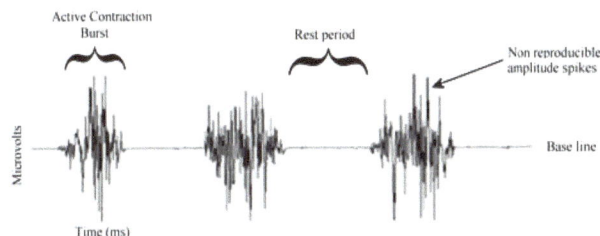

FIGURE 2.2.2. Raw EMG Recording of 3 contractions bursts of the Muscle Biceps br.[2]

In the relaxed state of muscle, one can observe an almost noise free zero potential signal on EMG electrodes. It can be called baseline noise. The raw EMG baseline noise depends on many factors, especially the quality of the EMG amplifier, the environment noise and the quality of the given detection condition. The healthy relaxed muscle shows no significant EMG activity due to lack of depolarization and action potentials. By its nature, raw EMG spikes are of random shape, which means one raw recording burst cannot be precisely reproduced in exact shape. This is due to the fact that the actual set of recruited motor units constantly changes within the matrix/diameter of available motor units. If occasionally two or more motor units fire at the same time and they are located near the electrodes, they produce a strong superposition spike. By applying a smoothing algorithm (e.g. moving average) or selecting a proper amplitude parameter (e.g. area under the rectified curve), the non- reproducible contents of the signal is eliminated or at least minimized. Raw Surface EMG can range between +/- 5000 microvolts (athletes!) and typically the frequency contents ranges between 6 and 500 Hz, showing most frequency power between 20 and 150 Hz [3].

The myoelectric prosthesis discussed in this chapter is designed towards detecting the peak of electromyograms to initiate a certain set of actions to complete the task.

2.3. Mechanical design and modeling

Robotic manipulators in general are open kinematic chains that form coupled mechanical systems. The dynamic equations of these manipulators play a very important role in the evaluation of its design, and in the design of suitable control laws. A prosthetic hand is also a form of robotic manipulator. A lot of research has been done in the past on the design of robotic fingers and arms. There is a wide applicability of robotic hands.

The design of fingers proposed in this chapter, uses five actuators for driving the four fingers and the thumb. It is a driven mechanism aimed to demonstrate only basic manipulation like grasping of cylindrical or spherical object. The force sensor at the tip of finger feeds back the grasping force to the controller.

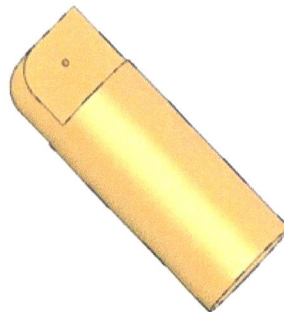

FIGURE 2.3.1. CAD model of one link

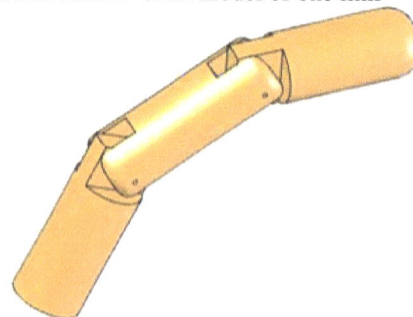

FIGURE 2.3.2. CAD model of a finger

Figure 2.3.1 and 2.3.2 shows the proposed design of a finger with three links connected by a hinge joint. One face of link at the joint is filleted to allow the motion in one direction, as present in original finger.

FIGURE 2.3.3. Image of developed finger

Figure 2.3.3 shows the actual finger with the driving mechanism. The servomotor is attached to the first link of the finger by a metallic spoke which in turn is connected to the next link by same mechanism. The desired amount of curvature is present in the finger in the relaxed position so as to grasp the object when actuated. The actuation of motor forces the connected links to move and consequently the hand closes and grips the object. Figure 2.3.4 shows the grasping of an object by the prosthetic hand. Figure 2.3.5 shows the assembly of hand with elbow.

2.3.1. Mechanical properties of hardware designed. Table 1 shows the mechanical properties of the links of a finger. The properties include link mass, length, and inertia about x, y and z-axis.

TABLE 1. The anthropomorphic data for human hand

Link	Mass (kg)	Link length (cm)	Inertia Ixx $(Kg - m^2)$	Inertia Iyy $(Kg - m^2)$	Inertia Izz $(Kg - m^2)$
Link1	0.024	0.045	$2e^{-6}$	$6.2e^{-6}$	$6.2e^{-6}$
Link 2	0.013	0.035	$0.75e^{-6}$	$1.2\,e^{-6}$	$1.2e^{-6}$
Link 3	0.006	0.025	$0.32e^{-6}$	$0.52e^{-6}$	$0.5e^{-6}$

Denavit-Hartenberg notation has been used to compute all the kinematics and dynamics of finger. Also Newton-Euler formula has been used to compute the torques. To compute the kinematics and dynamics of joints of finger, forward cases were considered. Based upon the anthropometric data available, the structural characteristics of the finger are computed and tabulated as shown in Table 1. The Robotics Toolbox 7.1 for MATLAB is used for simulation of the finger and inverse dynamics calculations. For our simulation, the torque function uses precomputed joint trajectories, and the trajectory is interpolated for the specified time length. The trajectory is computed for a destination angle of 30, 40 and 20 degrees or 0.5236, 0.6981 and 0.3491 radians of each link of finger. The torques were calculated for its own weight. The Servo Motor HS-55 MICRO LITE SERVO having a weight of 8 grams and torque of 0.1 N-m is suited for our purposes.

2.4. Electronics design

The electronics of the system comprises of set of sensors a central controller and actuators. The electronics are EMG electrode, its amplifier and a comparator circuit, proximity sensor electronics, force sensor and its amplifier and a Micro- controller breakout board. Figure 2.4.1 shows the different electronic components involved.

In the following section, electronics involved in the developed prosthetic hand is discussed in detail.

2.4.1. EMG Amplification. In general, signals resulting from physiological activity have very small amplitudes and must therefore be amplified before their processing and display can be accomplished [4]. Bio-potential measurements are very sensitive to electromagnetic interference (EMI) [5]. The frequency bandwidth of a bio-potential amplifier should be such as to amplify, without attenuation, all frequencies present in the electrophysiological signal of interest [4]. Another critical issue is that of the medical safety of the patient being studied and it must be the one of the primary objectives throughout the development effort [4].

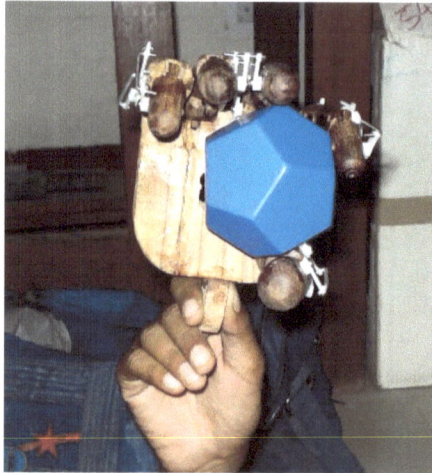

FIGURE 2.3.4. An object grasped by
the prosthetic hand

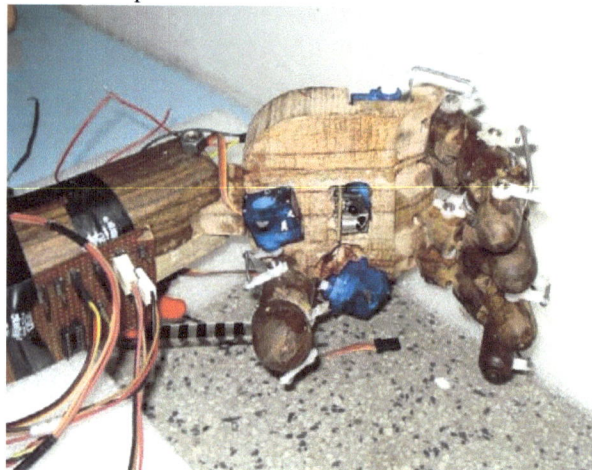

FIGURE 2.3.5. The prosthetic hand and
elbow assembly

In this section development of amplifier cum filter for the EMG signals is discussed. Firstly the sample electromyogram is studied from an already existing amplifier developed by Recorders & Medicare Systems (P) Ltd. Chandigarh, India. For having an electromyogram, stick two electrodes on the skin over the corresponding muscle as shown in Figure 2.4.2. These electrodes are connected to a bio-instrumentation amplifier that measures the electrical potential difference between both electrodes, then that potential is filtered and the resulting signal is visualized on a computer.

The amplifier is serially interfaced to the computer. The sampling rate for the EMG data collection is 24000 samples per second. The response obtained can be seen on an interface developed by Recorders & Medicare Systems (P) Ltd. Chandigarh, India [6]. The responses for carpi radialis muscle in closed and open hand positions are shown in Figures 2.4.3 and 2.4.4 respectively. It can be observed that in open hand position, carpi radialis is in extension state and hence the EMG potential is higher than as for closed position.

These signals can be processed to obtain desired digital values. However for high sampling rate the conventional serial ports cannot be used in this case. Furthermore, there is a need for on board hardware signal processing, as the computer cannot be carried with the prosthetic hand. To address this issue an amplifier circuit is developed which detects the peak when there is a muscle activity. Here a comparator has been used to detect the peak. Once the peak

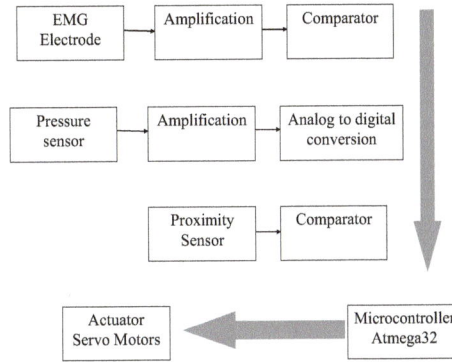

FIGURE 2.4.1. Signal flow diagram through different electronic components

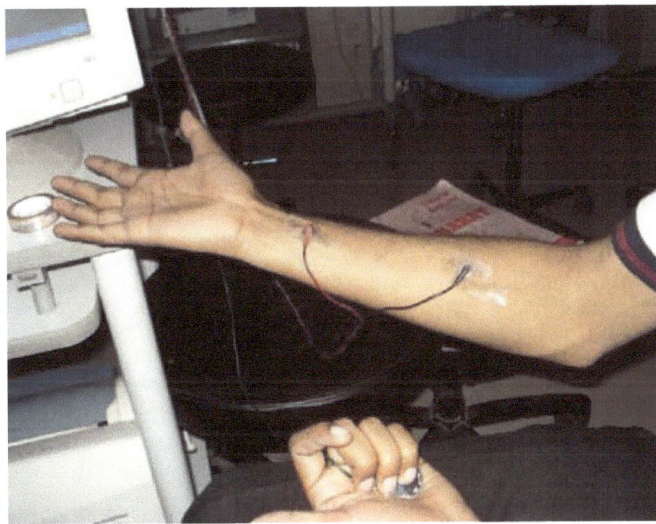

FIGURE 2.4.2. Application of electrodes for acquisition of EMG signals

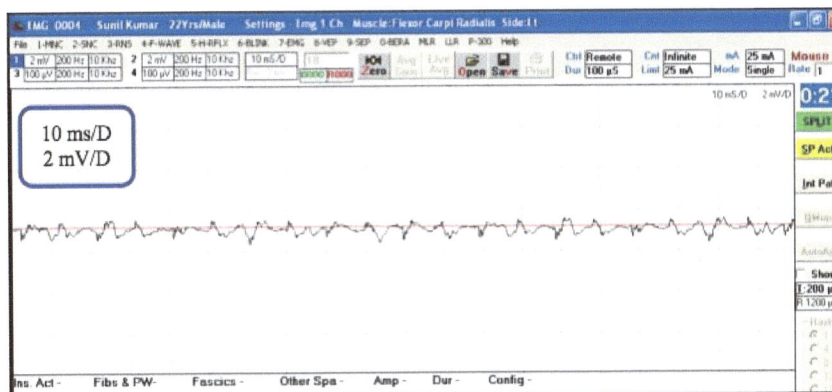

FIGURE 2.4.3. EMG Signal for flexor carpi radialis in closed hand position

is detected it is used as an interrupt to controller. The amplifier actually consists of pre-amplification phase, followed by a filtering phase (notch filtering and band pass filtering) and another amplification phase. The outline of EMG amplifier is shown in Figure 2.4.1.

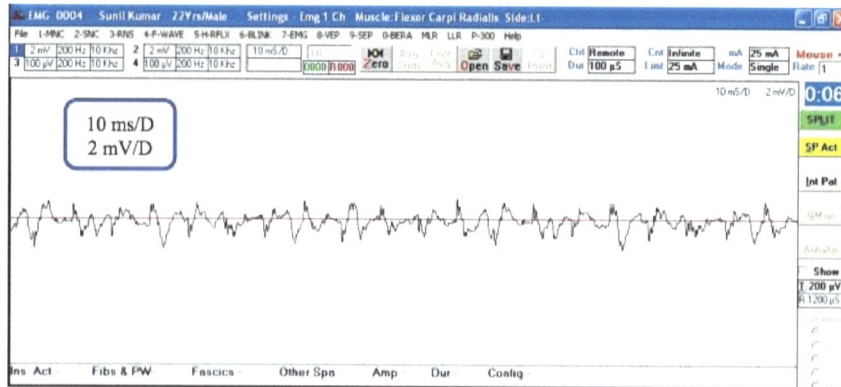

FIGURE 2.4.4. EMG Signal for flexor carpi radialis in open hand position

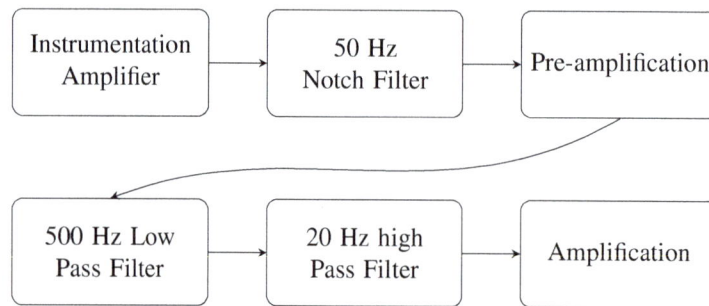

FIGURE 2.4.5. EMG Signal for flexor carpi radialis in closed hand position

The salient features of EMG are tabulated in Table 2 and the amplifier is built around these salient features.

TABLE 2. The salient features of EMG Amplifier

Frequency	20-500 Hz
Common Mode Rejection Ratio	120 db
Noise Range	0-20 Hz
Notch Filter	50 Hz
Amplitude Range	100 microvolts to 90 milli volts
Total Gain Obtained	17420

Total amplification achieved is17420 times. EEG (Electroencephalography) electrodes were used because of their high conductivity.

The basic elements of EMG amplifier are:

(1) *Superficial Electrodes :* they collect the electrical activity of the muscle. With these electrodes it is possible to obtain an idea of the global electro genesis of the muscle, but low amplitude or high frequency potentials are not detected. However adhesive EEG electrodes can be used which are most appropriate for superficial EMG.

(2) Instrumentation Amplifier (AD620): an instrumentation amplifier is commonly used in bio-signal amplification. The Bio-signal typically requires several stages of amplification and instrumentation amplification is the first stage. Care is taken while amplifying EMG signal, as the idea is to reject the noise and amplify only

the signal.

The first stage of amplifier, instrumentation amplifier is connected in differential mode because of the benefits of common-mode noise reduction and improvement in signal to noise ratio. The Common Mode Rejection Ratio (CMRR) measures the accuracy with which the differential amplifier can subtract the signals. A perfect differential amplifier would have a CMRR of infinity but CMRR of 32,000 or 90 dB is generally sufficient to suppress extraneous electrical noises.

It is also outfitted with input buffers, which eliminate the need for input impedance matching and thus makes the amplifier particularly suitable for use in measurement and test equipment. So the characteristics include very low DC offset, low drift, low noise, very high open-loop gain, very high common-mode rejection ratio, and very high input impedances. Instrumentation amplifiers are used where great accuracy and stability of the circuit for both short and long-term are required.

The instrumentation amplifier used here is AD620. The AD620 is a low cost, high accuracy instrumentation amplifier that requires only one external resistor to set gains of 1 to 1000. Some specifications of instrumentation amplifier are shown in Table 3.

TABLE 3. Specifications of instrumentation amplifier [7]

Input Impedance (Differential)	CMRR (min)	Quiescent I
10‖2 GOhm‖pF	100 db	0.9 mA

The Figure 2.4.6 show the typical connections of AD620 Integrated Circuit(IC) available commercially.

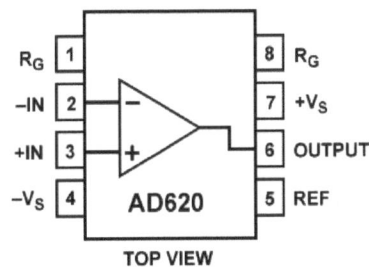

FIGURE 2.4.6. Pin connection of AD620[7]

This IC consists of an Instrumentation amplifier, which is represented in Figure 2.4.7.
The gain equation, as given in data sheet for AD620 is

$$R = \frac{49.4 kohm}{G - 1}$$
(2.4.1)

Where R is resistance and G is gain. For gain of 500 the resistance comes out to be 99 ohm.

(3) *Notch Filter :* A notch filter is a band-stop filter with a narrow stop band (high Q factor), to filter out the DC component at 50 Hz. This is the second stage of amplification. Figure 2.4.8 shows the design of notch filter. Values for a 50 Hz notch would be: C1, C2 = 47 nF, R1, R2 = 10 k ohm, R3, R4 = 68k ohm.

$$F_{notch} = \frac{1}{2pi RC}$$
(2.4.2)

R = R3 = R4, C = C1 = C2
Where F_{notch} is center frequency of notch in Hertz.

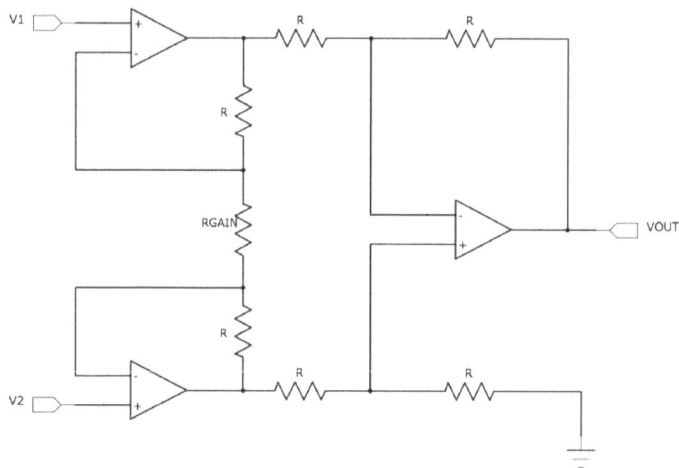

FIGURE 2.4.7. Instrumentation Amplifier

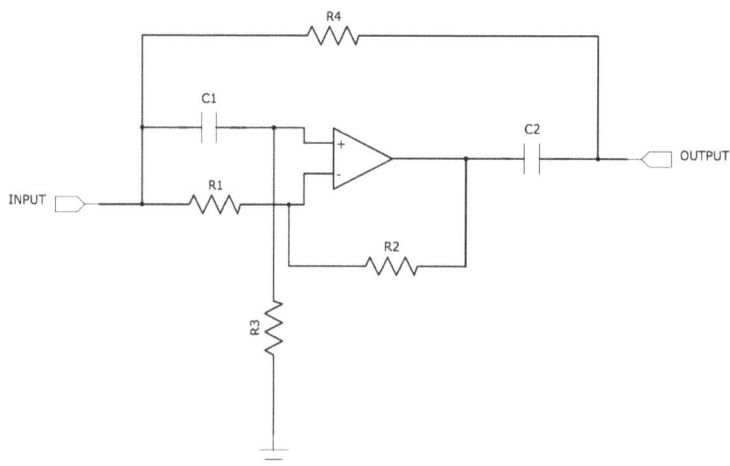

FIGURE 2.4.8. Notch filter

(4) *Band Pass Filter :* A band-pass filter is a device that passes frequencies within a certain range and rejects (attenuates) frequencies outside that range. The band pass filter used here is made from low pass filter and high pass filter in series. The frequency range of interest is in between 20 Hz to 500 Hz. For the low pass filter and the high pass filter the selected value of capacitance is 0.1 micro Farad. The low pass filter is set at 500 Hz and high pass is set at 20 Hz. The design of filters are shown in Figures 2.4.9 and 2.4.10.

Figure 2.4.11 shows the EMG amplifier fabricated. The electrodes shown are EEG electrodes.

Preparation of skin must be done before using this sensor. Skin should be clean in order to get good signal. Alcohol or water can be used to clean the skin. Removal of hair will also help in getting better adhesion of electrodes to skin.

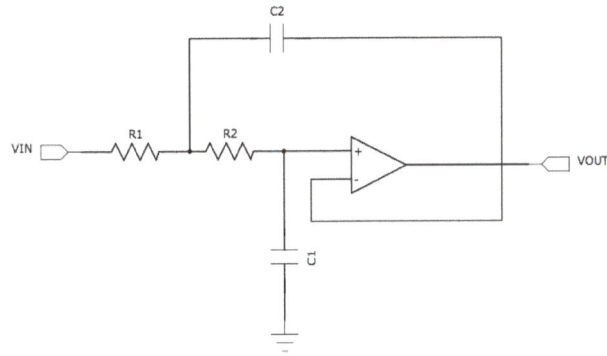

FIGURE 2.4.9. Low pass filter

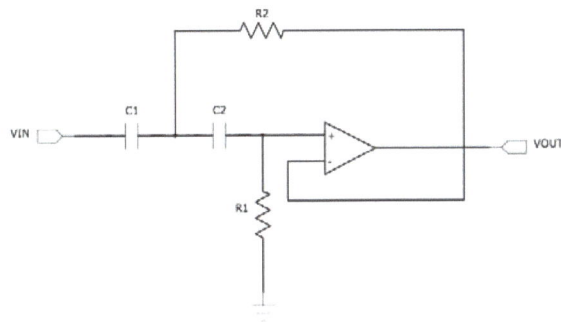

FIGURE 2.4.10. High pass filter

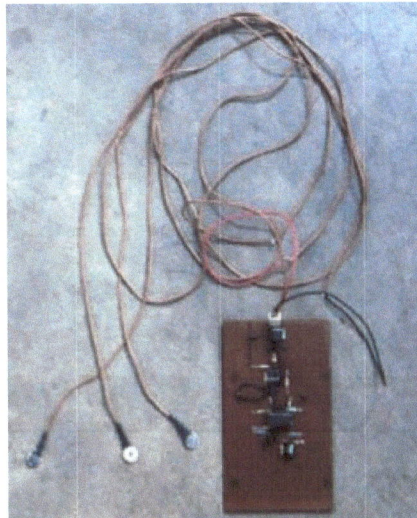

FIGURE 2.4.11. EMG Amplifier

2.4.2. Proximity Sensor. The obstacle-detecting sensor is used as a proximity sensor. It is used to find the object in the near grasp. It can activate an interrupt. It has an infrared led as emitter and TSOP1738[**9**] as receiver of modulated signal. The infrared light emitted by led is modulated at 38 Khz for TSOP1738 to detect when it is reflected by obstacle. The distance for detection can be adjusted by a potentiometer.

Figure 2.4.12 shows the proximity sensor used in prosthetic hand. It is place at the center of palm.

FIGURE 2.4.12. Proximity sensor [8]

Figure 2.4.13 shows the block diagram of receiver. Incoming signal (reflected infrared light in this case) goes to the amplifier with automatic controllable gain (AGC) then band-pass filter passes only specific frequency, and finally it goes to demodulator, which opens output transistor when it receives signal input after band-pass filter.

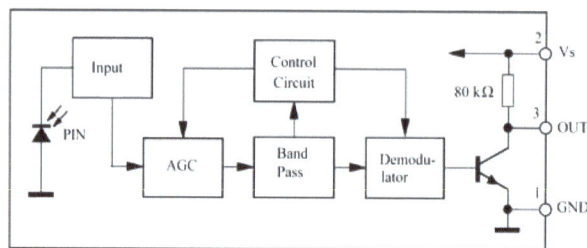

FIGURE 2.4.13. Block diagram of Proximity sensor [8]

If this receiver detects 38Khz modulated light from infrared led it turns on the output transistor and an interrupt is generated for microcontroller.

2.4.3. Force Sensor. Flexi Force A201 force sensor is used as force sensor. In this sensor, the silver circle on top of the pressure-sensitive region defines the active force sensing area. Silver extends from the sensing area to the connectors at the other end of the sensor, forming the conductive leads. A201 sensors are terminated with male square pins, allowing them to be easily incorporated into a circuit. The two outer pins of the connector are active and the center pin is inactive [10].

The Flexi force single element force sensor acts as a force-sensing resistor in an electrical circuit. When the force sensor is unloaded, its resistance is very high. When a force is applied to the sensor, this resistance decreases. The resistance can be read by probing a multimeter to the outer two pins (center pin is inactive) , then applying a force to the sensing area. Figure 2.4.16 shows both the force versus resistance and force versus conductance (1/R) of flexi force sensor.

By putting incremental weight from 1 to 5 kilograms, the calibration of sensor is done. The voltage is recorded from the amplified circuit with sensor as shown in figure 2.4.17. The calibration result from this circuit is shown in Table 4.

2.4.4. Microcontroller. Atmega32 from Atmel AVR is used as the controller. Atmega32 is a low cost 8-bit micro-controller with 32k bytes In-System programmable flash [11]. More details about it are discussed in next section. The pin diagram of the controller is shown in Figure 2.4.18.

A breakout board from this controller is fabricated to facilitate the connection to various other modules.

FIGURE 2.4.14. FlexiForceš sensor [**10**]

FIGURE 2.4.15. Flexi force on pros-thetic hand

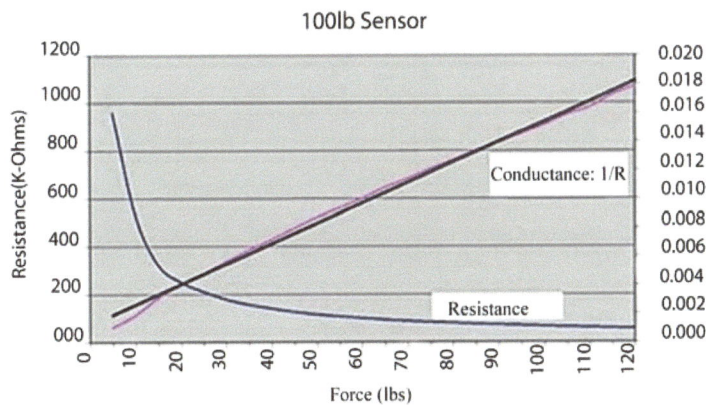

FIGURE 2.4.16. Force versus resistance and force versus conductance (1/R) of flexi force sensor[**10**]

2.5. Control

The control forms the most important aspect of any dynamical system. The control schematic of prosthetic hand developed is shown in Figure 2.5.1. There is a central controller, which takes the data at various ports/pins from EMG sensor, proximity sensor and Force sensor.

The output from EMG sensors after being passed through comparator is connected to INT1 (External Interrupt Pin) of Atmega32 while from proximity sensor is connected at INT0. A rising edge on INT1 triggers an interrupt, which initiates the actuation of servomotor. The peak of EMG is detected and amplified. The peak in EMG signal comes when the user of prosthetic hand tries to contract the muscles to which electrodes are connected in order to

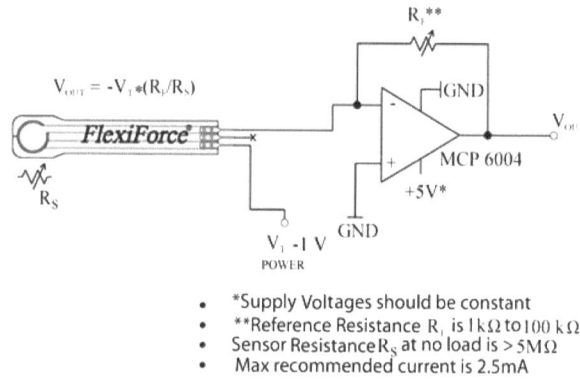

- *Supply Voltages should be constant
- **Reference Resistance R_i is 1 kΩ to 100 kΩ
- Sensor Resistance R_S at no load is > 5MΩ
- Max recommended current is 2.5mA

FIGURE 2.4.17. Recommended amplification circuit for Tekscan for flexi force sensor[10]

TABLE 4. Calibration Result of Flexi force sensor

Sl. No.	Force Applied (N)	Practical Vout (V)	Theoretical Vout(V)
1	9.8	0.60	0.83
2	19.6	1.95	2.08
3	29.4	2.60	2.50
4	39.2	3.10	2.94
5	49	4.00	4.16
6	58.8	4.60	4.56
7	68.6	4.80	4.76

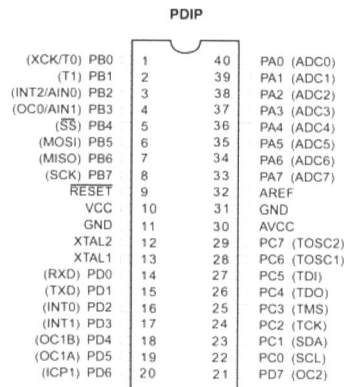

FIGURE 2.4.18. Pin diagram of Atmega32 from its datasheet [11]

accomplish a task. The comparator value is tunable as it is connected to a potentiometer. The proximity sensor is there which is also connected to external interrupt pin. In case of EMG peak and null proximity the actuation is performed as half closing and opening of grasp but once the proximity sensor detects an object, closing action is performed. The force sensor tries to control the force exerted so as to not crush the object. The force sensor signal is connected to the ADC. The algorithm is shown in Figure 2.5.2.

2.5.1. Microcontroller. Atmega32 is used as the microcontroller for controlling the prosthetic hand. It is embedded in a surface mounted board whose peripherals are connected to interfacing connections for servomotor, force sensor, proximity sensor and the EMG amplifier. The Atmega32 is a low-power CMOS 8-bit microcontroller based on the AVR enhanced RISC (Reduced Instruction Set Computing) architecture[11]. By executing powerful instructions in a single clock cycle, the Atmega32 achieves throughputs approaching 1 MIPS per MHz, which allows the

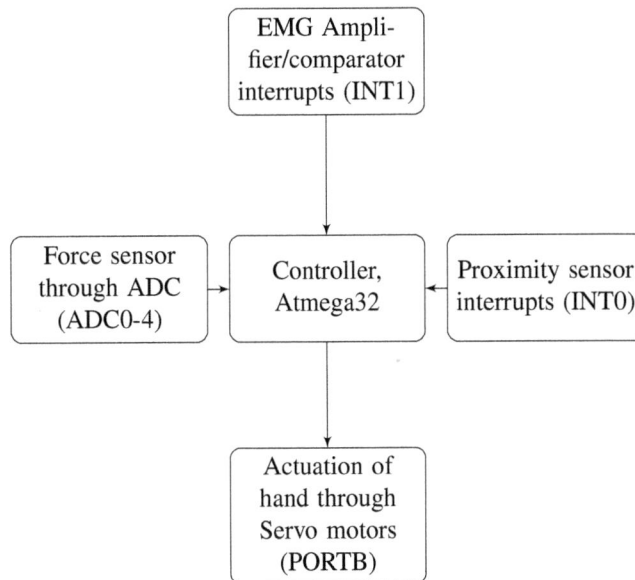

FIGURE 2.5.1. Pin connections of various components of prosthetic hand to microcontroller

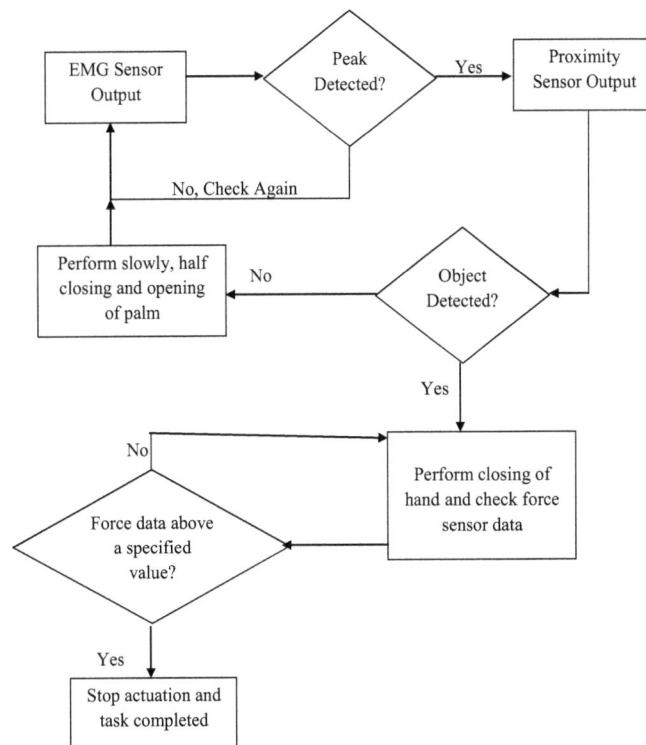

FIGURE 2.5.2. Algorithm for control of prosthetic hand.

system designer to optimize power consumption versus processing speed[**11**]. Out of the various capabilities of the microcontroller following were used:

- External Interrupt
- Analog to digital converter
- Timer interrupts for servo drive

2.5.1.1. *External Interrupt.* The INT pins (the pin-out of Atmega32 shown in Figure 2.4.18 shows the locations of these pins) trigger the External Interrupts typically. This feature provides a way of generating a software interrupt. In Atmega32, a falling or rising edge or a low level can trigger the external interrupts. When the controller receives an interrupt, it ceases its normal polling and accomplishes the interrupt task. The recognition of falling or rising edge interrupts on INT0 and INT1 requires the presence of an I/O clock. This implies that these interrupts can be used for waking the part also from sleep modes other than idle mode.

2.5.1.2. *Analog to Digital Converter.* The main characteristics of ADC in Atmega32 are as follows [**11**]:

- 10-bit resolution
- Up to 15 kilo samples per second at maximum resolution
- 8 Multiplexed Single Ended Input Channels

The other features are 7 Differential Input Channels, Optional Left adjustment for ADC Result Readout, 0 - VCC ADC Input Voltage Range, Selectable 2.56V ADC Reference Voltage, Free Running or Single Conversion Mode. The Atmega32 features a 10-bit successive approximation ADC. The ADC is connected to an 8-channel analog multiplexer, which allows 8 single-ended voltage inputs constructed from the pins of Port A. The single-ended voltage inputs refer to 0V (GND). The device also supports 16 differential voltage input combinations. The ADC contains a sample and hold circuit, which ensures that the input voltage to the ADC is held at a constant level during conversion. The ADC has a separate analog supply voltage pin, AVCC. AVCC must not differ more than ± 0.3 V from VCC. Internal reference voltages of nominally 2.56V or AVCC are provided On-chip[**11**].

2.5.1.3. *Timer.* The Timer of Atmega32 can be used in various modes like output compare mode, overflow mode, etc. Here timer has been configured in overflow mode and one can control many servomotors. It is not used as an interrupt.

2.6. Actuators

The actuators that are used in this prosthetic hand is servo motor. Servomotor is an electromechanical device in which an electrical input determines the position of the armature of a motor. Sending a pulse width modulated or PWM signal to the PWM input pin does servo control. As long as the coded signal exists on the input line, the servo will maintain the angular position of the shaft. Typically servo motors from Futaba and Hitec expects a signal with pulse width between 1 ms to 2 ms after every 20-22 ms. The signal that is given to the servo is one that is high (5V) for 1-2ms and low (0V) for the remaining of the 20ms period. The duration of the high signal determines the position that the servo attempts to maintain. The servo must continually receive this signal in order to maintain its position. The positions of shaft of servomotors are decided by the pulse length linearly as below.

1.0 ms = full left

1.5 ms = middle

2.0ms = full right

In this prosthetic hand, a servo that has 180 degrees of rotation has been used. To generate a 50Hz signal with a digital high whose width varies between 1 and 2ms, the phase and frequency correct mode of the timer of Atmega32 is used. In phase and frequency correct mode the timer starts at zero, counts up to a user defined value called ICRn(n is the timer number, 0 for timer0), an 8-bit register in Atmega32, and then counts back down to zero. The counting up and down process should take 20ms in order to generate the 50Hz signal. The clock frequency is an externally connected crystal oscillator of 16 Mhz. There are prescalers that can be selected to achieve the 50 Hz signal from 16 Mhz clock.

(2.6.1)
$$PWM = \frac{Fclock}{2.N.TOP}$$

where Fclock in this case is 16 Mhz, N is prescaler, TOP is the value of ICR0 register. With Values of N as 1024 and TOP as 156, 50 Hz frequency for pulse width modulation is achieved. OCRn (n as zero in the discussed case) is the register responsible for pulse width. If OCR0 is 156 then pulse is 20 ms (1/50 Hz). Value of OCR as 8 gives a 1ms pulse approximately.

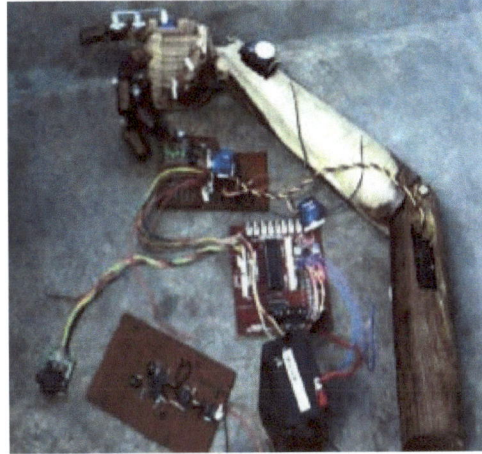

FIGURE 2.6.1. Final Assembly.

2.7. Conclusion

The prosthetic hand fabricated from wood takes feedback from EMG sensor, force sensor and proximity sensor and provide an excellent demonstration of one of the application of surface EMG signal. The developed hand is although made from rigid material but the presented work has the scope of implementing the use of flexible material like Polyvinyl Chloride or other polymer. The flexibility in this under actuated model can help in achieving basic manipulation. Also fixed torque servomotors are used, but a different force can be applied through this as the distance between point of action and axis of torque changes. This chapter provides appropriate data for EMG signal processing. The work enlightens the path towards the development of cheap bio-instrumentation amplifier. The commercially available EMG sensors are so costly that they are not viable for the presented cause. The use of proximity sensor adds to the feature of hand as it helps in the position feedback. The proximity sensor gives the proximal value with respect to object to the processing unit. The force sensor gives the value of force. Both the data are fed to processing unit and the object grasp and handling is done.

Bibliography

[1] C.J. De Luca, M.Knaflitz. Surface electromyography: : Detection and Recording, Delsys, Inc, 2002.

[2] R.M. Enoka, Neuromechanical Basis of Kinesiology, Human Kinetics Pub; 2nd edition (October 1994).

[3] Peter Konrad, A Practical Introduction to Kinesiological Electromyography, Version 1.0 April 2005, © Noraxon.

[4] D.Prutchi and M.Norris. Design and Development of Medical Electronic Instrumentation, Hoboken, New Jersey. John Wiley & Sons Inc., 2005.

[5] Spinelli, E.M., Martinez, N.H. and Mayosky, M.A., A Transconductance Driven-Right-Leg Circuit, IEEE Transactions on Biomedical Engineering, December 1999. 46(12): pp. 1466-1470

[6] Recorders & Medicare Systems(P) Ltd, Aleron series, http://www.rmsindia.com/brochures/Aleron.pdf.

[7] AD620 datasheet. Available from: http://www.analog.com/static/imported-files/data_sheets/AD620.pdf

[8] Proximity sensor, Tri India, Mumbai, India Available from: http://www.thinklabs.in/shop/

[9] TSOP 1738 Available from: http://www.datasheetcatalog.org/datasheets/208/301092_DS.pdf

[10] Tekscan, FlexiForceš, Available from: http://www.tekscan.com/pdf/A201-force-sensor.pdf

[11] Atlem AVR Atmega32 Data sheet. Available from: www.atmel.com/dyn/resources/prod_documents/doc2503.pdf.

CHAPTER 3

Control of Free and Constrained Motion of a C5 Parallel Robot

Brahim Achili [1]
Computer Science Lab. LIASD
University of Paris 8
2, rue de la liberté, 93526 St Denis, France
achili@ai.univ-paris8.fr

Boubaker Daachi,
Images, Signals and Intelligent Systems Lab. LISSI University of Paris
Est Creteil (UPEC)
94400 Vitry/Seine, France
daachi@u-pec.fr

Arab Ali chérif
Computer Science Lab. LIASD
University of Paris 8
2, rue de la liberté, 93526 St Denis, France
aa@ai.univ-paris8.fr

Yacine Amirat,
Images, Signals and Intelligent Systems Lab. LISSI University of Paris
Est Creteil (UPEC)
94400 Vitry/Seine, France
amirat@u-pec.fr

ABSTRACT. The work presented in this chapter is twofold. Firstly, the robust adaptive position control of a 6 degree of freedom (6dof) parallel robot called C5 is addressed. Coupling of the sliding modes and Multi-Layer Perceptron (MLP) neural networks form the basis of the proposed approach. This means that for the derivation of the control law, there is no requirement for the inverse dynamic model. The MLP neural network is integrated into the control scheme for the estimation of both gravity and friction forces alongside the dynamic effects, which do not constitute part of the model. The non-linearity problem present in neural networks is resolved using Taylor series expansion. The proposed approach permits the adjustment of the neural network parameters and sliding mode control terms by considering a reference model and the closed-loop stability, in the Lyapunov sense. Secondly, the force control of a C5 parallel robot is also addressed. It is based on the multi-layer perceptron (MLP), without any ex-ante knowledge of the dynamic model of the robot. Otherwise, the control type that we propose to apply is a black box one. The neural network corrective is used adaptively in the goal to ensure the system stability in the Lyapunov sense. We implemented the proposed approaches on a C5 parallel robot and carried tests of robustness against external influences in order to check the originality of our work.

Keywords: Adaptive Control, Multi-Layer Perceptron (MLP)neural networks, Taylor series expansion, stability analysis, Lyapunov method, Parallel robot, experimental results, sliding mode technique,robustness against external influences.

---

[1]Corresponding author
Part of information included in this chapter has been previously published in International Journal of Control Volume 83, Issue 10, 2010.

3.1. Introduction

A parallel robot is defined as a closed-loop mechanism wherein the mobile platform is linked by least two independent kinematic chain to the base. Such types of robots present numerous advantages in terms of compliance, fast execution and movement, precision, and the load that they can handle [3]. Owing to their low inertia, the effects of dynamic coupling are lowered, thus resulting in an increased dynamic performance. The application and use of parallel robots is seen in diverse fields including Pick and Place [31], flight simulator [2], surgical medicine [1] and electronic industry [4]. These robots have a complex design structure which makes their control less obvious and robust using the current control approaches [5, 9, 6, 7]. The literature reveals a set of control approaches that are strictly reserved to precision and speed of motion. This heightens the need for robust adaptive control techniques which have not been investigated in sufficient depth [10]. The controller ensures the precise and robust tracking of the trajectory, hence its importance. Furthermore, a number of techniques have emerged. They include proportional, integration, derivative (PID) control, and Computed Torque Control [11], adaptive control [12], neural networks control [13] and fuzzy adaptive control [14].

Good knowledge of the inverse dynamic model is a requirement for computed torque control. Theoretically, it ensures the linearization and decoupling of the equations that govern the robot's motion, hence resulting in a uniform response for a given configuration. This technique is relatively effective in terms of accuracy for quick motion than both PD and PID controls [15]. However, it does present a degree of sensitivity to changes in the parameters of the system and external influences[16]. In practice, it is difficult to know the dynamic model of the robot with relative precision. Therefore, to deal with the problem of dynamic model uncertainties, an a compensation method is required. In the area of artificial intelligence techniques for system control, a variety of paradigms are proposed, they include neural network control [13, 17] and fuzzy control [18, 19, 20, 21]. Research in relation to neural networks and the learning paradigm is conducted in order to improve trajectory tracking. In the literature, two types of neural control approaches are proposed. The first consists of specifying the neural controller parameters which are transformed into a shortcoming due to the controller inability to adapt to parametric variations in the system. To overcome this shortcoming another approach provided adaptation in the controller parameters. In comparison, the time complexity of the second control type is shown to be higher than the first control type. The suggested neural methods are able to be differentiated through the alternative to incorporate an ex-ante knowledge in the model, the number of measured joint variables, the guaranty of the closed-loop stability as well as the number of parameters. In [17], only one neural network is used for tracking position-velocity trajectories. In [22], the model structure is known ex-ante and the functions of the model are identified by a neural network at run-time. The adaptation of the neural parameters is established by the system's closed-loop stability in the Lyapunov sense. Another set of adaptive neural control techniques using the system model (nominal or identified) were also studied [8, 23].

Another approach [24, 25] based on sliding mode control was suggested for the purpose of robust control of robotic systems. This is characterized by resilience against external perturbations and parametric changes. It is a practically adequate means for the nonlinear control of robotic systems. However, it does present a major drawback which is reflected in the discontinuity of the control signal. To overcome this problem of discontinuity, several solutions were proposed in the literature. In [26], the *sign* function is substituted with the sat (saturation) function. Other methods [27, 28] resort to neural networks in conjunction with sliding modes to decrease the chattering phenomenon.

All methods mentioned above deal with the position control of the nonlinear systems. When the system comes in contact with its environment, the force control is required. Besides, several approaches have been proposed in the literature. The problem to control the constrained motion of the robot has received considerable attention in the literature due to its complexity [32, 33, 35]. For these approaches, the problem of force and position regulation has been successfully solved. On the other hand, various approaches have been proposed in the literature, which require a knowledge of the robot model. These methods are not sufficient because of parametric variations of the robot model. In order to overcome this problem, an adaptive approach is necessary [36, 37, 38].

A second category of force control has been proposed such as impedance control [39]. Latter achieve indirect force control by means of closed-loop position. Furthermore, an impedance control has been studied in [40], where the impedance parameters like mass, damping and spring are exactly known. However, in practice these parameters change, then this method is not interesting because it does not take into account the parametric variations.

Some research has been directed towards development of neural networks-based approaches for control of complex processes, and satisfactory results have been obtained in [41, 42, 43, 49]. These approaches deal with neural adaptive force control; the structure of robot dynamic model is considered uncertain. Nevertheless, the major short-coming of the existing methods is the possibility of undesired oscillation which may occur until the controller adapts to the environment dynamics. Another method has been proposed in [44] would not need ex-ante information about the surface elastic model linearity. However, the disadvantage of this approach is the used linear neural network approximator to estimate the stiffness parameters.

In addition, there are other methods where both the dynamics as well as the inverse kinematics models are required to calculate the control law. In this case, the control signal must be calculated in real time from the complex dynamic equations. However, this inversion of the kinematics model can cause instability of the system in closed loop. In order to avoid the online calculation of the inverse kinematics model, a new interest research in neural networks has been conducted. Thus, various methods based on neural networks have been widely studied by researchers to model the inverse kinematics of robot [45, 46]. An approach based on neural networks has been proposed in [47] to control an assembly cell. This approach realizes, simultaneously, an identification and control of systems, and it is implemented on a C5 parallel robot. However, in this technique the neural network parameters are not adaptive in Lyapunov sense but they are calculated by using the dynamic back propagation method. Thus, the stability study of system in Lyapunov sense is not taken into account.

We present in this chapter a synthesis of our research developed in [48, 34]. Full theoretical formulations of the adaptive position and force control approaches are presented. The stability of the two control systems (position control and force control) in the Lyapunov sense is investigated with the objective to work out the control laws. Regarding the proposed position control approach, it is based on the coupling of sliding modes and MLP neural networks. It does not require the inverse dynamic model for deriving the control law which is unlike to [19, 29] approaches. The basis of the developed control law is the reference model. The generally bounded inertia and Coriolis terms are compensated by the sliding mode control terms. In addition, the forces of gravity, the frictions and the non- modeled dynamic effects are estimated by a neural network in view of exploiting the advantages of each technique via sliding modes and neural network. The force control approach also developed in this chapter is based on the multi-layer perceptron (MLP), without an ex-ante knowledge of the robot's dynamic model. Otherwise, the control type that we propose to apply is a black box one. The neural network corrective is used adaptively in the goal to ensure the system stability in the Lyapunov sense, unlike the previous approach cited in [47]. We implemented on a C5 parallel robot and carried out tests of robustness against external influences/perturbations in order to check for contributions in this research stream.

This chapter is organized as follows: Section 2 describes the mechanical architecture of the C5 parallel robot and the nomenclature of notations used in the chapter. Section 3 is dedicated to the position controller design and stability analysis. The experimental results are detailed and discussed in subsection 3.4. Section 4 is dedicated to the force controller design. Experimental results are shown in subsection 4.5. Finally, in this section a conclusion is drawn and future perspectives are presented.

3.2. Description of the C5 parallel robot and nomenclature

3.2.1. Description of the C5 parallel robot. The experimental setup shown in Figure 3.2.1 is composed of a 2D Cartesian robot linked to a 6 DOF parallel robot (also called C5 parallel robot) as shown in figure 3.2.2. The 2D Cartesian robot allows moving parts in order to carry them from a given position towards the operational area. It can perform small corrective displacements around the nominal trajectory. The 2D Cartesian robot allows movement of the parts so as to carry them from a specific position to the operational area. It is also able to perform relatively small corrective displacements around the nominal trajectory. The C5 parallel robot consists of a static part and a mobile part both connected through six actuated links. Each segment is embedded to the static part at point A_i and connected to the mobile part at point B_i via a spherical joint attached to two crossed sliding plates (Figures 3.2.2 and 3.2.3).

A theoretical research effort concerning this type of design architecture was presented in [30]. The C5 parallel robot is furnished with six linear actuators; each driven by a DC motor which drives a ball and screw arrangement. The measurements of positions are obtained from six incremental encoders, connected to the DC motors.

3.2.2. Nomenclature of notations.

- J_p: The robot's Jacobian matrix.
- Γ: The torque vector.
- q: The joint positions' vector.
- $\dot{q}$: The joint velocities vector.

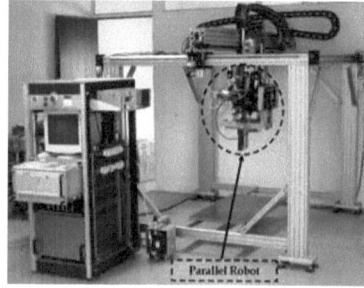

FIGURE 3.2.1. Experimental setup.

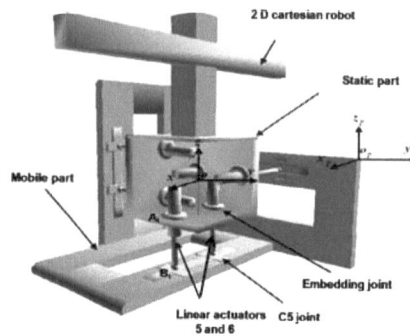

FIGURE 3.2.2. C5 parallel robot.

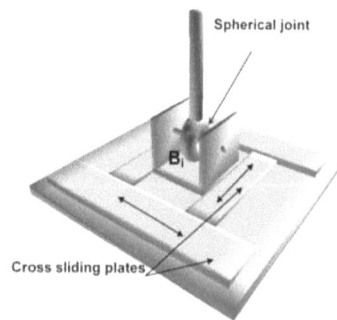

FIGURE 3.2.3. Detail of the C5 joint

- $\ddot{q}$: The joint accelerations vector.
- $M(q)$: The inertia matrix which is symmetric and positive definite.
- $C(q, \dot{q})\dot{q}$: The centrifugal and Coriolis vectors.
- $\varpi(q, \dot{q})$: The friction, gravitational, and the non-modeled dynamic effects vector.
- $I_{n \times n} \in R^{n \times n}$: The identity matrix.
- $x_m = \begin{bmatrix} q_m \\ \dot{q}_m \end{bmatrix}$: The state vector of the reference model.
- $A_m \in R^{2n \times 2n}$: The state matrix.
- $B_m \in R^{2n \times n}$: The input matrix.
- $c \in R^n$: The input vector.

- w_1^*: The weights between the inputs and the hidden layer.
- w_2^*: The weights between the hidden layer and the output.
- ε The approximation error, such that ($\|\varepsilon\| \leq \bar{\varepsilon}$), with a sufficiently small $\bar{\varepsilon}$.
- $\hat{\varpi}$ An estimate of ϖ (unknown).
- w_1 and w_2 are the estimates of w_1^* and w_2^* respectively.
- ϵ_φ represent the error due to the first order Taylor-Young series approximation.
- c_1: The positive constant calculated from the derivative $\varphi\prime$.
- $\|\cdot\|_F$ represents the Frobenius norm.
- $\Lambda \in R^{n \times n}$: The square matrix.
- k_v: The positive-definite matrix.
- p: The number of hidden neurons.
- m: The dimension of x.
- n: The neural network output number.
- $e_f = f_d - f$; f_d and f are the desired and measured joint forces respectively.
- $\dot{e}_f = \dot{f}_d - \dot{f}$; $\dot{e}_f$ is the derivative of force error e_f.
- k_p and k_v are positive definite diagonal matrices.
- k_r: The positive definite diagonal matrix.
- $sign(.)$: The sign function.
- $k_r sign(e_f)$: The robustness term with respect to external. disturbances.
- η_1, η_2: The positive matrices of adaptation gain.

3.3. Adaptive position controller design

The purpose of this section is to design a robust adaptive position controller for the C5 robot. An adaptive multi-layer perceptron (MLP) neural network and sliding mode technique is used to work out the control laws which are the state vector of the equation (3.3.3) tracks precisely the state vector of the reference model given in equation (3.3.5) (reference trajectory). Indeed, the role of the MLP neural network is to estimate gravitational and frictional forces and the non- modeled dynamic effects represented by equation (3.3.1). The sliding mode terms are used to compensate for inertia, Coriolis and centrifugal forces.

3.3.1. Problem formulation. The inverse dynamic model of the C5 parallel robot can be expressed in the joint space in the following standard form:

$$(3.3.1) \qquad M(q)\ddot{q} + C(q,\dot{q})\dot{q} + \varpi(q,\dot{q}) = \Gamma$$

Let $x \in R^{2n}$ be the state vector defined as:

$$(3.3.2) \qquad x = \begin{bmatrix} q \\ \dot{q} \end{bmatrix}$$

In order to work out the control laws, the dynamic model (3.3.1) should be re-written in the form of a state equation:

$$(3.3.3) \qquad \dot{x} = \underbrace{\begin{bmatrix} 0_{n \times n} & I_{n \times n} \\ A_1 & A_2 \end{bmatrix}}_{A(q)} x + \underbrace{\begin{bmatrix} 0_{n \times n} \\ B_1 \end{bmatrix}}_{B(q)} u - \underbrace{\begin{bmatrix} 0_n \\ H_1 \end{bmatrix}}_{H(q,\dot{q})}$$

where

$$(3.3.4) \qquad \begin{cases} A_1 = 0_{n \times n} \\ A_2(q,\dot{q}) = -M^{-1}(q)C(q,\dot{q}) \in R^{n \times n} \\ B_1(q) = M^{-1}(q) \in R^{n \times n} \\ H_1(q,\dot{q}) = M^{-1}(q)\varpi(q,\dot{q}) \in R^n \end{cases}$$

Consider the following reference model:

$$(3.3.5) \qquad \dot{x}_m = \underbrace{\begin{bmatrix} 0_{n \times n} & I_{n \times n} \\ A_{m1} & A_{m2} \end{bmatrix}}_{A_m} x_m + \underbrace{\begin{bmatrix} 0_n \\ B_{m1} \end{bmatrix}}_{B_m} c$$

For the undertaken experiments, the reference trajectory is computed by choosing the input signal c with a quasi-sinusoidal shape. The output of the reference model denoted by x_m is then used as desired trajectory in the control scheme.

3.3.2. MLP neural network approximation. Our function approximation ϖ is a Multi-Layer Perceptron (MLP) neural network with one hidden layer and a linear output. The approximation is structured as follows: $w_2^{*t} \varphi(w_1^{*t} x)$ where $\varphi : R^p \rightarrow R^p$ is the activation functions vector and $w_1^{*t} \in R^{p, b}$, $w_2^{*t} \in R^{n, p}$ are the optimal parameters to be approximated in an adaptive way (b, is input size of the neural network x, n is the number of outputs of the neural network and p the number of hidden neurons).

For any x, we have:

$$\| \varpi(x) - \widehat{\varpi}(w_1^*, w_2^*, x) \| < \varepsilon$$

with:

$$\widehat{\varpi}(w_1^*, w_2^*, x) = w_2^{*t} \varphi(w_1^{*t} x)$$

The neural approximation ϖ can be written in the following manner:

$$\widehat{\varpi} = w_2^t \varphi(w_1^t x)$$

First order Taylor series expansion is used in order to deal with the non-linearity of the function φ,.

$$(3.3.6) \qquad \varphi(w_1^{*t} x) = \varphi(w_1^t x) - \varphi\prime(w_1^t x) \tilde{w}_1^t x - O(w_1^{*t} x)$$

with:

•

$$\tilde{w}_1 = w_1 - w_1^*$$

We remark that $O(w_1^{*t} x)$ tends to 0 if $\tilde{w}_1$ tends to 0

To simplify the problem formulation, the following notations are introduced:

$$\varphi = \varphi(w_1^t x)$$
$$\varphi^* = \varphi(w_1^{*t} x)$$
$$\varphi\prime = \varphi\prime(w_1^t x)$$

$$\varphi(w_1^t x) - \varphi(w_1^{*t} x) = \tilde{\varphi}$$

Let us consider the following equation:

$$(3.3.7) \qquad \tilde{\varpi} = \varpi(w_1^*, w_2^*, x) - \widehat{\varpi}(w_1, w_2, x)$$
$$(3.3.8) \qquad = w_2^{*t} \varphi^* - w_2^t \varphi + \varepsilon$$
$$(3.3.9) \qquad = \left(w_2^t - \tilde{w}_2^t \right) (\varphi - \tilde{\varphi}) - w_2^t \varphi + \varepsilon$$
$$(3.3.10) \qquad = w_2^t \varphi - w_2^t \tilde{\varphi} - \tilde{w}_2^t \varphi + \tilde{w}_2^t \tilde{\varphi} - w_2^t \varphi + \varepsilon$$
$$(3.3.11) \qquad = \tilde{w}_2^t \tilde{\varphi} - w_2^t \tilde{\varphi} - \tilde{w}_2^t \varphi + \varepsilon$$

By using equation (3.3.6), we get:

$$(3.3.12) \qquad \tilde{\varpi} = -w_2^t \varphi\prime \tilde{w}_1^t x - \tilde{w}_2^t \varphi + \epsilon_\varphi + \varepsilon$$
$$\epsilon_\varphi = \tilde{w}_2^t \varphi\prime \tilde{w}_1^t x - w_2^{*t} O(w_1^{*t} x)$$

If $\tilde{w}_1^t$ and $O(w_1^{*t} x)$ tend to 0 and owing that $\tilde{w}_2^t$ is bounded then ϵ_φ tends to 0.
Recall that:

$$(3.3.13) \qquad \| \epsilon_\varphi \| \le c_1 \| \tilde{w}_2 \|_F \| \tilde{w}_1 \|_F \| x \| + \| w_2^* \|_F \| O \|$$
$$\bar{\epsilon}_\varphi = c_1 \| \tilde{w}_2 \|_F \| \tilde{w}_1 \|_F \| x \| + \| w_2^* \|_F \| O \|$$

with:

(3.3.14)
$$\|w_1^*\|_F \le \|w_1^*\|_{max}$$
$$\|w_2^*\|_F \le \|w_2^*\|_{max}$$

The equation (3.3.12) will be used in the subsection 3.4 for the stability analysis.

3.3.3. Control laws design.
The tracking error is defined as follows:

(3.3.15)
$$e = x_m - x$$

(3.3.16)
$$x_m = x + e$$

The rationale behind our approach can be summarized in the following points:

  (1) determines a stable sliding surface $s = 0_n$, so that in ideal sliding mode, the state vector defined in (3.3.3) tracks the reference trajectory x_m, i.e the tracking error e tends to 0 when t tends to ∞.
  (2) determines the control laws thus facilitating the attraction of the system's state to the sliding surface $s = 0_n$.

Let $s \in R^n$ be the vector defined as follows:

(3.3.17)
$$s = Ue$$

(3.3.18)
$$U = [\Lambda, I_{n\times n}] \in R^{n \times 2n}$$

Let $e = \left[\epsilon^T, \dot{\epsilon}^T\right]^T$, with $\epsilon = (q_m - q) \in R^n$. In ideal sliding mode ($s = 0_n$), the sliding error dynamics is given by:

(3.3.19)
$$\dot{\epsilon} = -\Lambda\epsilon$$

Therefore, the stability of the sliding surface $s = 0_n$ is guaranteed if and only if Λ is a positive-definite matrix.
The proposed control law is given by:

(3.3.20)
$$u = K \cdot x + L \cdot e + N \cdot c + \widehat{\varpi}(q, \dot{q}) + k_v s$$

with:

(3.3.21)
$$\begin{cases} K = \bar{W}_1 + W_1 \operatorname{sgn}(x)^T \in R^{n \times 2n} \\ L = \bar{W}_2 + W_2 \operatorname{sgn}(e)^T \in R^{n \times 2n} \\ N = \bar{W}_3 + W_3 \operatorname{sgn}(c)^T \in R^{n \times n} \end{cases}$$

$\bar{W}_1 \in R^{n \times 2n}, \bar{W}_2 \in R^{n \times 2n}$ and $\bar{W}_3 \in R^{n \times n}$ are constants; $W_i \in R^n$ are variable terms defined as follows:

(3.3.22)
$$W_i = r_i \left(\alpha_i s + \operatorname{sgn}(s)\right) \text{ for } i = 1, 2, 3$$

where $r_{\{.\}} \in R^+$ are adaptive positive gains, which are calculated from an adaptation algorithm whose basis is closed-loop system stability (in the Lyapunov sense):

(3.3.23)
$$\begin{cases} \dot{r}_1 = \eta_1(\|s\|_1 - r_1 \|s\|_2^2) \|x\|_1 \\ \dot{r}_2 = \eta_2(\|s\|_1 - r_2 \|s\|_2^2) \|e\|_1 \\ \dot{r}_3 = \eta_3(\|s\|_1 - r_3 \|s\|_2^2) \|c\|_1 \end{cases}$$

$\eta_{\{.\}} \in R^{*+}$ and $\alpha_{\{.\}} \in R^{*+}$ are arbitrary positive adaptation gains.
  α_k is selected such that:

$$\rho_1 \alpha_k \ge \beta_k \qquad \forall k \in \{1, 2, 3\}$$

where ρ_1 and β_k are two bounded strictly positive constants.
  Assumption : *To design control laws, matrices $M(q)$ and A must be bounded:*

(3.3.24)
$$\begin{aligned} \left\|M(q)U(A_m - A) - \bar{W}_1\right\|_2 &\le \beta_1 \\ \left\|M(q)U(A_m + I_{2n\times 2n}) - \bar{W}_2\right\|_2 &\le \beta_2 \\ \left\|M(q)B_{m1} - \bar{W}_3\right\|_2 &\le \beta_3 \end{aligned}$$

The neural network parameters used for estimation $\varpi(q, \dot{q})$ are calculated from the adaptation law given in equation (3.3.25), which is worked out from the stability analysis presented in the next section

(3.3.25)
$$\begin{cases} \dot{w}_1 = \gamma_2 x s^t w_2^t \varphi' \\ \dot{w}_2 = \gamma_3 \varphi s^t \end{cases}$$

where γ_2 and γ_3 are two positive gain matrices.

Concerning the stability analysis, we have used the Lyapunov principle (for more details see [**34**]).

3.3.4. Experimental results and discussion. In order to demonstrate the relevance of the proposed approach, it was implemented and tested for the position control of the C5 parallel robot in Matlab/Simulink environment with a Real-Time dSPACE DS1103. The desired trajectory is a quasi-sinusoidal trajectory, in these experiments. This study consists of two cases: 1) the adaptive control law given in equation (3.3.20) with a multi-layer perceptron neural network (MLP-NN) is implemented; 2) the MLP-NN is not implemented. Numerous experiments were carried out in order to select the neuron number in the hidden layer. For the purpose of obtaining an improved trajectory tracking, this number is set to the value of 3. The characteristics of the neural network and the gain of control laws are given below:

TABLE 1. Parameters of the proposed approach.

Number of NN inputs	12
Number of NN outputs	06
Number of hidden layer neurons	03
$diag(k_v)$	150
$\alpha_i,\ (i = 1, 2, 3)$	200
$\eta_1 = \eta_2 = \eta_3$	5
$diag(\gamma_2)$	20
$diag(\gamma_3)$	20
A_{m1}	$-4\,I_{6\times 6}$
A_{m2}	$-4\,I_{6\times 6}$
B_{m1}	$4\,I_{6\times 6}$
Identity matrix $I_{n\times n}$	$I_{6\times 6}$

A study was carried in similar experimental conditions. Its aims were to do a comparison between the results obtained for the above two cases i.e. with and without MLP-NN. The trajectory tracking in relation to the control law with MLP-NN is represented by a solid line. The dotted line represents the control law without MLP-NN. The reference trajectory is generated from a reference model which corresponds to a second-order system that is relatively stable. According to 3.3.1, the control law presented in this study ensures a good performance trajectory tracking. It is observed that trajectory tracking is greatly reduced in the absence of the neural network.

To validate the effectiveness of the proposed control scheme against external disturbances, another experiment was also done. A mass was attached to the end-effector of robot in its rest position and retired at time t=7000 ms. As shown in Figure 3.3.2, the disturbance is completely absorbed at t=9000 ms.

During the disturbance, the tracking error is very important but the disturbance is quickly rejected as shown in Figure 3.3.2. After the rejection of disturbance, the robot continues to satisfactorily track the reference model states.

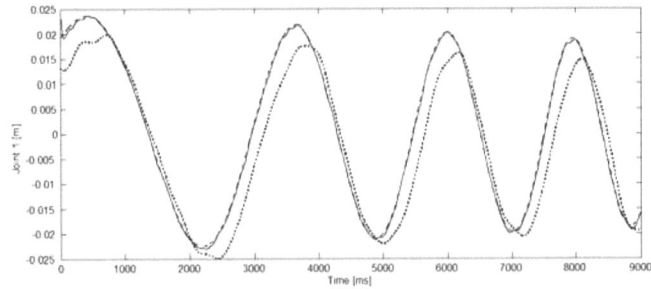

FIGURE 3.3.1. Tracking errors for the first axis (NN in solid line, without NN in doted line)

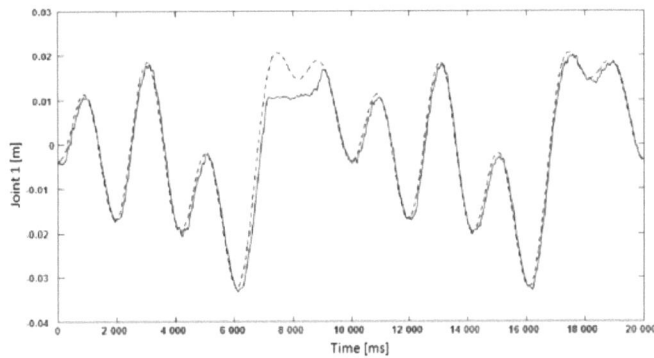

FIGURE 3.3.2. Control robustness test for axis 1

3.4. Adaptive force controller design

In this section, the adaptive force control of the parallel robot is developed. The force control law has been developed in order to compensate the dynamic interaction between the end-effector motion and the environment induced force. It is based on the multi-layer perceptron (MLP), without an ex-ante knowledge of the robot's dynamic model. Otherwise, the control type that we propose to apply, is a black box one. The neural network corrective is used adaptively in the goal to ensure the stability of the system in Lyapunov sense.

3.4.1. Problem formulation. We consider in this study, a C5 parallel robot whose dynamic model is unknown. It can be presented under the following black box formulation:

$$(3.4.1) \qquad\qquad\qquad\qquad H(z) = u$$

where H is a nonlinear function whose expression is unknown and u is the control input. z is an input vector of the neural network (MLP) whose content is fixed thereafter by a practice manner. In fact, the content of the vector z can be fixed by a priori training on a data base relative to the studied system. This solution allows us to avoid the noisy accelerations data in our adaptive control law given in subsection 4.3 (Control laws design). Consequently, the content of z contains only the joint positions, velocities and forces $(q, \dot{q}, f)$. All detail of the determination of the content of z has been presented in our previous work [**48**].

3.4.2. Neural network non linearity treatment.
Using the development in Taylor-Young series, to treat adaptively the non linearities, the function φ can be written:

$$(3.4.2) \qquad \varphi(w_1^{*T} z) = \varphi(w_1^T z) - \varphi'(w_1^T z)\widetilde{w}_1^T z - O(w_1^{*T} z)\widetilde{w}_1^T z$$

with $\widetilde{w} = w_1 - w_1^*$ and $O(w_1^{*T} z)$ converges to zero if $\widetilde{w}_1$ converge to zero, φ' represent the bounded derivative of φ. To simplify we use the following notations:

$$(3.4.3) \qquad \begin{cases} \varphi = \varphi(w_1^T z) \\ \varphi^* = \varphi(w_1^{*T} z) \\ \varphi' = \varphi'(w_1^T z) \\ \widetilde{\varphi} = \varphi(w_1^T z) - \varphi(w_1^{*T} z) = \varphi - \varphi^* \end{cases}$$

It comes that:

$$(3.4.4) \qquad H - \widehat{H} = w_2^{*T}\varphi^* - w_2^T\varphi + \epsilon$$

where ϵ is the neural approximation error such that $(\|\epsilon\| \leq \bar{\epsilon})$, with a known and relatively small $\bar{\epsilon}$

Adding and subtracting $w_2^{*T}\varphi$ in equation $(3.4.4)$:

$$(3.4.5) \qquad H - \widehat{H} = -w_2^{*T}\widetilde{\varphi} - \widetilde{w}_2^T\varphi + \epsilon$$

Adding and subtracting $w_2^T\widetilde{\varphi}$ into equation $(3.4.5)$ we obtain:

$$(3.4.6) \qquad H - \widehat{H} = \widetilde{w}_2^T\widetilde{\varphi} - w_2^T\widetilde{\varphi} - \widetilde{w}_2^T\varphi + \epsilon$$

Using Taylor-Young approximation we have:

$$(3.4.7) \qquad H - \widehat{H} = -w_2^T\varphi'\widetilde{w}_1^T z - \widetilde{w}_2^T\varphi + \epsilon + \sigma$$

where σ represents the approximation errors due to Taylor-Young series of the first order:

$$(3.4.8) \qquad \sigma = (\widetilde{w}_2^T\varphi' - w_2^{*T}O(w_1^{*T} z))\widetilde{w}_1^T z$$

with

$$\sigma \to 0$$
$$if\ \widetilde{w}_1 \to 0\ and\ \widetilde{w}_2\ is\ bounded$$

Recall that:

$$(3.4.9) \qquad \|\sigma\| \leq c_1 \|\widetilde{w}_2\|_F \|\widetilde{w}_1\|_F \|z\| + \|w_2^*\|_F \|O\| \|\widetilde{w}_1\|_F \|z\|$$
$$\bar{\sigma} = c_1 \|\widetilde{w}_2\|_F \|\widetilde{w}_1\|_F \|z\| + \|w_2^*\|_F \|O\| \|\widetilde{w}_1\|_F \|z\|$$

where c_1 is the positive constants computed from the derivative φ'. $\|\cdot\|_F$ represents the Frobenius norm. with

$$(3.4.10) \qquad \|w_1^*\|_F \leq \|w_1^*\|_{max}$$
$$\|w_2^*\|_F \leq \|w_2^*\|_{max}$$

The expression $(3.4.7)$, serves us to determine the adaptation laws of the neural network parameters.

3.4.3. Control laws design. We define the desired contact force F_d in task space, and then to develop our controller it is necessary to calculate the desired forces f_d in joint space by using the inverse force model (IFM).
Let us consider the following control law:

$$(3.4.11) \qquad u = \widehat{H} + k_p e_f + k_v \dot{e}_f + k_r sign(e_f)$$

with

From equations (10.2.1) and (3.4.11), it comes that:

$$(3.4.12) \qquad H - \widehat{H} = \widetilde{H} = k_p e_f + k_v \dot{e}_f + k_r sign(e_f)$$

we obtain:

$$(3.4.13) \qquad \widetilde{H} = k_p e_f + k_v \dot{e}_f + k_r sign(e_f)$$

Substituting equation (3.4.7) into equation (3.4.13), we obtain:

$$
\begin{aligned}
k_p e_f + k_v \dot{e}_f + k_r sign(e_f) &= -w_2^T \varphi' \widetilde{w}_1^T z \\
&\quad -\widetilde{w}_2^T \varphi + \epsilon + \sigma \\
(3.4.14) \qquad k_v \dot{e}_f &= -k_p e_f - k_r sign(e_f) \\
&\quad -w_2^T \varphi' \widetilde{w}_1^T z - \widetilde{w}_2^T \varphi + \epsilon + \sigma
\end{aligned}
$$

The equation (3.4.14) will be used in the next subsection for the stability analysis.

3.4.4. Stability study. In order to calculate the adaptation laws of neural network, it is necessary to demonstrate the stability of our control system in closed loop.
Let us consider the following adaptation laws:

$$(3.4.15) \qquad \begin{cases} \dot{w}_1 = \eta_1 z e_f^T w_2^T \varphi' \\ \dot{w}_2 = \eta_2 \varphi e_f^T \end{cases}$$

Theorem 1. The system defined by (10.2.1) in closed loop with the control law given by (3.4.11) as well as the adaptation laws (3.4.15) is stable.

PROOF. Let the Lyapunov function:

$$(3.4.16) \qquad V = \frac{1}{2} e_f^T k_v e_f + \frac{1}{2} tr(\widetilde{w}_1^T \eta_1^{-1} \widetilde{w}_1) + \frac{1}{2} tr(\widetilde{w}_2^T \eta_2^{-1} \widetilde{w}_2)$$

Differentiating the equation (3.4.16), we get.

$$
\begin{aligned}
\dot{V} &= e_f^T k_v \dot{e}_f + tr(\widetilde{w}_1^T \eta_1^{-1} \dot{w}_1) \\
(3.4.17) \qquad &\quad + tr(\widetilde{w}_2^T \eta_2^{-1} \dot{w}_2)
\end{aligned}
$$

Using equation (3.4.14) into (3.4.17), we obtain:

$$
\begin{aligned}
\dot{V} &= e_f^T(-k_p e_f - k_r sign(e_f) - w_2^T \varphi' \widetilde{w}_1^T z - \widetilde{w}_2^T \varphi + \epsilon + \sigma) \\
(3.4.18) \qquad &\quad + tr(\widetilde{w}_1^T \eta_1^{-1} \dot{w}_1) + tr(\widetilde{w}_2^T \eta_2^{-1} \dot{w}_2)
\end{aligned}
$$

we develop equation (3.4.18), we get:

$$
\begin{aligned}
(3.4.19) \qquad \dot{V} &= -e_f^T k_p e_f - e_f^T w_2^T \varphi' \widetilde{w}_1^T z - e_f^T \widetilde{w}_2^T \varphi \\
&\quad + tr(\widetilde{w}_1^T \eta_1^{-1} \dot{w}_1) + tr(\widetilde{w}_2^T \eta_2^{-1} \dot{w}_2) \\
&\quad + e_f^T(\epsilon + \sigma) - e_f^T k_r sign(e_f) \\
(3.4.20) \qquad \dot{V} &= \dot{V}_1 + \dot{V}_2 + \dot{V}_3 + \dot{V}_4
\end{aligned}
$$

with

$$(3.4.21) \qquad \dot{V}_1 = -e_f^T k_p e_f + e_f^T(\epsilon + \sigma)$$

$$(3.4.22) \qquad \dot{V}_2 = -e_f^T w_2^T \varphi' \widetilde{w}_1^T z + tr(\widetilde{w}_1^T \eta_1^{-1} \dot{w}_1)$$

$$(3.4.23) \qquad \dot{V}_3 = -e_f^T \widetilde{w}_2^T \varphi + tr(\widetilde{w}_2^T \eta_2^{-1} \dot{w}_2)$$

$$(3.4.24) \qquad \dot{V}_4 = -e_f^T k_r sign(e_f)$$

According to equation (3.4.15), it comes that:

(3.4.25)
$$\begin{cases} \dot{V}_2 = 0 \\ \quad et \\ \dot{V}_3 = 0 \end{cases}$$

From the equation (3.4.24), we obtain:

$$\dot{V}_4 = -e_f{}^T k_r sign(e_f) \leq -\lambda_{\min}(k_r) \underbrace{e_f{}^T sign(e_f)}_{\|e_f\|_1} \leq 0$$

$\lambda_{\min}(k_r)$ is the smallest eigen value of k_r.
From the equation (3.4.21), we get:

(3.4.26)
$$\dot{V}_1 \leq -\lambda_{\min}(k_p) \|e_f\|^2 + \|e_f\| (\bar{\varepsilon} + \bar{\sigma})$$

$\lambda_{\min}(k_p)$ is the smallest eigen value of k_p.
$\dot{V}_1 \leq 0$ if:

(3.4.27)
$$\|e_f\|_2 \geq \frac{(\bar{\varepsilon} + \bar{\sigma})}{\lambda_{\min}(k_p)}$$

Consequently: $\dot{V} \leq 0$, and the system (10.2.1) is stable. $\qquad\qquad\qquad\qquad\qquad\qquad$ □

3.4.5. Experimental Results. We have experimentally validated the proposed approach in order to demonstrate its effectiveness. In this experiment, we define the desired trajectory F_d in task space. The trajectory has been chosen so that the robot comes into contact with its environment, only along the x axis previously described in section (Experimental cell description). Thus, in this case, the only component of the contact force vector F to control is F_x. We note that the sinusoidal trajectory is used like a desired force trajectory.

This study addressed two cases: the first case, we have tested a PID force control, in the second case; the proposed adaptive neural approach has been implemented. The role of the neural network is produce an estimation of gravitational forces, inertia, frictions and other dynamics of robot. The characteristics of this network are given as: input number = 12, output number = 6 and the hidden layer contains 3 neurons; the neural adaptation gains are: $\eta_1=\eta_2=0.5$. Here, the same proportional and derivative gains have been chosen in our controller and PID; they are designed as follows: $diag(k_p = 6)$, $diag(k_v = 0.01)$, $diag(k_r = 0.1)$.

A comparative study was conducted in similar experimental conditions. This study compared the results of two cases (PID control and neural adaptive control). All results given here are filtered in the same conditions.

According to Figure 3.4.1, the results obtained by the PID controller are insufficient. However we obtained a good force tracking when the neural adaptive control is used, as shown in Figure 3.4.2; in this situation, the measured force is much closer to the desired trajectory unlike the previous PID control.

3.5. Conclusion

This work has presented a robust adaptive position control approach which tracks the trajectory of the six degree of freedom C5 robot. To derive the control law, the adaptive artificial neural network and sliding mode techniques have been combined due to the advantages the approach presents, mainly that the ex-ante knowledge of the model parameters of the robot are not required. Both the neural parameters and sliding mode terms were worked out from the stability analysis in Lyapunov sense. Furthermore, in order to demonstrate the tracking performance of the proposed controller, it was implemented on the C5 parallel robot. The results produced with this approach confirm the precise tracking. However, the robustness of the control system was checked against external disturbances and to that end, a variety of tests have been performed. The experimental results demonstrate that the controller is robust. Secondly, the adaptive force control of a C5 parallel robot has also been proposed. By an a priori learning of the inverse dynamics of the considered system, we were able to determine the best content of the nonlinear function that represents this system. Using these results and one hidden layer perceptron to perform an adaptive estimation of all unknown dynamics, we established a stable adaptive force control law. The control scheme has then been implemented on a C5 parallel robot hence obtaining good performances. The proposed approach may be extended with a particular focus on hybrid force/position control of the C5 robot in prospective work. The dynamic interaction between the end-effector motion

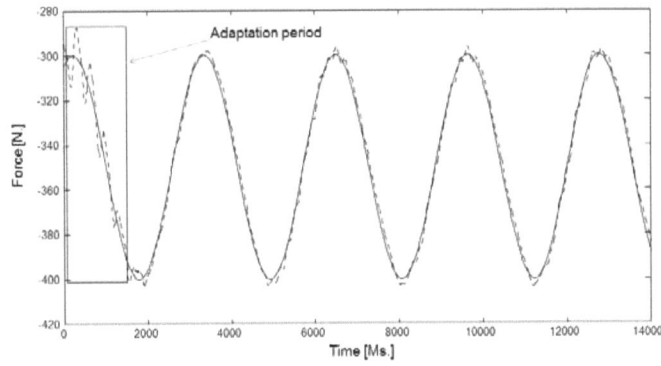

FIGURE 3.4.1. Trajectory tracking with PID controller

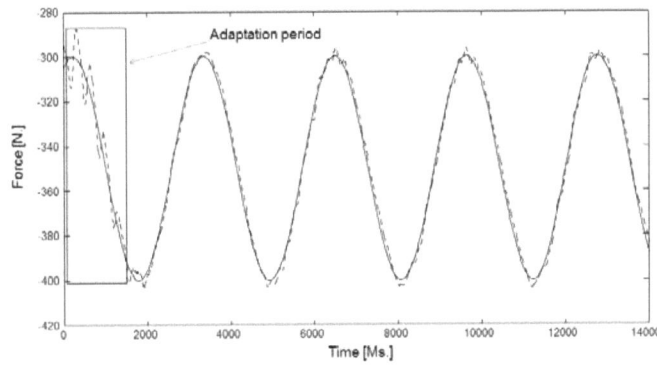

FIGURE 3.4.2. Trajectory tracking with neural controller

and the environment induced force will be compensated by the hybrid force/position control law. Thus, a parameter-adaptation algorithm which uses the stability criteria may be considered, and that would depend on both the position and the force tracking errors.

Bibliography

[1] Merlet, J.P. (2002), "Optimal design for the micro parallel robot mips," *In IEEE Int. Conf. on Robotics and Automation*, pp. 1149–1154.

[2] V. E. Gough. "Contribution to discussion of papers on research in automobile stability, control and tyre performance", Proceedings of the Auto Div. Inst. Mech. Eng, (1956-1957) pp. 392-394.

[3] Khalil, W., and Dombre, E. (2002), "Modeling, identification and control of robots," *Hermès penton Ltd.*

[4] Thornton, G.S. (2005), "The GEC Tetrabot-a new serial-parallel assembly robot," *In IEEE Int. Conf. on Robotics and Automation*,pp. 437-439, Philadelphia.

[5] Vivas, A., and Poignet, P. (2005), "Predictive functional control of a parallel robot," *Control Engineering Practice, Elsevier Science*, vol. 13, N. 7, pp. 863-874.

[6] Guegan, S. (2003), *Contribution à la modélisation et l'identification dynamique des robots parallèles,* PhD thesis, Ecole Centrale de Nantes.

[7] Vivas, A., Poignet, P., and pierrot, F. (2003), "Predictive functional control of a parallel robot," *Proceedings of the IEEE/RSJ International Conference on Intelligent Robots and Systems,*pp. 2785-2790.

[8] Achili, B., Daachi, B., Ali Cherif, A. and Y. Amirat (2008), "Compensateur neuronal adaptatif pour la commande d'un robot parallèle à liaison C5," *CIFA'2008, Bucarest, Roumanie.*

[9] Achili, B., Daachi, B., Ali Cherif, A. and Y. Amirat (2010), "A C5 Parallel Robot Identification and Control," *Intenational Journal of control, automation and systems,*vol.8, no.2, pp.369-377.

[10] Achili, B., Daachi, B., Ali Cherif, A., and Y. Amirat (2009), "Combined Multi-Layer Perceptron Neural Network and Sliding Mode Technique for Parallel Robots Control: An Adaptive Approach,"*IEEE International Joint Conference on Neural Networks,* pp. 3121-3128, Atlanta - Georgia.

[11] Middleton, R.H., and Goodwin, G.C. (1988), "Adaptive computed torque control for rigid link manipulators," *System Control Lett*, 10, pp. 9-16.

[12] Slotine, J.J.E., and Li, W. (1988), "Adaptive manipulator control: a case study," *IEEE Trans. Automat. Control*, 33, pp. 995-1003.

[13] Sun, F., Sun, Z., and Woo, P.Y. (2001), "Neural network-based adaptive controller design of robotic manipulators with an observer," *IEEE Trans. Neural Networks*, 12, pp. 54-67.

[14] Goléa, N., and Goléa, A. (2008), "Observer-based adaptive control of robot manipulators: Fuzzy systems approach," *Applied soft computing*, vol. 8, Issue 1, pp. 778-787.

[15] Luh, J.Y.S. (1983), "Conventional controller design for Industrial robots-a tutorial," *IEEE Trans. Systems Man Cybernet*, 13, pp. 298-316.

[16] Zhiyong, Y., Jiang, W., and Jiangping, M. (2007), "Motor-mecanism dynamic model based neural network optimized computed torque control of a high speed parallel manipulator," *Mechatronics*, 17, pp. 381-390.

[17] Lewis, F. L., Liu, K., and Yesildirek, A. (1995), "Neural net robot controller with guaranteed tracking performance," *IEEE Trans. on Neural Net.*, vol. 6, N. 3, pp.703-715.

[18] Chen, B.S., Uang, H.J., and Tseng, C.S. (1998), "Robust tracking enhancement of robot systems including motor dynamics: a fuzzy-based dynamic game approach," *IEEE Trans. Fuzzy Systems*, 6, pp. 538-552.

[19] Song, Z., Yi, J., Zhao, D., and Li, X. (2005), "A computed torque controller for uncertain robotic manipulator system: Fuzzy approach," *Fuzzy Sets and Systems*, 154, pp. 208-226.

[20] Touati, Y., and Amirat, Y. (2006), "Fuzzy logic controller design methodology for Cartesian robot control," *International Journal of Computer Applications in Technology*, vol.27, n.2/3, pp.85-96.

[21] Touati, Y., Amirat, Y., and Ali Cherif, A. (2007), "Fuzzy logic based approach for robotics systems control. stability analysis," *IROS 2007*, pp.3968-3973.

[22] Daachi, B., and benallegue, A. (2006), "A neural network controller for end-effector tracking of redundant robot manipulators," *J Intell Robot Syst*, 46, pp 245-262.

[23] Tso, S.K., Fung, Y.H., and Lin, N.L. (2000), "Analysis and real-time implementation of a radial-basis-function neural-network compensator for high-performance robot manipulators," *Mechatronics*, 10, pp. 265-287.

[24] Madani, A., and Benallegue, T. (2006), "Backstepping Sliding Mode Control Applied to a Miniature Quadrotor Flying Robot," *IEEE Industrial Electronics, IECON 2006 - 32nd Annual Conference on*, pp 700-705.

[25] HongBo, G., YongGuang, L., GuiRong, L., and HongRen, L. (2008), "Cascade control of a hydraulically driven 6-DOF parallel robot manipulator based on a sliding mode," *Control Engineering Practice*, 16, pp. 1055-1068.

[26] Slotine, J.J.E., and LI, W. (1988), "Adaptive manipulator control: a case study," *IEEE Trans. Automat. Control*, 33, pp. 995–1003.

[27] Ertugrul, M., and Kaynak, O. (2000), "Neuro sliding mode control of robotic manipulators," *Mechatronics*, Vol. 10.

[28] AV Topalov, and Kaynak, O. (2007), "Neuro-adaptive SM tracking control of robot manipulators," *Int. J. of adaptive control and signal processing*, Vol. 21, pp. 674-691.

[29] Lin, F.J., and Wai, R.J. (2002), "hybrid computed torque controlled motor-toggle servomechanism using fuzzy neural network uncertainty observer," *Neurocomputing*, 48, pp. 403-422.

[30] Dafaoui, M., Amirat, Y., and Pontnau, J. (1998), "Analysis and design of a six dof parallel robot. Modeling, singular configurations and workspace," *IEEE Transactions on Robotics and Automation*, Vol 14, N 1, pp. 78-92.

[31] Khalil, W., and Ibrahim, O. (2004), "General Solution for the Dynamic Modeling of Parallel Robots," *International Conference on Robotics and Automation*, New Orleans.

[32] Minami, M., and Xu, W. (2008) "Shape-grinding by Direct Position / Force Control with On-line Constraint Estimation," *International Conference on Intelligent Robots and Systems*, pp. 943-948 Nice, France, Sept, 22-26.

[33] Zhao, Y., and Cheah, C. (2004) "Position and force control of robot manipulators using neural networks," *in Proceedings of the IEEE 2004 conference on robotics, automation and mechatronics*, pp. 300-305.

[34] Achili, B., Daachi, Amirat, Y. and Ali-Cherif A. (2010), "Robust Adaptive Control For a Parallel Robot, *International Journal of Control*, Vol. 83, Issue 10, pp. 2107-2119.

[35] Osumi, H., and Tomiyama, T. (2008) "Development of Force Control Device with High Power and High Resolution," *International Conference on Intelligent Robots and Systems*, pp. 943-948, Nice, France, Sept, 22-26.

[36] Farooq, M., and Wang, D. B. (2008) "hybrid force/position control scheme for flexible joint robot with friction between the end-effector and the environment," *International journal of engineering science* vol. 46, pp. 1266-1278.

[37] Chiu, C. S., Lian, K. Y., and Wu, T. C. (2004) "Robust adaptive motion/force tracking control design for uncertain constrained robot manipulators," *Automatica*, 40(12), pp. 2111-2119.

[38] Jung, S., and Hsia, T. C. (2000) "Robust neural force control scheme under uncertainties in robot dynamics and unknown environment," *IEEE Transactions on Industrial Electronics*, 47(2), pp. 403-412.

[39] Hogan, N. (1985) "Impedance control: an approach to manipulation," *Parts I-III, ASME J. Dyn. Systems Measurement Control*, 107, pp. 1-24.

[40] Khalil, W., and Dombre, E. (1999) *Modélisation, identification et commande des robots*, Hermès Science Publications, Paris.

[41] Karayiannidis, Y., and Doulgeri, Z. (2006) "An adaptive law for slope identification and force position regulation using motion variables,". *Proceedings of IEEE 2006 international conference on robotics and automation*, pp. 3538-3543.

[42] Lewis, F. L. (1998) "adaptive neural network control of robotic manipulators," *World Scientific Series in robotics and Intelligent Systems*.

[43] Daachi, B., Benallegue, A., and M'Sirdi, N. K. (2001) "A Stable Neural Adaptive Force Controller for a Hydraulic Actuator," *In proc. of IEEE ICRA*, pp 3465-3470.

[44] Karayiannidis, Y., Rovithakis, G., and Doulgeri, Z. (2007) "Force/position tracking for a robotic manipulator in compliant contact with a surface using neuro-adaptive control," *Automatica* 43, pp. 1281-1288.

[45] Zhang, T., Nakamura, M., Goto, S., and Kyura, N. (2005) "High accurate contour control of an articulated robot manipulator using a Gaussian neural network," *Int. J. Industrial Robot* 32 (5), pp. 408-418.

[46] Temurtas, F., Temurtas, H., and Yumusak, N. (2006) "Application of neural generalized predictive control to robotic manipulators with a cubic trajectory and random disturbances," *J. Robot. Auto. Syst.* 54 (1), pp. 74-83.

[47] Touati, Y., Amirat, Y., Saadia, N., and Ali-Cherif, A. (2008) "A neural network-based approach for an assembly cell control," *Elsevier Science Publishers* , Vol 14, pp. 1335-1343.

[48] B. Achili, B. Daachi, A. Ali-Cherif and Y. Amirat,(2009) "Robust neural adaptive force controller for a C5 parallel robot", *IEEE International conference on advanced robotics ICAR*, pp. 1-6, munich.

[49] A. Benallegue, D. Y. Meddah and B. Daachi, "Neural Network Identication and Control of a Class of Non linear Systems".16th IMACS WORLD CONGRESS 2000 on Scientific Computation, Applied Mathematics and Simulation, Lausanne (Switzerland), August 21-25, 2000.

CHAPTER 4

Optimal Mechatronics for Driving Simulator Design

Lamri Nehaoua[1], Hichem Arioui, Nicolas Séguy,
Informatique, Biologie Intégrative et Systèmes Complexes
Evry University
Evry, 91020, France
{nehaoua,arioui,seguy}@ibisc.univ-evry.fr

ABSTRACT. This chapter discusses the necessary conditions for the successful design of a driving simulator. This success is assessed, among other factors, by the quality of the rendered motions, when driving the simulator.

We discuss throughout this chapter, the philosophy of driving simulation to better explain the challenges of these applications. Particular attention will be paid to the interaction between the various entities, making up a simulation system operational. This interaction is intended to highlight the importance of mechatronics in the successful design of such a simulator.

Keywords: Driving Simulator, Mechatronics, Optimal Design.

4.1. Introduction

The concept of driving simulators existed as far back as the early 1970s. At that time, considerable research on safety was performed focusing on vehicle dynamics, in which stability issues were important. A dynamically correct simulator should be something that could be used to study how different vehicle parameters influence stability.

Simulators were mainly used in aviation and then primarily for training in the use of cockpit instruments. In a driving simulator, the driver does not rely on instruments to the same extent. Here, the surroundings and dynamic forces are more important. It soon became clear that a high fidelity driving simulator requires a sophisticated motion simulation, a detailed model of the vehicle's dynamics and a visual description of the road environment.

The achievement of a driving simulator is a serious multidisciplinary challenge, because each driving simulator is a single prototype and there is no specific standard design. The success of such implementation can be reached with the cooperation of different skills and experts, from the designer to the final user. Therefore, the realization of such a platform requires a design at various levels of abstraction in order to address the best different specifications and constraints. It is obvious that the achievement of each level will involve limited choices in finding the good balance between the various hardware and software components.

A driving simulator is considered as a complex mechatronics system. It is composed, among other components, of a mechanical platform and an embedded electronics for its functioning. Its design should follow the approaches of modeling and simulation of multi-physics to extract an optimal solution. Previously, the choice of the optimal architecture was done through seeking the most appropriate solution for each component. The performances of the whole are checked by a juxtaposition of the different blocks. Otherwise, and following a mechatronics approach, the optimal design of a driving simulator requires a multi-physic modeling and optimization of the system in its entirety. However, in practice, such an approach is still not feasible. Thus, the problem is divided into sub-problems that can be optimized separately. Then, each solution is evaluated and adjusted to provide a sub-optimal architecture.

The quality of a driving simulator is assessed by its fidelity level, from a perceptive point of view; to better feed-back the car motions into the driver. This restitution is largely dependent on the platform of mechatronics (mechanics, embedded electronics, actuation's technologies and control laws). Thus an optimized mechatronics strongly helps the

---

[1]Corresponding author

FIGURE 4.2.1. University of Padua Riding Simulator

system to provide optimum motions, conversely, a non-synchronized mechatronics induce perceptive errors to users or can even make them feel very uncomfortable.

4.2. Overview on existing simulators

The literature reveals that numerous driving simulators exist in the world. Whether academic, industrial or commercial, many institutions have built their own vehicle prototypes for different purposes [**1, 2, 3**]. However, the bibliography on two-wheeled riding simulators is sparse. In fact, Japanese and Italian industrial institutions did most works. In 1988, HONDA Corporation started to develop series of motorcycle simulators. The first prototype of a dynamic platform was designed to test the feasibility of driving simulation to reproduce the basic maneuvers of a motorcycle dynamics. The mobile platform has 7 actuated axes to simulate 4 Degrees of Freedom (DoF) including roll, yaw, pitch, and handlebar steer. A cradle mechanism was developed to simulate the feeling of sustained acceleration. Next, a second prototype was developed and installed in the center of traffic education at Suzuka since 1991 to assert the simulator's effectiveness as an approved training tool [**4, 5**]. The architecture of the new platform was completely modified. The cradle system was removed and only 3 DoF were retained: pitch ($\pm 10°$), roll ($\pm 15°$) and steer ($\pm 30°$).

Outside Japan, a simulator was born from the collaboration between the PERCRO laboratory and the motorcycle manufacturer Piaggio. Appointed as a rapid prototyping tool, it is based on a 6 DoF mechanical parallel platform, hydraulically actuated. A real scooter chassis is mounted on the mobile platform [**6**]. In the same way, a bicycle simulator was built at the Korean Advanced Institute of Sciences and Technologies (KAIST). The motion generation is also ensured by a 6 DoF mechanical platform electrically actuated on which a bicycle frame is fixed. The handlebar and the pedal are respectively equipped with active and passive haptic devices [**7**]. Based on a serial mechanical architecture, a riding simulator is designed at the department of mechanical engineering at Padua University [**8**], Figure 4.2.1. This simulator allows the simulation of 5 DoF including roll, pitch, yaw, lateral displacement and steer angle.

Within the framework of the French project SIMACOM , a riding motorcycle simulator was conceived in collaboration between the National Institute of Research in Transportation and Safety (INRETS) and the laboratory of Informatics, Integrative Biology and Complex Systems (IBISC) France [**9**]. It allows the simulation of 5 DoF, namely, roll, pitch, yaw, steer and arms displacement, Figure 4.2.2.

4.3. Single/Double track dynamics

For a driving simulator, the dynamic model of the vehicle is responsible for generating reference trajectories to move the mechanical platform. Accordingly, the platform computes the new states of the virtual vehicle in response to the various actions of the driver.

In theory, a vehicle has six degrees of freedom (DoF) consisting of three translations and three rotations, Figure 4.3.1. The translation on the X axis denotes the longitudinal displacement, along the Y axis is the lateral displacement and vertical translation is done along the Z axis which reflects the movement of the chassis. The rotation around the Z axis is the yaw ψ of the vehicle that determines its trajectory, a second rotation called roll φ around the X axis defines the inclination of the body when taking a turn or lane change. Finally, the rotation ψ around the Y axis describes the pitching of the vehicle encountered during acceleration and braking phases.

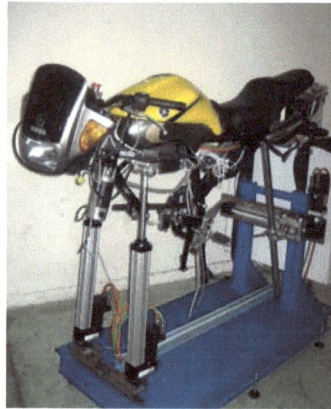

FIGURE 4.2.2. IBISC-INRETS Riding Simulator

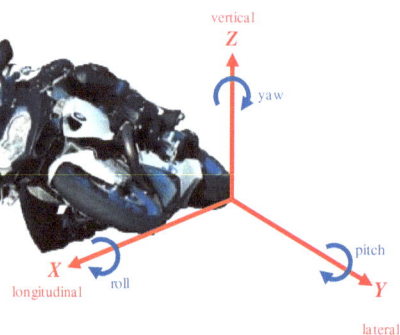

FIGURE 4.3.1. Single track vehicle motions and frames

Over its movement, most of the external forces acting on a vehicle are generated at tire-ground interface, Figure 4.3.2. Initially, we can identify the lateral guidance force, longitudinal force of traction / braking, the vertical one and their corresponding moments [10]. Although these forces are strongly linked, they are generally processed separately. On the other hand, aerodynamic actions consist of the longitudinal force, lateral thrust and vertical lift applied at the center of gravity. Beyond these efforts, of external origin, others are of intrinsic nature and result primarily from various connections and mechanical knowledge of the vehicle such as: suspension, steering, etc.

The complexity of the dynamic model depends on: (1) the mechanical architecture of the driving simulator and (2) the nature of the application to carry out. For example, if the simulator is used for automotive applications, the dynamic model must be dynamically rich. Otherwise, representations able to transcribe the lateral and the longitudinal dynamics are sufficient.

Considering the motorcycle as a set of rigid bodies connected by simple joints, multi-body system mechanics theory offers a convenient framework to derive its motion model. In [11], authors have adopted Lagrange formalism to derive their motorcycle dynamic model. A direct application of this approach, leads to unattractive performances in terms of the number of operations and implementation facilities. In Khalil [12] and Hollerbach [13], Recursive Newton-Euler Algorithm (RENA) was shown to be effective, Figure 4.3.3. Originally, this technique was developed to solve the inverse dynamic model of open chain manipulators with a fixed base for control purposes. By projecting the dynamics of each body in its attached frame, the acceleration of the joint variable can be easily derived without the need to inverse the whole system inertia matrix. In addition, this technique, named Articulated Body Algorithm (ABA), is more numerically stable than the inertia matrix inversion method.

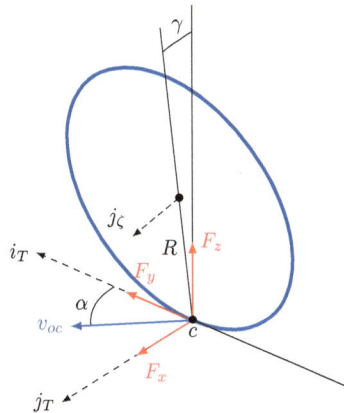

FIGURE 4.3.2. Efforts at tire-road interface

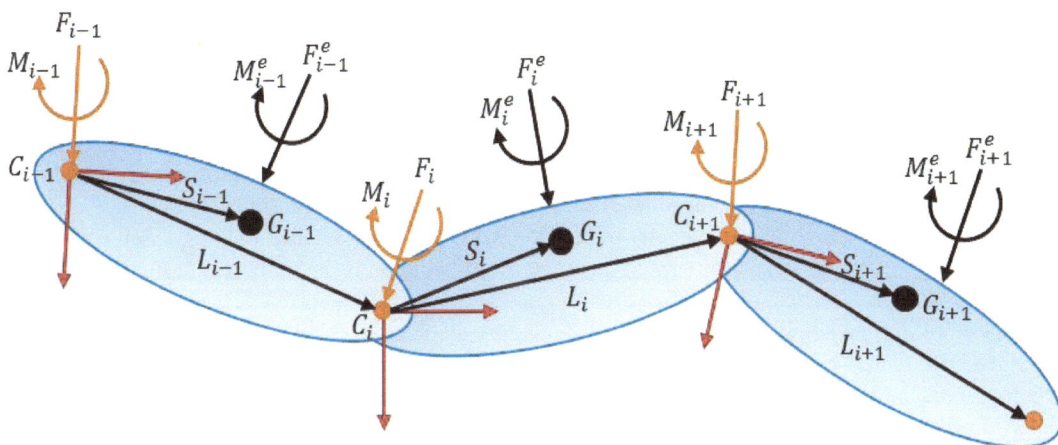

FIGURE 4.3.3. Open chain multi-body system

As an alternative to this approach, a more flexible algorithm has been presented in [**14**]. In this model, the motorcycle is considered as the saddle body, the front upper body (handlebar and upper part of suspension), the front lower part (lower part of the suspension), the swinging arm and the two tires. The handlebar and the swinging arm are attached to the saddle body by a simple revolute joint. The front lower part is linked to the front upper part by a prismatic joint. In addition, the rear suspension is connected to the saddle with a revolute joint from one side and to the swinging arm with a revolute joint also on the other side. This creates a closed kinematic loop leading to solving difficulties. Finally, the rear and front wheels are respectively connected to the other tips of the swinging arm and front lower part, as sketched in Figure 4.3.4.

Furthermore, an engine model is often very useful, even indispensable. Its goal is to compute the engine power transmitted to the vehicle based on the driver actions on the pedals: throttle, clutch, brakes and gearbox selector. To establish a model close to a real behavior, one should take into account the thermodynamics equations related to flows and combustion phenomena. Furthermore, although the engine seems to have a continuous functioning, it is actually a hybrid system with a succession of cycles almost independent (compression, expansion, etc.). Some engine models already exist in the literature, but they can be implemented on powerful computers. Therefore, they are not suitable for driving simulation applications. Nevertheless, some simple models can be adjusted for real-time operations.

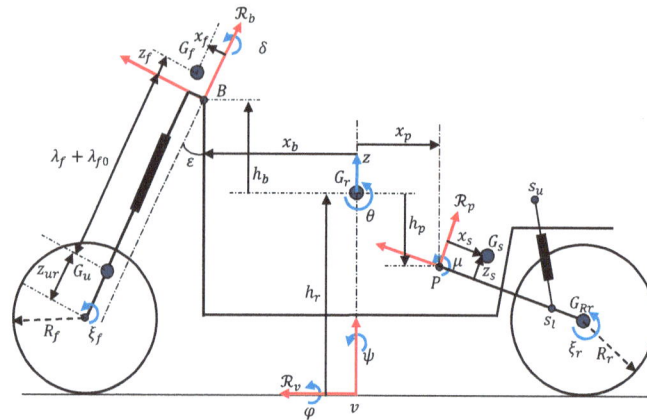

FIGURE 4.3.4. motorcycle kinematic configuration

4.4. Design and mechanical aspects

During the design phase, a first consideration should be conducted for all sub-systems with a central question: "what do we need to reproduce to the driver?" according to the planned tests and the application for which the simulator is designed. For example, for the straight line driving case, a vibrating table is largely sufficient. For other situations, a more sophisticated mobile platform may be required.

First, simulator design requires the definition of an appropriate mechanical architecture. An intuitive choice is to opt for a parallel platform like Gough-Stewart's. The advantage of these platforms is that they cover the 6 DoF with the possibility to choose the instant center of rotation, which position relative to the system of driver perception (vestibular system), seems important. However, the price of this solution is far from being affordable to be supported by the users.

The choice of driving simulator architecture is guided by the need to ensure a sufficient perception level, previously defined. The purpose of driving simulator applications relates generally two distinct frameworks that are: (1) risk training to make drivers aware of dangerous situations and (2) drivers' behavioral observation in normal driving, especially in urban situations. Simulator's users can as well be training centers or road safety agencies. The system must, therefore, be of an acceptable cost and easily transportable. Thus, it is necessary for every application to identify driving situations, which should be considered. Risk training intends to increase the awareness of new drivers to road risks. By motorcycle , some situations are unassailable, as the skidding of the front wheel; the simulator must still allow the driver to adjust himself to this situation. In this case, the transit time stability to instability is very short; therefore, the reproduction of this behavior involves the performance of an actuation system. While the purpose of behavior observation is to enable researchers to understand the cues perceived by the driver, in order to develop assistance system or to test the interaction with a special infrastructure. Extreme cases should therefore be produced as well as cases of normal driving.

For both previous scenarios, the learning context should be preferred to the faithful reproduction of the entire motions or risks. The proposed platform architecture (workspace, dimensioning of actuators, etc.) should promote this approach. This is consistent with the constraints of developing a low-cost tool. Under these considerations, the degrees of freedom are fixed. For a single track vehicle, three motions should be reproduced. The roll motion is considered to be the most important one. This degree of freedom is essential for stabilizing and leading the motorcycle . It is also involved in cornering maneuvers, slalom and lane changing. Then, the pitch motion is reproduced to feedback rear longitudinal acceleration, braking phases and fork dynamics. Finally, the yaw motion may be specifically used in order to reproduce skidding phenomena. The skidding of the front wheel is not interesting because it has fatal consequences.

In addition, we know that the multiplication of perceptual stimuli can strongly increase riding simulation sensations. Based on this idea, one can add passive/active device so that the driver is well "'immerse'" in his virtual environment. For example, a force feedback system can be implemented on the handlebars, Figure 4.4.1. The aim is to create other transitory phenomena to feed back to the user: inertial cause on bust during acceleration and braking. Thus, an effort is created in the arms of the rider by varying the distance between the saddle and the handlebar. Another

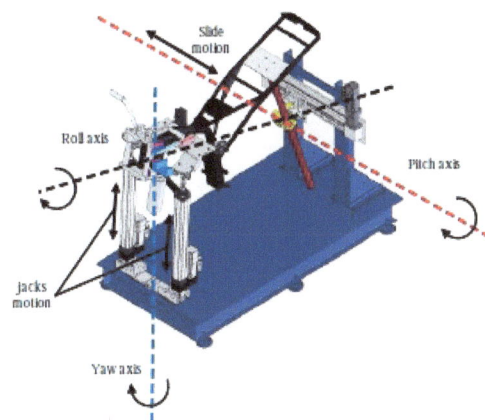

FIGURE 4.4.1. CAD model of the mechanical platform with its different rotation axes

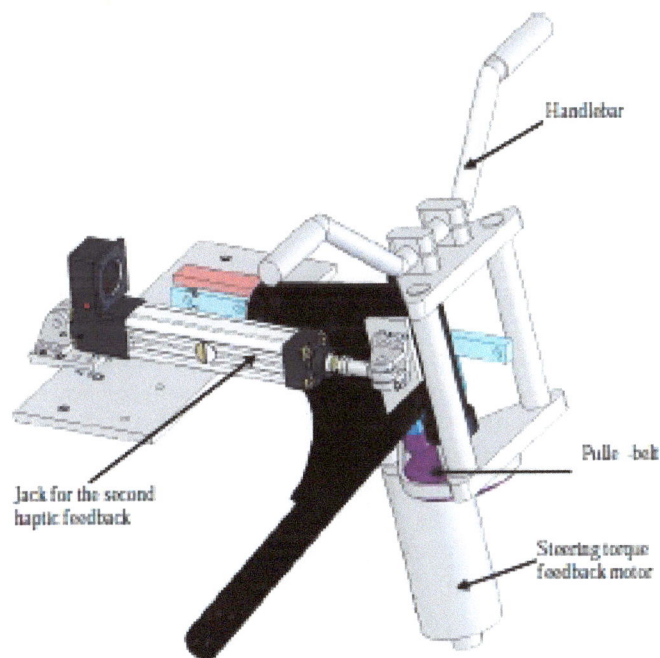

FIGURE 4.4.2. Haptic device for torque feedback on handlebar

force feedback can be developed to render the resulting torque of the tire-ground on the steering axis of the motorcycle
.

At the kinematic level, the position of the different axes of possible rotations is crucial. At the authors' knowledge, no psychophysical study has been conducted, except in some special cases [**5**].Therefore, these axes are defined from the kinematics of a real motorcycle . In [**9**], the authors reproduce the necessary yaw in order to feel the rear wheel skidding, a slide system is placed on the back of the motorcycle frame. The roll axis is placed in the motorcycle symmetry plane with an adjustable height in order to test various configurations and to achieve the best perception results. Lastly, for the pitch axis, it is the displacement of the front fork in the acceleration and braking phases which were privileged, therefore this axis goes through the back of the motorcycle frame, Figure 4.4.2.

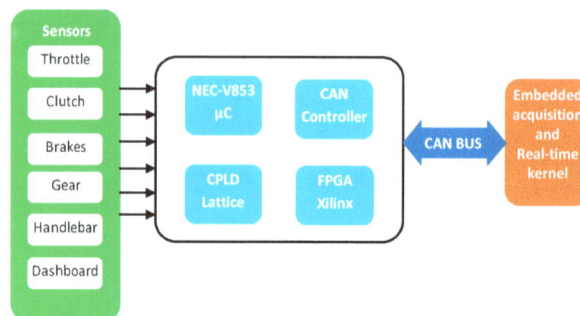

FIGURE 4.5.1. Rider's action instrumentation. Sensors, acquisition board and communication

4.5. Platform instrumentation

Actuating the simulator's mechanical platform amounts to setting its states of motion in reply to driver's actions. This results in the generation of so-called reference trajectories that are in adequacy with the user's actions, actuators performance and the platform mechanics. All inputs to this mechatronic system should be sent from either form of sensory transducers and sub-blocs communication. To that effect, the simulator is equipped with conventional motorcycle controls, with associated sensors, Figure 4.5.1. Information from these sensors is also considered as inputs to the virtual motorcycle and summarized as following:

- The position of the throttle and clutch lever: these inputs serve for computing the engine output power. Throttle and clutch actions are instrumented by using simple linear resistive sensors known as potentiometers. This last consists of three terminals where the second one is connected to a sliding wiper attached to each level. Analog electrical information, image of the lever position, is delivered.
- The braking power of the front brake lever is returned through a pressure sensor that is mounted on the hydraulic assembly. This sensor delivers an analog electrical signal, image of the rider's torque exerted on the lever's handle. The braking power of the rear brake pedal is provided with the same device as well.
- Gearbox information: the gearbox signal translates the shift state used to update the engine speed. For this, a mechanical switch (SPDT: single pole double throw) is mounted beside the speed selector, which returns binary information. A typical position switch with a mounted lever creates a mechanical contact when a shift is engaged which closes the electrical circuit within the switch. A logic circuit based on Flip-flops, is included to prevent the signal from bouncing resulting in lever micro-impacts.
- The handlebar position: theoretically, two-wheeled vehicles are controlled mainly by the steering torque applied by the rider on the handlebar.

However, the cost generated by the implementation of a torque sensor is in contradiction with a low-cost design philosophy. Recently, control theory has provided engineers with powerful techniques allowing implementation of virtual sensors, namely, unknown inputs observers. These observers can perform, starting from some measurements, the estimation of an unknown information signal like the rider torque exerted on the motorcycle handlebar. To achieve this, the handlebar's position is an important measure that should be acquired by a maximum of precision. To that end, a common sensor for position measurement is the optical incremental encoder providing digital information with a given precision, 1024 points in the current simulator. Nevertheless, whatever the simulated maneuvers, the steering angle is infinitely small and therefore one must choose a high resolution sensor. Another simple solution is to trade on the available sensor dynamics by using a pulley-belt system. This solution is more practical and advantageous to also implement a torque feedback on the handlebar.

For the present motorcycle simulator, the acquisition of the rider actions is ensured by a set of two plugged electronic cards forming a compact acquisition board. The first provides an I/O interface for sensor signal routing, while the second deals with the signal low-level management. This board consists of a micro-controller V853 from NEC Corporation with 12 analog inputs (± 10Volt), 16 analog outputs (0-5Volt) as well as multiple digital I/O. Binary buffered inputs dedicated to the acquisition of the optical encoder channels are managed by an FPGA (Xilinx XCS30XL). Intercommunication among host computers is proposed in various flexible configurations ranging from the parallel port to CAN (Controller Area Network) communication frame.

The overall architecture of an embedded system (sensors, acquisition, data processing, decision making, and actuation) requires special attention with respect to the data exchange flow and requirements on the data reliability level. Signal alteration, sampling and protection against noises must be taken into consideration. According to the number of users and mechatronics subsystems, point-to-point or point-to-multipoint communications are possible. A distributed vision of such architecture constitutes an initial guess to reduce its complexity. Indeed, a multiplexed bus, type CAN, reduces the level of coupling between different components. Thus, for the present simulator, the acquisition is accomplished by establishing communication via the CAN bus by an integrated controller.

4.6. Actuator selection and driving

The choice of actuators is in line with the expected performance in terms of perception and ability to put the rider into a risk situation. This is done by full study of the stability of a two-wheeled vehicle. Indeed, the analysis of motorcycle dynamics gives rise to three significant, well separated, modes of instability [15]. The "Capsize" is related to the roll motion where gyroscopic effects turn out to be negligible to stabilize the vehicle. It is a non-oscillatory, well damped mode at low speed, but, beyond a speed of 20 (m/s), this mode becomes unstable and can be controlled by a steering torque applied on the motorcycle handlebar. The "wobble" is a fast oscillating mode involving the steering motion of the front fork, whose frequency range into 4-10 (Hz). Lastly, the "Weave", is side-to-side motions of the entire motorcycle involving yaw and roll oscillations with significant lateral displacement. Weave is an unstable mode at very low speed (less than 5 (m/ s)) and well stable at high speeds over 30 (m/s). Its frequency ranges between 0 and 4 (Hz), and it affects the entire two-wheeled vehicle which makes it difficult to be controlled by the motorcycle rider. These modes are at the limit of the vehicle stability and beyond, the rider lands with a falling and risk situations. Moreover, the two-wheeled vehicle is more powerful than a car vehicle with an important power/mass ratio leading to considerable accelerations.

To achieve the expected performance as prescribed in the simulator specifications, it is necessary to choose the platform mechanics (number of DoF) and the corresponding drive system. Advances in industrial computing and power electronics together have promoted the use of electrical machines over hydraulic ones. Indeed, hydraulic actuators present higher strength characteristic where the density of energy is almost 100 times greater than that of an ideal electric machine. But, this factor should not alone dictate the choice of the platform actuation, the versatility use, installation speed and high speeds of the electrical power with a better value for money; make the electrical actuators a convenient choice. Finally, the control problem is much simpler to lay with electrical drives. Hydraulic actuators exhibit strong nonlinearities marked by hysteresis. This fact is important and impacts the performance, precision and deployment.

If the possibility of a hydraulic actuator is ruled out, the platform actuation is to choose from a wide range of electric motors, Figure 4.6.1 [16]. They can be classified based on functionality, rotor design and induced electromotive force (emf). The direct current (DC) brush motor consists of a stator and a rotor with coil windings (armature). The stator, formed by electro-magnets, creates a fixed magnetic field, thereby generating a very regular torque. The rotor winding is driven by a DC current which can cause the rotor rotation under the action of the well-known Lorentz's electromagnetic force. Its simple nature gives him the advantage of being easy to control.

Synchronous brush motor looks like the previously described DC brush motor. Its rotor, powered by DC current through its brushes, creates a magnetic field that is in phase with the stator field. So, the rotor rotates at the same speed as the rotating field, hence its name "'Synchronous motor'". Furthermore, the permanent magnet synchronous motor (PMSM) is a synchronous motor where the conventional electromagnetic field poles in the rotor are replaced by permanent magnet poles yielding to a brushless motor. In this type of actuator, slip ring and brush assembly are dispensed with the help of an electronic switch enabling self-drive.

PMSM can be broadly classified according to the different ways the magnets can be arranged on the rotor. Magnet disposition has a direct impact on the flux density, winding inductance, reluctance torque and the shape of the induced emf, i.e., sinusoidal and trapezoidal, so:

- If the rotor magnet poles (also called smooth poles) are disposed, the induced emf is sinusoidal. In the literature, the sinusoidal type is shortly known as PMSM,
- If the rotor is of concentrated coil-wound (also called salient poles), the induced emf is trapezoidal. The trapezoidal type is called PM DC brushless motor,

The PM DC brushless motor with induced trapezoidal emf is driven by the six-step switching method where, the rotor phases are excited by 120° wide currents. The drawback of this method is that it produces large torque

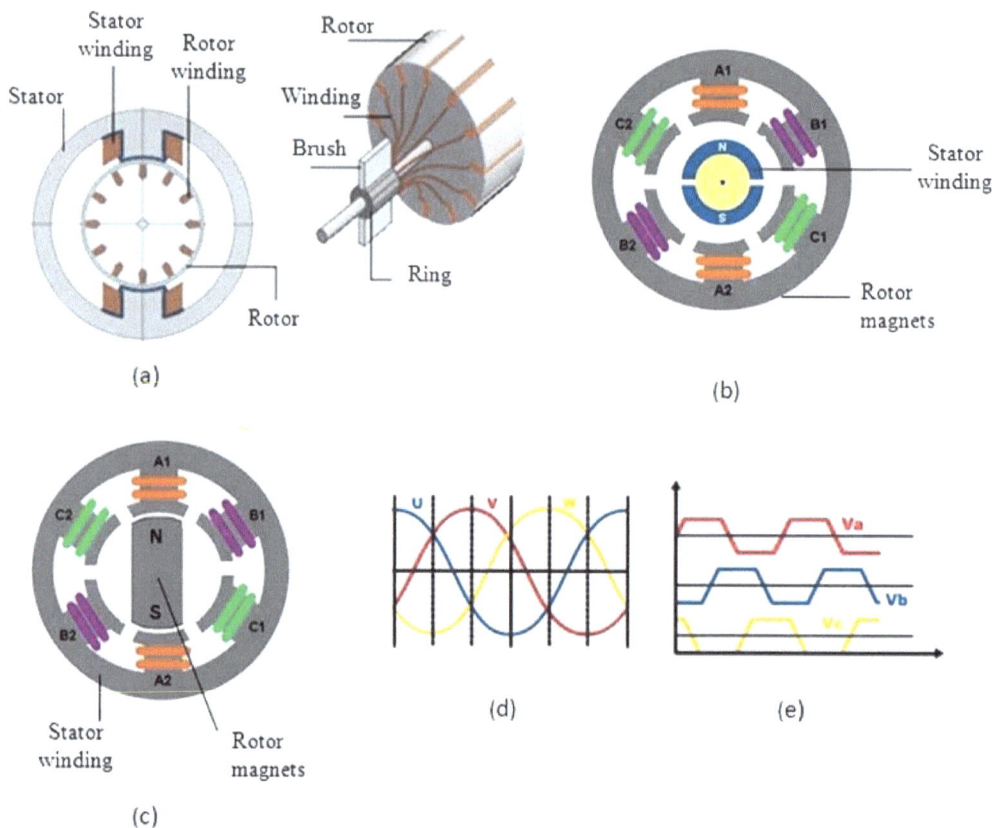

FIGURE 4.6.1. Electric machine classification : (a) direct current (DC) brush motor, (b) permanent magnet synchronous motor (PMSM) with one pair of smooth poles, (c) PM DC brushless motor with a pair of salient poles, (d) trapezoidal induced electromotive force, (e) sinusoidal induced electromotive force

ripple. However, its main advantage is that the commutation signals need to be generated only six times for every electrical cycle. So, implementing a position sensor like Hall Effect is very trivial. In addition, for a motor with sinusoidal emf, the controller aims to drive independently each phase current depending on the rotor position. This electronic switching method allows minimizing the produced torque ripple and improving the efficiency and power factor. However, controlling this type of motor is more difficult. Indeed, the drive system of the PMSM with sinusoidal emf needs the position information continuously in order to construct winding currents. Consequently, the precision of the current reference signal depends mainly on the position sensor resolution (optical encoder or resolver), which makes the cost of a PMSM higher against a PM DC brushless motor.

Finally, the asynchronous motor is the most widespread in industrial applications like ventilating and pumping, where there is a need for constant speed driving. Following the restrictions imposed by the simulator specifications and in a spirit of a low cost development, the choice of the platform actuation seems to be obvious. To move the different axis of motion of the simulator's mechanics, a PM DC brushless motor with a trapezoidal emf is adopted. On the one hand, this actuator is well adapted for low positioning performance, has about 15% more density than the PMSM with a lower cost. On the other hand, for the handlebar steering torque feedback, the ease of integration and the absence of torque ripple are the two mandatory criteria to be taken into account and hence, a DC brush motor is selected.

The use of brushless motors is the widespread norm, thanks to the development of power electronic converters. In an electromechanical system, variable frequency of the stator winding current is needed to achieve variable speed. For this, the designer uses inverters to convert the reference signals as delivered by the controller to an electric current that produces the actuation torque. So, it is essential to develop the associated drive electronics. This drive includes several subsystems ranging from alternative current (AC) supply rectifier to low-level current servoing. Thus, a major question arises; do we use a commercial drive or a custom home solution?

It is difficult to answer this question and a compromise must be reached. A commercial drive provides a fast and efficient way for prototyping and control. Nowadays, in addition to the inverter and the switching circuit, drives are allocated with embedded devices allowing position, speed and torque control. However, a drive is not a flexible solution for the implementation of advanced control laws. As an example, for torque control, one should make sure that the drive allows the measurement of phase currents and the absolute position of the rotor. Nevertheless, if a conventional proportional, integral and, derivative (PID) control scheme should be implemented, a commercial drive is an appropriate choice.

A custom home solution provides a flexible, powerful and an open control device. However, torque control requires high acquisition speeds (over 500 (Hz)) and the current control loop should be performed at a frequency higher than 8 (KHz) (depending on the power components that constitute the inverter). Moreover, in the industrial domain, the use of three-phase AC current supply is predominant. It follows that a custom solution requires designing a rectifier, inverter, switching circuit, current sensor, signal conditioning, protection, insulation and implementation of the control algorithms as close as possible to the switching devices (the use of a DSP is mandatory). This solution, being more flexible, requires a greater investment and resources.

4.7. Real-time monitoring, sequencing and synchronization

The simulator is a set of mechanical, electronic and software components that must communicate with each other with respect to prescribed temporal constraints, Figure 4.7.1. All tasks must be performed in a binding manner, real-time and delay free. Therefore, synchronization is a key element for the simulator's operation. The problem of delays may lead to a loss of controllability of the simulator and a poor motion rendering.

The need for real-time is essential and refers to two main aspects which are event management and control of the simulator. These two aspects should be performed in a predefined time entity and hence, rethinking and reconceptualizing a dedicated platform is inevitable. Indeed, in the initial conceptual design, most of the development is carried out using a simulation tool like Simulink under Microsoft Windows operating system or a similar platform. Such an approach is well suited for rapid prototyping and requires much less time. However, Simulink uses block diagram representations and a sequential time vector to evaluate the different model states and outputs. Since this vector is not connected to the system clock, the outputs are determined in a non real-time way following the performance of the used computer. So, a real-time implementation requires the programmer to have control of all events; semaphore resources and can assign a priority to each process defined as the most critical. This commitment cannot be managed without an appropriate operating system which facilitates access to CPU resources, management processes, memory, interrupts and so on.

Recently, many professional and free license operating systems with a real-time kernel have been proposed. However, the major difficulty lies in the development of drivers used to interface the hardware components to the real-time kernel. Therefore, the use of a professional real-time manager is recommended for rapid development. In this perspective, according to the suitable final deployment, one can cite for example:

- xPC Target: a toolbox from Mathworks intended for MATLAB/Simulink developers. The advantage of xPC Target is its user-friendly and its trivial configuration. File generation, compilation and loading are done in an implicit manner with a minimum of intervention from the user. An equivalent version for Windows, called Real-Time Windows Target, also exists [17].
- LabVIEW Real-Time Module Development: A real-time package developed by National Instruments.
- Dspace: widely used in the automotive field, it consists of an acquisition host computer and a real-time kernel manager. All models to be simulated must be converted into C (C++) language. The main advantage is that dSpace is compatible with the Matlab/Simulink. All Simulink models can be converted into C language using an appropriate package like Target-Link or Real-time workshop toolbox. The major drawback of this tool is its price.

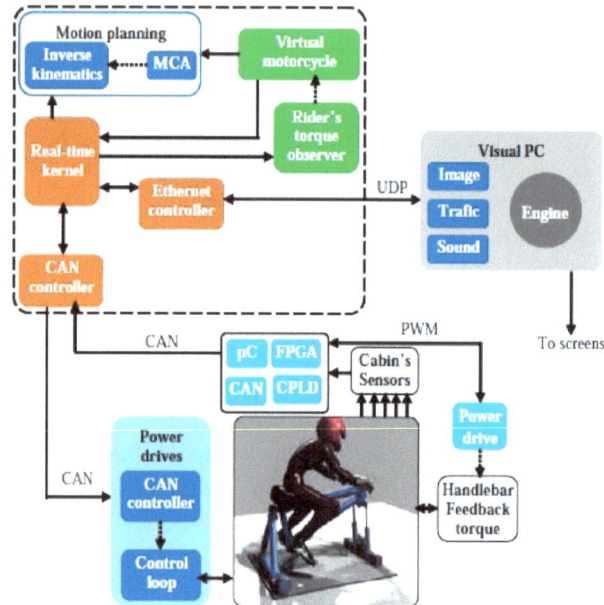

FIGURE 4.7.1. Overall motorcycle simulator mechatronic architecture

- Linux Kernel: Thanks to its stability and reliability, Linux is becoming more and more used in industry and application domains. Several real-time kernels have emerged as result of continuous efforts of a growing open source community. Thus, numerous solutions have emerged in the free license world such as Open RTLinux and RTCoreBSD. Professional versions, such as VxWorks, offer an adequate working interface and a wider portability.

Furthermore, the control task is a central element in the actuation of the platform mechanics. It consists of the trajectory generation, loop control and supervision. So, these three control steps will convert the states of the virtual vehicle, images of the rider's actions, in an actuator signals that drive the various motion axes. Otherwise, supervision is a task that includes, among system initialization, the handling of different events, the fault detection and diagnosis. In this sense, each motion axis (jacks and slide) is instrumented with two Hall Effect switch sensors. These sensors provide binary output indicating that a joint has reached the limit of its allowable travel. In this way, it is possible to scrutinize the possible excursion and to avoid any unforeseen damage and preserving the platform mechanics. In addition, supervision task allows a stable connection between the visual rendering computer and the real-time kernel computer. Therefore, planning semi-static data exchange in the embedded network can provide a common level of dependability. In this context, the use of a multiplexed field bus is entirely justified.

According to this perspective, a monitoring interface based on a state machine is developed, Figure 4.7.2. This machine contains the essential processes to be implemented, as summarized in the following points:

- At the beginning of the simulation, the actuators are set to state 1 "Login". A CAN request, with the given identifier, is sent to each drive to set it into a control mode.
- The second step involves positioning the simulator's mobile platform to the neutral position (jacks and slide at mid-travel). This position will be considered as a reference to any future movement of the platform. At this stage, essential information such as: position switch information and the maximum allowable workspace, will be available. Positioning is achieved simply by scrutinizing the displacement of each motion axis. By detecting a given limit position, every drive returns an error frame designated by an identifier of 1B9.
- Being at one or at the other limit switch (state 2 or 3), the maximum excursion of each actuator is scanned to determine the mid-point travel for the platform positioning (state 5). In this state, a CAN frame is sent in writing mode to force the initialization of the position value. Henceforth, the neutral position will be considered as the new reference.
- Once the platform is positioned at the neutral position, an UDP (Universal Datagram Protocol) request with identifier ID: 1 is sent to establish a direct connection with the visual rendering computer. If it responds with

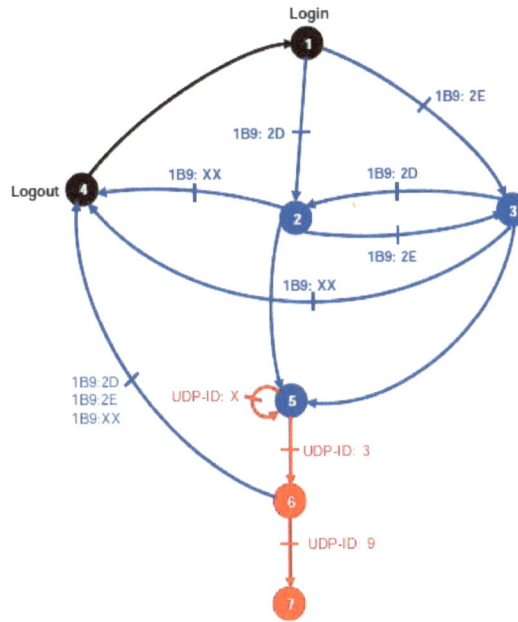

FIGURE 4.7.2. State machine for the simulator monitoring and fault diagnosis

an ID: 3, the connection is established (state 6), otherwise a second request is returned. From that moment, the driving simulation can start.

- If an error frame with an identifier of 1B9: XX is detected, a corresponding action must be taken. This frame error is the result of an event more or less critical. Indeed, during the simulation, several unexpected malfunctions may occur. These may be, limit switch reaching, current peak consumption and, overheating. Depending on the severity of the error event, a resolution procedure may be implemented. For example, if the movement has reached a given limit switch, re-scheduling is introduced to stop the simulation smoothly and hence ensuring the simulator's user safety. In the worst case, the drives are switched to control-off and an UDP request with ID: 9 allow the disconnection from the visual environment.

4.8. Motion planning and control

Generation of reference trajectories requires a planning of a motion profile compatible with the mechanics of the simulator and the capabilities of its actuators. In this context, reproducing the full scale vehicle dynamics is impossible even with the more recent mechatronic systems. Inertial forces present in a real driving situation, vary from one vehicle to another and are characterized mainly by accelerations' bandwidth. Designing a system which is able to simulate the frequency content of these accelerations goes inevitably through two different modules: one for rendering high frequency accelerations (also called transient) and the second for low frequency acceleration (also called sustained). For the transient components, a vibration table is an efficient way to reproduce the vehicle linear speed and the road irregularities. The rendering of sustained acceleration is highly dependent on the platform mechanics and on the simulator objectives. If the simulator aims to carry out a behavior study in normal traffic, some experts believe that for maneuvering below 0.3g (1g= 9.81 m/s^2), a simulator fixed base is sufficient. Otherwise, to simulate a dangerous or critical driving situation, the inertial forces restitution is a crucial element, and therefore, a mobile platform is necessary. In the latter case, the first challenge is to accomplish the platform movement in the allowable simulator's workspace. For this, geometric and kinematics constraints should be accounted for and the use of specific planning algorithms commonly known as Motion Cueing Algorithm (MCA) is essential.

MCA has the task of reproducing a part of the inertial forces present in a real driving situation to achieve a sufficient perception level of the simulation [18]. These algorithms are based on a simple frequency separation of the various accelerations to be restored through two main strategies. The first approach uses the longitudinal motion of the simulator to directly restore the transient acceleration. Moreover, the second one consists on tilting the rider on

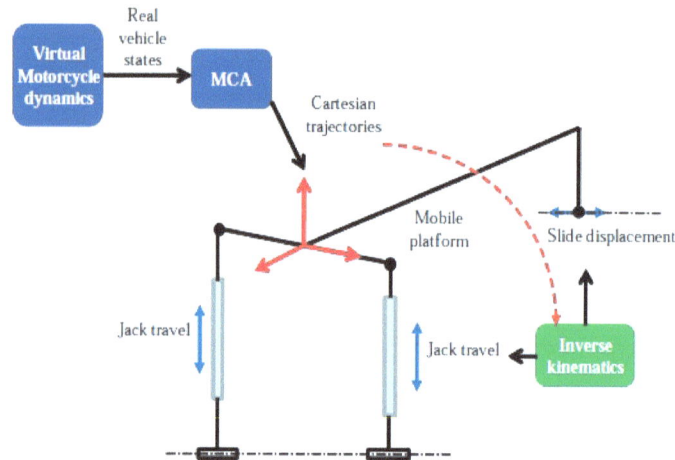

FIGURE 4.8.1. Motion planning by using MCA and inverse kinematics

the mobile platform to take advantage of the gravity vector in order to simulate the sustained component of the linear acceleration. In conclusion, when designing a system for motion cueing , it is important to take into consideration the targeted tests and manipulations. If a vibrating table is sufficient, or must have a large linear motion. Is it better to have no movement than to have a bad one?

However, the planned trajectory, as obtained with the motion cueing algorithm cannot be directly used to actuate the mechanical platform. According to the mechanical architecture of the simulator's platform, it is necessary to transform these trajectories, generated in the Cartesian coordinate space in trajectories expressed in actuators operational space or joint space, Figure 4.8.1. It is the role of geometric/kinematic models widely developed in robotics theory [**19**]. In the case of the present simulator, these amounts are used to calculate the jack travels and the displacement of the rear slide according to the desired orientation of the mobile platform.

Once the reference joint trajectories are defined, it is essential to develop a control scheme for the tracking task to drive the mechanical platform and imparting desired transient and steady-state performance. In general, electromechanical systems exhibit a dynamic behavior similar to that of a second order model and therefore can be reasonably controlled by PID controllers (Proportional-Integral-Derivative). Nevertheless, several control schemes intended for complex robotic manipulators like the computed torque method, robust control and sliding mode control, are known in the literature. Since the objective is to achieve a position tracking, all these commands need to implement an external control loop. This solution creates additional time delay in the overall simulation loop and imposes other constraints, especially if commercial drives are adopted.

However, the primary purpose of driving simulation is the creation of an appropriate illusion to lure the simulator user. The exact reproduction of the real movement is of secondary priority, thus, the controller synthesis must move in that direction. Simple strategies favoring compensation of inertial delays have shown their effectiveness. Nowadays, the servo power monitoring devices are featured with a sufficient intelligence based on axis by axis control with several nested control loops (position, speed and torque). The use of this solution provides significant time savings with a satisfactory performance level.

On the basis of a PID approach, the final control scheme consists of three nested control loops namely, the position loop, the speed loop and the current control loop. The position and velocity are estimated by an observer, from the resolver information of the brushless motor. These estimates are considered as measure variables and are used to compute the tracking error. However, to improve the performance of the servo system, compensation terms of speed and acceleration are expected. These terms, introduced in open loop, do not affect the stability are intended to offset the effect of friction and inertia.

The diagram of Figure 4.8.2 shows a commonly PID approach implemented in situ of a commercial drive. In this figure, the current loop does not appear. Indeed, this loop is intended for the motor current tracking according to the reference speed (PI block output) and hence to setup the motor input voltage. This voltage is then chopped and modulated by an appropriate switching scheme for commutating power semiconductor devices and ensures proper

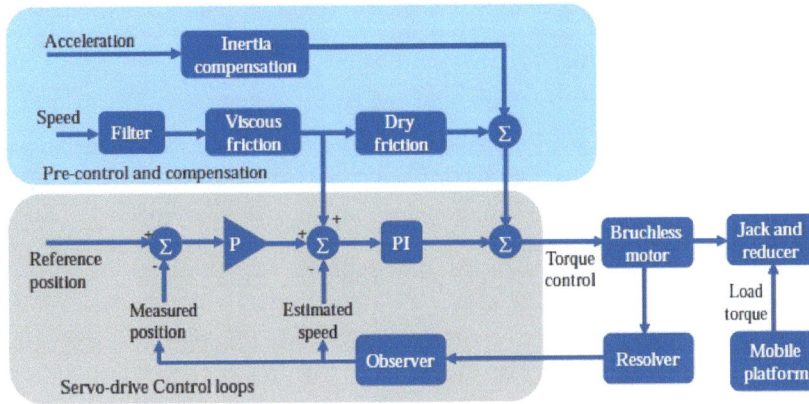

FIGURE 4.8.2. Servo-drive embedded control loops

monitoring of torque. In a direct current brush motor, the generated torque is generally proportional to the armature current. However, in a three phase PMSM, the torque depends on the phase current amplitude and frequency and therefore the switching algorithm should be very complicated. That is why these motors come with associated drives where the switching and the current loop is already implemented and pre-tuned.

Otherwise, controlling PM brushless DC motor is quite simpler than its counterpart PMSM for low performance applications. The design of a current controller is straightforward, due to the great similarity with DC machines, i.e., PM brushless DC can be modeled as three symmetric DC motors and the inner current control loop can be easily tuned by using a standard linear control approach.

Tuning the different control parameters can be achieved in several ways. As a demonstration, for a position control method, the controller can be expressed by the following equation:

$$(4.8.1) \qquad \tau_c = k_p e_p + \left(k_v + \frac{k_i}{s}\right)(e_v + k_{fvv}\dot{x}_{ref}) + k_{fa}\ddot{x}_{ref} + k_{fvs}\text{sgn}(\dot{x}_{ref})$$

where, τ_c is the control torque, e_p and e_v are position and speed tracking errors, x_{ref} is the reference position trajectory, k_p, k_v and k_i are the PID controller gains, k_{fvv}, k_{fvs} and k_{fva} are parameters related to the feed-forward compensation terms for viscous and dry friction and inertia, respectively. Next, the dynamic equations of each axis motion must be formulated. To simplify this demonstration, the model of electric motor is not considered because the electrical dynamic is much faster than the mechanical one; therefore, we can write the equation of motion of the motor- jack-reducer as:

$$(4.8.2) \qquad J_{eq}\ddot{x}_v + f_{eq}\dot{x}_v + f_s\text{sgn}(\dot{x}_v) = N\tau_c - \tau_l$$

where J_{eq}, f_{eq} and f_s are respectively inertia, viscous and dry friction of the equivalent assembly, N is the reducer ratio and τ_l is load torque applied by the simulator mobile platform on the motion axis (like jacks). The load torque can be found from the mobile platform dynamics expressed, by using a simple Newton-Euler modeling approach, by the following equation:

$$(4.8.3) \qquad \mathcal{M}\ddot{\bar{X}} + \bar{C} + \bar{G} = \mathcal{J}_{-1}^T \tau_l$$

Here, $\mathcal{M}$, $\bar{C}$ and $\bar{G}$ are mass matrix, vector of non-linearity and related gravity terms vector, respectively. $\mathcal{J}_{-1}^T$ is the transpose of the inverse Jacobian matrix allowing the transformation between Cartesian velocities and joint space velocities. Finally, $\bar{X}$ is the vector of the joint coordinates, namely jacks and the rear slide displacements.

Equations (10.2.1-10.2.3) can be combined and processed in order to tune the respective parameters of the final controller loops. Nowadays, with the rise of numerical optimization methods, it is possible to introduce constraints for performance and disturbance rejection, while imposing a defined structure of the desired controller. Thus, optimized gains can be used to refine the performance of the inner low-level servo controller and avoiding the need for external control loops. Consequently, platform mechanics is actuated directly by sending reference position trajectories to the motor drives via CAN bus .

4.9. Visual, sound and traffic systems

A visual system consists of a 3D graphics generator and projection devices. Generally, images are projected onto one or more curved or flat screens to ensure a front, side and (if necessary) back vision. In fact, driving is primarily a visual task and it is therefore evident that this information must be carefully produced. Indeed, the image quality is measured by its energy properties (brightness, contrast, resolution, color), its spatial (vision fields and depth) and temporal characteristics (transport delay and refresh rate). Therefore, producing a visual scene with great realism depends on the efforts made to satisfy these factors.

On the other hand, delays are a major problem in applications like driving simulation. Often, these delays are divided into two types: the refresh rate and transportation time. The former is described by the frequency at which the screen is updated or redrawn to give an impression of continuous image animation. The latter, the transportation delay is the time interval between the rider action and the visual projection of the virtual environment. This factor presents a real problem in driving simulation because it is an integral part of the overall simulator mechatronic (virtual motorcycle dynamics, control, acquisition, actuation, image generation, traffic computing, and projection of the visual scene and refresh rate). Singhal and Cheriton [20] showed that subjects can detect latencies of 100 (ms), with a tolerance of up 200 (ms). Nevertheless, it was reported that the overall delay in a driving simulator should not exceed 50 (ms) [21]. Thus, if the refresh rate is 60Hz (equivalent to 17 ms), it is evident, with a simple subtraction, that all other sub-systems should operate and communicate at delays less than 33 (ms).

Moreover, depriving drivers of acoustic cues leads to a systematic increase in the speed of the vehicle. To prevent this issue, a 3D sound system is used. Indeed, the sound system allows the enhancement of the driving simulation realism. The main characteristics of this element are: the number, quality and location of speakers, and also the diversity of sounds. Indeed, the dominant sound frequency band in a vehicle is about 20 to 500 (Hz), induced mainly by the engine [22]. Some sounds from tire/road interaction have major components at high frequencies. Therefore, the sound system should cover a few thousand hertz. Consequently, to achieve a realistic sound illusion, it is paramount to generate a spatial multi-channel sound. If a vehicle passes nearby, the direction of the sound should follow the projected image to avoid the rider disorientation.

Finally, traffic and scenario are important elements as they add some interactivity to the visual environment. Starting from the individual driver's behavior, visual objects like vehicles and pedestrians are imparted with a defined level of intelligence to immerse the simulator's user in realistic traffic conditions.

4.10. Rider safety versus existing security systems

The automotive industry has greatly evolved in recent years. Today, cars are considered as complex mechatronic systems. Each feature is seen with a dedicated processor for managing multiple sensors and processing a considerable flow of data. This complexity is the result of a massive spreading of security features, support, comfort, and power management. In particular, driving safety has become the unifying theme among car manufacturers, official institutions and customers. The present challenge consists in developing security assistance and support, more efficiently and less costly.

In this context, the number of deaths in road accidents has experienced a huge reduction (about 50%) during the last decade. However, analysis of accident statistics shows that the number of deaths when a motorcycle is involved, has increased. In fact, the motorcycle remains a particularly dangerous mode of transport: the number of deaths is still very high, and if one takes account of the number of traveled kilometers, the risk of death for a motorcycle rider is 21 times higher than that of other transportation modes. In additions, motorcycle and scooters are becoming more and more popular for urban an suburban travel in European countries. The associated traveled kilometers and the number of vehicles sold is continuously on the rise.

During the last *twenty* years, passenger cars experienced several advances in passive and active safety systems. Nowadays, cars are all equipped with one or several airbags and ABS (Anti-lock Braking System) systems. More powerful systems such as ESP (Electronic Stability System), brake assist systems and traction control systems or belt tensioners are becoming increasingly widespread. Conversely, during the same period, the backwardness taken by motorcycle continues to grow. For example, ABS , which has been around for over 15 years, is still reserved to few top range motorcycle models. The braking distributor is becoming less expensive but still remains marginal in the world of motorcycles. The use of motorcycle airbags, whose development seems problematic, did not become widespread. The recent braking amplifier seems to be promising, but its diffusion remains, very restricted.

4.11. Conclusion

The design of each driving simulator is a very difficult multidisciplinary challenge. Each component is essential (vision, motion restitution, communication, etc.) for the success of the driving simulation and to increase the immersion level. One of these important aspects is mechatronics , including mechanics, embedded electronics, actuation technologies and control laws, all in a single system).

The design of a driving simulator is based, among others, on a mechanical platform and an embedded electronics for its achievement. The performances of the final architecture are checked by a juxtaposition of the different components. Therefore, an optimal design of such systems requires a multi-physic modeling and optimization of the global system.

The quality of a driving simulator is rated by its fidelity level. The quality of motion restitution is largely dependent on the platform's mechatronics . Without this condition, perceptive errors are transmitted to users and can make them quite uncomfortable. Thus, as stated before, an optimized mechatronics strongly helps the system to provide optimum motions.

Bibliography

[1] Dagdelen M, Reymond G, Kemeny A, Bordier M, Maiza N. MPC based motion cueing algorithm : Development and application to the ULTIMATE driving simulator. Driving Simulation Conference DSC Europe 2004; 221-233.

[2] Nehaoua L, Mohellebi H, Amouri A, Arioui H, Espié S, Kheddar A. Design and control of a small-clearance driving simulator. IEEE Transactions on Vehicular Technology 2008 ; 57(1) :736-746.

[3] Arioui H, Hima S, Nehaoua L, Bertin RJV, Espié S. From Design to Experiments of a 2-DOF Vehicle Driving Simulator. IEEE Transactions on Vehicular Technology 2011 ; 60(2) :357-368.

[4] Miyamaru Y, Yamasaky G, Aoky K. Development of motorcycle riding simulator and its prehistory. JSME Review 2000; 50.

[5] Yamasaky G, Aoky K, Miyamaru Y, Ohnuma K. Development of motorcycle training simulator. JSAE Review 1998; 19: 81-85.

[6] Ferrazzin D, Barbagli F, Avizzano CA, Pietro DD, Bergamasco M, Designing new commercial motorcycles through a highly reconfigurable virtual reality-based simulator. Advanced Robotics 2003; 17(4): 293-318.

[7] Kwon DS. Kaist interactive bicycle simulator. Proceedings of IEEE International Conference on Robotics and Automation (ICRA01) ; 2313-2318.

[8] Cossalter V, Doria A, Lot R. Development and validation of a motorcycle riding simulator. World Automotive Congress FISITA 2004, Barcelona

[9] Arioui H, Nehaoua L, Hima S, Séguy N, Espié S. Mechatronics, design and modeling of a motorcycle riding simulator. IEEE/ASME Transactions on Mechatronics 2010; 15(5): 805-818.

[10] Pacejka HB, Sharp RS. Shear force development by pneumatic tyres in steady state conditions : A review of modeling aspects. Vehicle System Dynamics 1991; 20 :121-176.

[11] Cossalter V, Lot R. A motorcycle multibody model for real time simulation based on the natural coordinates approach. Vehicle System Dynamics 2002 ; 37(6) :423-447.

[12] Khalil W, Dombre E. Modélisation, identification et commande des robots. Hermes science publications, Paris, 2nd edition,1999.

[13] Hollerbach JM. A recursive lagrangian formulation of manipulator dynamics and a comparative study of dynamics formulation complexity. IEEE Transaction on systems, man and cybernetics 1980; 10(11): 730-736.

[14] Hima S, Nehaoua L, Séguy N, Arioui H. Suitable two wheeled vehicle dynamics synthesis for interactive motorcycle simulator. Proceedings of the 17th IFAC World Congress 2008; 96-101.

[15] Sharp RS. The stability and control of motorcycles. Journal of Mechanical Engineering Science 1971; 13:316-329.

[16] Bishop RH. Mechatronic systems, sensors, and actuators: fundamentals and modeling. CRC Press Inc 2007.

[17] xPC Target for use with real-time workshop. User's guide. The MathWorks.

[18] Nahon MA, Reid LD. Simulator motion-drive algorithms : A designers perspective. Journal of Aircraft 1990; 13(2): 356-362.

[19] Nehaoua L, Hima S, Arioui H, Séguy N, Espié S. Design and modeling of a new motorcycle riding simulator. Proceedings of the 2007 American Control Conference. 176-181.

[20] Singhal SK, Cheriton DR. Exploiting position history for efficient remote rendering in networked virtual reality. Teleoperation and Virtual Environment, Presence 1995; 4(2) :169-19.

[21] Franck LH, Casalli JG, Wierville WW. Effects of visual display and motion system delays on operator performance and uneasiness in a driving simulator. Human Factors 1988; 30 :201-217.

[22] Genuit K, Bray W. A virtual car : Prediction of sound and vibration in an interactive simulation environment. Proceedings of SAE Noise & Vibration Conference & Exposition, Grand Traverse 2001, USA.

[23] Larnaudie B, Bouaziz S, Maurin T, Espié S, Reynaud R. Motorcycle platform for dynamics model extraction. Proceedings of the IEEE Intelligent Vehicle Symposium 2006.

CHAPTER 5

Robust Monitoring of an Omnidirectionnal Mobile Robot

Youcef TOUATI[1],
Polytech-Lille, LAGIS, UMR CNRS 8219
Avenue Paul Langevin, 59655 Villeneuve D'Ascq, France
youcef.touati@polytech-lille.fr

Rochdi MERZOUKI
Polytech-Lille, LAGIS, UMR CNRS 8219
Avenue Paul Langevin, 59655 Villeneuve D'Ascq, France
rochedi.merzouki@polytech-lille.fr

Belkacem OULD BOUAMAMA
Polytech-Lille, LAGIS, UMR CNRS 8219
Avenue Paul Langevin, 59655 Villeneuve D'Ascq, France
Belkacem.Ouldbouamama@polytech-lille.fr

ABSTRACT. This chapter deals with robust fault detection and isolation using the bond graph (BG) approach and linear filters. The bond graph tool is used to model the dynamic system and the uncertainties on the sensors and actuators. The same model is used to generate systematically the analytical redundancy relations and the thresholds. A specific form of digital linear filter is used to evaluate the residuals and to ameliorate the robustness and the detectability of the faults. the developed procedure is applied to experimental data of an electromechanical system which is a subsystem of a mobile robot named Robotino.

Keywords: Fault detection and isolation; Mobile Robot; uncertainty modeling.

5.1. Introduction

The diagnosis system is very interesting to the functioning of the autonomous dynamic systems. It is used in order to determine the best way to control normal and faulty situations to ensure the safety of both the system and the environment. The last three decades have seen the emergence of numerous approaches on fault detection and isolation such as observer [1, 2], parity space [4], and graphical approaches [5]. These are based on analytical or graphical system models of the system. The problem of these classical approaches is the non-consideration of uncertainties and modeling errors. However, in real applications, the uncertainties and the modeling errors always exist and cause false alarms and non-detections. The latter reduce confidence on the diagnosis system. To solve this problem, several robust methods of fault detection and isolation (RFDI) are proposed in the literature [6, 7, 8]. Most of these methods try to eliminate or minimize the effect of the uncertainties on the residuals, while others deal with the decoupling of the uncertain part of the residuals from the nominal part, and try to generate the threshold of the residuals using the maximum of the uncertain part. The robust diagnosis with BG approach is based essentially on the generation of analytical redundancy relations (ARRs) directly and systematically from the graphical model in derivative causality. The generated ARRs from a bond graph model that considers the uncertainties are composed of two parts, the nominal and the uncertain part. The residuals are obtained from the nominal parts while the thresholds are obtained from the uncertain parts.

---

[1]Corresponding author

In general, the ARRs generated from an over-constrained BG model can contain 1 or 2 order derivative of the output signals, which amplify the measurement errors and eventually the thresholds. This can be cause the non detection of certain faults. The objective of this work is to solve the problem caused by the signal differentiation when the residuals are computed in real time. The main idea is to use linear filters to improve the detectability of the residuals to certain faults appearing with low frequencies and low values, and at the same time, trying to keep the sensitivity of these residues to high-frequency faults using a bank of filters on the same residual with the estimation of different thresholds to each resulting residual of each filter.

This chapter is organized as follows: In the first section, the procedure of ARRs generation in the presence of measurement uncertainties is presented. In the second section, a method of residual evaluation is proposed using linear filters, in order to solve the problems of fault detection due to the signal differentiation. Finally, an application to an electromechanical system that composes a mobile robot named Robotino is done in order to validate the proposed method.

5.2. Bond graph modeling

The BG is a unified graphical tool for multi-physical system modeling [5, 10]. Introduced by Paynter [9] in 1961, The bond graph theory has been developed and used for simulation, for structural analysis and for diagnosis.

Bond Graph modeling is based on the phenomena of power exchange between different components of a system. This means that each component is represented by a bond graph element according to its physical characteristics. These components are connected by bonds which represent the direction of the power $P(t)$ which is equal to the product of the effort $e(t)$ and the flow $f(t)$. Named Te power variables, the effort and flow represent the different physical quantities depending on the physical domains, such as the voltage and the current in electrical engineering, the torque and the angular speed in mechanical rotation, the pressure and the volumetric flow rate in hydraulics...etc

The BG tool can represent four levels of modeling: The technological, physical, mathematical, and algorithmic levels. The first level represents the architecture of the system where the different part and subsystems are clearly represented in the graph. The second level or the physical level represents the nature of the power exchange between the system components, using the basic physical concepts and laws, such as the dissipation, storage and transformation of energy. These elements are modeled by the graphical representations shown in Figure 5.2.1.

The C, I and R elements are called passive elements that do not produce energy, where the C-element is used to model the components which store the potential energy as capacitors, tanks and springs. I-elements can model the storage of the kinetic energy as the inertia; and R-element represent the components which dissipate energy such as electrical and mechanical resistances. Other elements called active elements are modeled by Se and Sf, and deliver energy to the system by imposing respectively effort and flow to the latter. The previous elements are interconnected with each other using junction elements 0 and 1, where the 0-junction is called the *common effort junction* (All elements connected to the 0-junction have the same effort), and the 1-junction is called the *common flow junction* (All elements connected to the 1-junction have the same flow). Moreover, there exist two other junction elements which represent transformers TF and gyrators GY. The active and passive elements must be connected to junctions to obtain the complete structure of the model. The detectors are represented by Df for flow detectors and De for effort detectors. The mathematical level represent the mathematical model that can be obtained from the different elements such as the junction equations ($\sum e_i = 0$ for 1-junctions and $\sum f_i = 0$ for 0-junctions) which are called the structural equations; and the behavioral equations (eg. $e = Rf$ for R-element) that explain how the energy is transformed. The algorithmic level is an important propriety of the BG. It indicates the way the unknown variables are calculated. The elimination of unknown variables is performed using the causality affectation which appears on the graphical model as a perpendicular stroke at the end of each bond (eg. R-element in resistance and conductance causality Figure 5.2.2).

5.2.1. Integral and derivative causality. The term integral and derivative causality is used to indicate if the causal behavioral equations of the dynamic elements C and I are calculated using the integral or the derivative of the known variable. However, the integral causality is preferred in simulation to simplify the calculations. The C and I elements in integral causality give the behavioral equations (5.2.1), while the derivative causality is gives equations (5.2.2):

(5.2.1)
$$e_C(t) := \frac{1}{C} \int f_C(t)dt;$$
$$f_I(t) := \frac{1}{I} \int e_I(t)dt.$$

$\Phi_C\left(e(t),\int f(t)dt\right)=0$	$\dfrac{e(t)}{f(t)}\!\!\nearrow C$
$\Phi_I\left(f(t),\int e(t)dt\right)=0$	$\dfrac{e(t)}{f(t)}\!\!\nearrow I$
$\Phi_R\left(f(t),e(t)\right)=0$	$\dfrac{e(t)}{f(t)}\!\!\nearrow R$
$e(t)=Se$ $f(t)=Sf$	$\left(Se,\,Sf\right)\dfrac{e(t)}{f(t)}\nearrow$
$\Phi_{TF}\left(e_1(t),e_2(t)\right)=0$ $\Phi_{TF}\left(f_2(t),f_2(t)\right)=0$	$\dfrac{e_1(t)}{f_1(t)}\!\!\nearrow TF \dfrac{e_2(t)}{f_2(t)}\nearrow$
$\Phi_{TF}\left(e_1(t),f_2(t)\right)=0$ $\Phi_{TF}\left(f_1(t),e_2(t)\right)=0$	$\dfrac{e_1(t)}{f_1(t)}\!\!\nearrow GY \dfrac{e_2(t)}{f_2(t)}\nearrow$

FIGURE 5.2.1. Basic Bond Graph elements.

$\dfrac{e(t)}{f(t)}\!\!\nearrow R$	$f(t)\;\boxed{R}\;e(t)$	$e(t):=Rf(t)$
$\dfrac{e(t)}{f(t)}\!\!\rightarrow\!| R$	$e(t)\;\boxed{R}\;f(t)$	$f(t):=\dfrac{1}{R}e(t)$

FIGURE 5.2.2. R-element causalities.

$$(5.2.2)\qquad \begin{aligned} f_C(t) &:= C\frac{de_C(t)}{dt};\\ e_I(t) &:= I\frac{df_I(t)}{dt}. \end{aligned}$$

Let us consider the electrical system shown in Figure 5.2.3. Using the different notions of BG modeling, the BG model can be obtained as follows: the system is composed of three components: a voltage source E, resistance R and inductance I. The voltage source can be modeled by an effort source Se because it imposes the voltage (effort) to the system, the resistance dissipates energy and it can be represented by the R-element, and the inductance stores the electromagnetic energy, it can be modeled by the I-element. The system components have the same electric current so the BG elements must be interconnected by a 1-junction. Finally, the BG model in integral causality is obtained by connecting the different elements using the bonds.

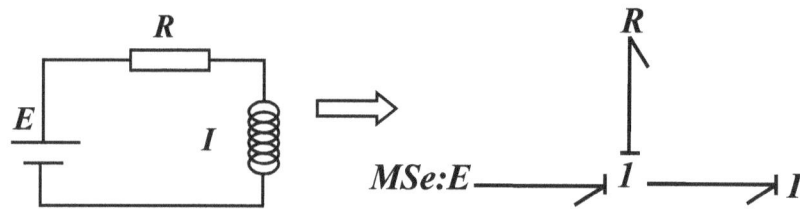

FIGURE 5.2.3. Bond Graph model in integral causality.

5.3. Fault detection and isolation using bond graph

Fault detection and isolation (FDI) using BG approach is based essentially on the generation of analytical redundancy relations (ARRs) using directly the graphical model. The method consists in using the structural and the causal properties of the BG to deduce the structural equations of *observed junctions* using only the known variables.

Definition The observed junction is the 0 or 1-junction which is connected to a dualized detector (eg. The 1-junction of the model shown in Figure 5.3.3 - (b)).

Definition Dualization of detectors is the replacing of the detectors De and Df (Figure 5.3.1-a) by a signal source SSe and SSf respectively (Figure 5.3.1-b).

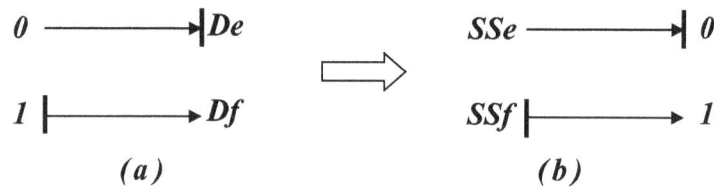

FIGURE 5.3.1. Dualisation of detectors.

Definition In a bond graph model, a causal path is an alternating of bonds and elements (R, C, I...) called nodes such that all nodes have a complete and correct causality. Depending on the causality, the passed variable is the effort or the flow. To change this variable, the causal path must pass through a junction element GY, or a passive element (I, C or R) (Figure 5.3.2).

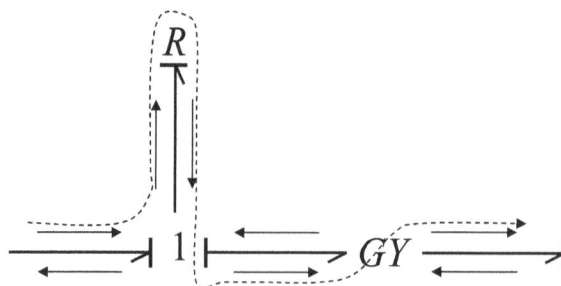

FIGURE 5.3.2. Causal path.

Definition A system is under-constrained if its dynamic BG elements can not accept the derivative causality when the detectors are dualized. However, in certain cases the initial conditions can be eliminated by using derivatives on the ARRs until the elimination of all integrals. This method is not preferred in the presence of noises.

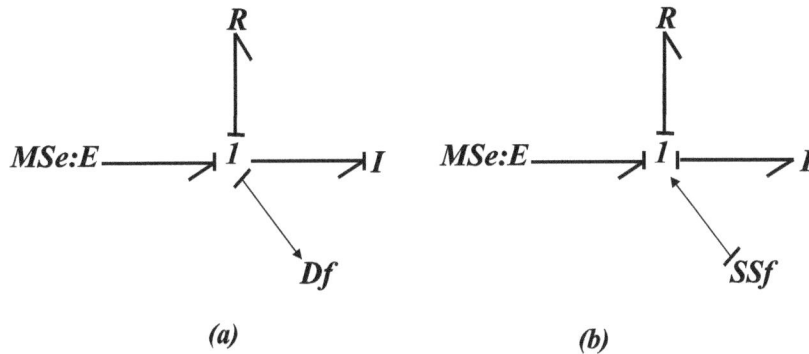

FIGURE 5.3.3. (a)model in integral causality,(b) model in derivative causality with the dualization of the detector.

In case of derivative causality, the detector must be dualized, it means that the detectors (De or Df) become sources of information (flow or effort). If the model in derivative causality is obtained without conflict of causality, the system is called over-constrained and it can be supervised; but if there is a conflict of causality the system is under-constrained, the system is not supervised. Noting that the flow sensor must be connected to one-junctions and the effort sensor to zero-junctions.

5.3.1. ARR generation. In BG approach, the ARR generation can be done using the following rules:

- Put the model in preferred derivative causality if possible and dualize the detectors when possible.
- Write the ARRs of the model using the structural equations of the observed junctions, and use the causal path to eliminate the unknown variables.
- Write the ARRs of detectors redundancy from the junctions that contain one or more non-dualized detectors by comparing its measurements with the other measurements of the other redundant detectors.

Example From the 1-junction of the bond graph model illustrated in Figure 5.3.3-(b), the following ARR can be generated:

$$ARR : E - R \cdot i - I \cdot \frac{di}{dt} = 0.$$

This can be obtained from the structural equation of the 1-junction (5.3.1). the unknown variables e_1, e_2, and e_3 are eliminated using the causal paths starting from the dualized detector SSf to the 1-junction (5.3.2).

(5.3.1)
$$\begin{cases} f_1 = f_2 = f_2 = i; \\ e_1 - e_2 - e_3 = 0; \end{cases}$$

(5.3.2)

Equation	Causal path
$e_1 := E;$	$MSe \to 1 - junction.$
$e_2 := R \cdot f_2 := R \cdot i;$	$SSf \to R \to 1 - junction.$
$e_3 := I \cdot \frac{df_2}{dt} := I \cdot \frac{di}{dt};$	$SSf \to I \to 1 - junction.$

5.3.2. Measurement uncertainties modeling using bond graph. The BG is a graphical tool used for physical modeling, simulation and ARR generation. Furthermore, it can be easily used to model the measurement uncertainties in order to generate both the ARRs and the thresholds. The latter is based on the structural and the causal properties of the BG model in derivative causality. The procedure of measurement uncertainties modeling is applied only on the dualized detectors because they give the information to the model from a real physical detector.

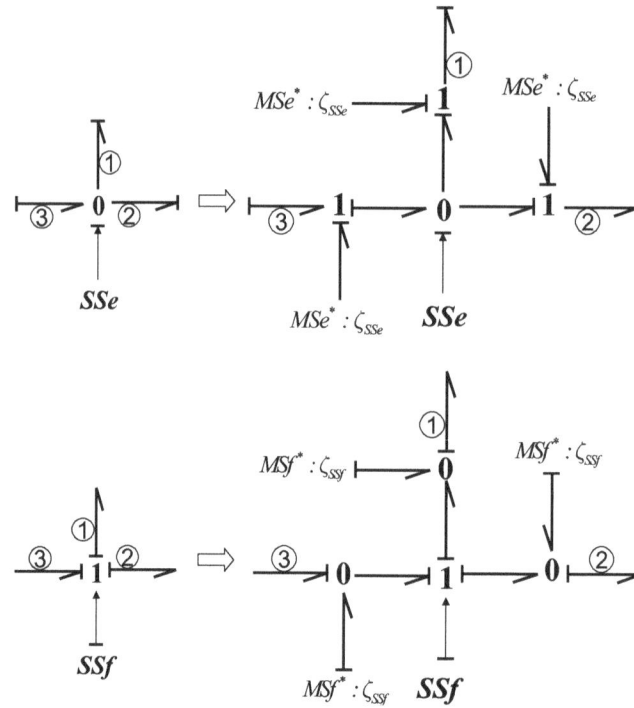

FIGURE 5.3.4. Measurement uncertainty modeling.

The dualized detectors on the BG model, used for diagnosis, in derivative causality impose the information signal to the observed junctions that are connected to these dualized detectors. Hence, the following equations can be obtained from Figure 5.3.4-(a,b), respectively:

(5.3.3)

$$(a) : \begin{cases} e_3 = SSe \\ e_1 = SSe \\ e_2 = SSe \end{cases}$$

$$(b) : \begin{cases} f_3 = SSf \\ f_1 = SSf \\ f_2 = SSf \end{cases}$$

The measurement uncertainties modeling can be done as shown in Figure 5.3.4-(c,d), this is based on the following equation (5.3.4):

(5.3.4)

$$(c) : \begin{cases} e_3 = e_6 - \zeta_{SSe} = SSe - \zeta_{SSe} \\ e_1 = e_4 - \zeta_{SSe} = SSe - \zeta_{SSe} \\ e_2 = e_5 - \zeta_{SSe} = SSe - \zeta_{SSe} \end{cases}$$

$$(d) : \begin{cases} f_3 = f_6 - \zeta_{SSf} = SSf - \zeta_{SSf} \\ f_1 = f_4 - \zeta_{SSf} = SSf - \zeta_{SSf} \\ f_2 = f_5 - \zeta_{SSf} = SSf - \zeta_{SSf} \end{cases}$$

Where SSf and SSe represent the measured signals ,ζ_{SSf} and ζ_{SSe} represent the measurement errors respectively on SSf and SSe.

From a BG point of view, the dualized flow (effort) detector imposes the flow (effort) to the 1-junction (0-junction) and consequently to all bonds connected to it. Based on this property, the measurement uncertainty is represented on these bonds by adding virtual sources modulated by the bounded measurement error.

5.3.3. Threshold generation in presence of measurement uncertainties. The thresholds generation in presence of measurement uncertainties using the BG model can be done using the following rules:

- The BG model of the system must be in preferred derivative causality.
- The measurement uncertainties must be represented on the graph using the procedure of measurement uncertainties modeling explained in Subsection 5.3.2.
- The uncertain part of the ARR must be generated using the causal paths which start from the virtual sources that represent the measurement uncertainties. All paths must be considered.
- The threshold is deduced using the sum of the maximum of the uncertain part.
- For the material redundancy equations, the thresholds are deduced by adding the maximum of uncertainties on the two detectors.

Example Let us consider the BG model shown in Figure 5.3.3. Two ARRs can be generated from this model, one

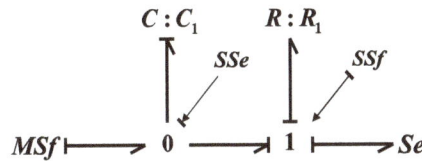

FIGURE 5.3.5. Example of BG model.

from the 0-junction and the other from the 1-junction. The two ARRs are used as residuals, and theoretically equal to zero in a normal situation, without considering the uncertainties and model errors. In presence of uncertainties on the sensors measurement, and if we know that the measurement error is an additive and bounded error, then the residual can be bounded by a two thresholds. The latter can be obtained using the BG model directly. The measurement error can be modeled by replacing the junction that contains the detector by a sub-model that corresponds to this junction as shown in Figure 5.3.4, where the measurement error is represented by a virtual source of flow or effort depending on the nature of the detector. This procedure of measurement uncertainties modeling is well presented in [**3**]. Applying the procedure of measurement uncertainties modeling, the model of Figure 5.3.3 becomes as shown in Figure 5.3.6. The model can be used to generate the uncertain part directly, by using the causal paths. For example in Figure 5.3.6:

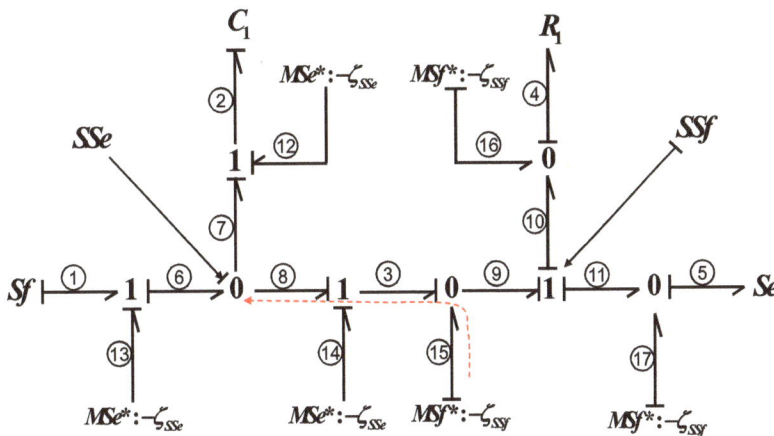

FIGURE 5.3.6. A BG model with measurement uncertainties.

$MSf* : \zeta_{SSf} \to 15 \to 3 \to 8 \to ARR_1$. The causal paths used to generate the uncertain part of the ARR must start from virtual sources that represent the measurement errors to the observed junctions (junctions which are connected to a detector). In Figure 5.3.6, We notice that the virtual source of effort connected with bond 13, and the virtual source of flow connected with bond 17 have no causal path to an observed junction, so the virtual sources that represent the measurement error do not appear in the residuals and can be removed.

The robust ARRs obtained from this model (Figure5.3.6) can be written as follows:

$$ARR_1 : MSf - C_1 \frac{dSSe}{dt} - SSf + \zeta_{SSf} + C_1 \frac{d\zeta_{SSe}}{dt} = 0$$
$$ARR_2 : SSe - R_1 SSf - Se + \zeta_{SSe} + R_1 \zeta_{SSf} = 0$$

These two ARR can be decomposed to two residuals r_1 and r_2(equation 5.3.5), and two thresholds a_1 and a_2 (equation 5.3.6).

(5.3.5)
$$r_1 = MSf - C_1 \frac{dSSe}{dt} - SSf$$
$$r_2 = SSe - R_1 SSf - Se$$

(5.3.6)
$$a_1 = \zeta_{SSf} + sC_1 \zeta_{SSe}$$
$$a_2 = \zeta_{SSe} + R_1 \zeta_{SSf}$$

where s is the differential operator. a_1 and a_2 can be obtained from the graphical model using the following causal paths.

(5.3.7)
$$a_1 : \begin{cases} MSf* : -\zeta_{SSf} \xrightarrow{15} 0 \xrightarrow{3} 1 \xrightarrow{8} 0_{ARR} = \zeta_{SSf}. \\ MSe* : -\zeta_{SSe} \xrightarrow{12} 1 \xrightarrow{2} C_1 \xrightarrow{2} 1 \xrightarrow{7} 0_{ARR} = sC_1 \zeta_{SSe}. \end{cases}$$

$$a_2 : \begin{cases} MSe* : -\zeta_{SSe} \xrightarrow{14} 1 \xrightarrow{3} 0 \xrightarrow{9} 1_{ARR} = \zeta_{SSe}. \\ MSf* : -\zeta_{SSf} \xrightarrow{16} 1 \xrightarrow{4} = R_1 \xrightarrow{4} 1 \xrightarrow{10} 0_{ARR} = R_1 \zeta_{SSf}. \end{cases}$$

The residuals r_1 and r_2 can be bounded by the maximum values of the a_1 and a_2, respectively. The following equations (5.3.8) can be written:

(5.3.8)
$$r_1 = MSf - sC_1 SSe - SSf \leq \max(\zeta_{SSf}) + \max(sC_1 \zeta_{SSe})$$
$$r_2 = SSe - R_1 SSf - Se \leq \max(\zeta_{SSe}) + \max(R_1 \zeta_{SSf})$$

The maximum values of a_1 and a_2 are calculated using the bounded measurement uncertainties Δ_{SSe} and Δ_{SSf}. $\max(sC_1 \zeta_{SSe})$ is given by equation (5.3.9).

(5.3.9)
$$\max(sC_1 \zeta_{SSe}) = C_1 \max \left(\frac{d\zeta_{SSe}}{dt} \right)$$

Generally, the derivative is calculated using the following formula:

$$\frac{d\zeta_{SSe}}{dt} \equiv \frac{\zeta_{SSe}^i - \zeta_{SSe}^{i-1}}{t_i - t_{i-1}}$$

where ζ_{SSe}^i is the measurement error at the time t_i. So we can write

$$\frac{\zeta_{SSe}^i - \zeta_{SSe}^{i-1}}{t_i - t_{i-1}} \leq \frac{2\Delta_{SSe}}{t_i - t_{i-1}}.$$

Figure 5.3.7 shows a bounded signal with its derivative where $t_i - t_{i-1} = 0.1$.

Sometimes, the variation of the measurement error is not big enough to be just enveloped by the threshold. In this case, we can say that we have an over-estimated threshold. This may cause non-detection of certain faults.

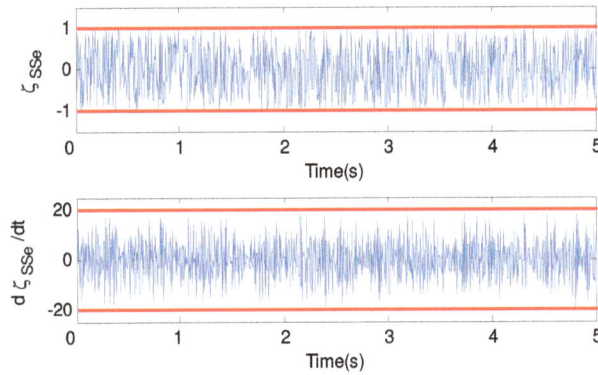

FIGURE 5.3.7. Residual evaluation with a bank of filters.

5.4. Residuals evaluation

The problem of over estimation of the thresholds can cause the non detection of certain faults that appear with low additive values. A residual evaluation method is presented in this section in order to improve the detectability of certain faults. The procedure consists of using a linear filter to solve the problem due to the signal differentiation, where the threshold can be very big and cause the non detectability of certain faults. We use a linear digital filter on the residual to minimize the threshold of the derivative of the measurement error. The filter is not applied only on the derivative signal but on the residual because it can cause non-synchronization between the signals used to calculate the same residual. The later is the sum of flows or efforts calculated from different signals. The non-synchronization between these signals can cause false alarms and non-detections. The used filter is linear, which allows the calculation of the thresholds after the filtering.

5.4.1. Linear filtering. To reduce the effect of the derivative of the measurement errors, a specific form of linear digital filters is used. Consequently, the linear filter is used to reduce the over estimation of the thresholds. The used filter is linear in order to be able to calculate the bound of the measurement error of the derivative signal. Let us consider the following form of digital filter (5.4.1):

(5.4.1)
$$F : Y_k = \sum_{l=0}^{p-1} h_l X_{k-l}$$

Where p is the filter order. This filter is applied to the residuals obtained from the BG model, as shown in Figure5.4.1.

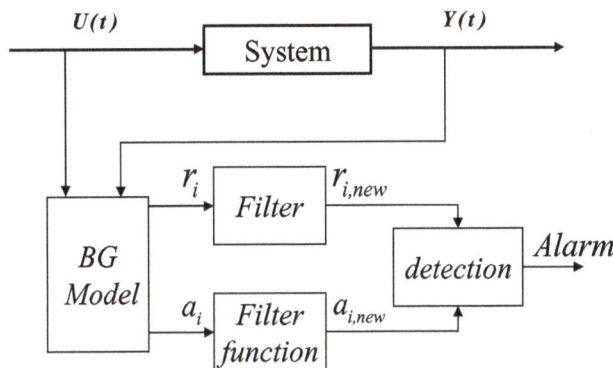

FIGURE 5.4.1. Thresholds and residuals generation

The filter F is applied to the residual r to obtain a new residual r_{new}, and a new threshold which can be calculated using the structural properties of the BG model, the parameters of the model, the bounds of the measurement errors, and the linear filter parameters.

This method is applied when the measurement error is bounded, without considering the distribution of the measurement error, because generally, it is unknown. In this case the threshold can be calculated as follows:

- The ARR is the sum of efforts or flows, so we can write:

$$
\begin{aligned}
r &= A_1(\dot{\theta}_1) + A_2(\theta_1) + \cdots + A_1(\dot{\zeta}_1) + A_2(\zeta_2) + \cdots \\
F(r) &= F\left(A_1(\dot{\theta}_1) + A_2(\theta_1) + \cdots + A_1(\dot{\zeta}_1) + A_2(\zeta_2) + \cdots\right) \\
&= F\left(A_1(\dot{\theta}_1) + A_2(\theta_1) + \cdots\right) + F\left(A_1(\dot{\zeta}_1) + A_2(\zeta_2) + \cdots\right) \\
&= F(r_m) + F\left(A_1(\dot{\zeta}_1)\right) + F\left(A_2(\zeta_2)\right) + \cdots
\end{aligned}
$$

Where r_m is the nominal part of the ARR, with $r_m = (A_1(\dot{\theta}_1) + A_2(\theta_1) + \cdots)$. A_i are linear functions that represent the linear relation between the sensor signals and the residuals. θ_i $(i = 1, 2, 3, ..., n)$ are the signals given by the detectors. ζ_i are the bounded measurement errors on θ_i by Δ_{θ_i}.

- Because the proprieties of the measurement error are unknown except that it is bounded. The terms which do not contain the derivative of the signal, can be bounded as follows:

$$
\begin{aligned}
\max\left(F\left(A_i(\zeta_i)\right)\right) &= F\left(\max\left(A_i(\zeta_i)\right)\right) \\
&= A_i\left(\max\left(F(\zeta_i)\right)\right) \\
\max\left(F(\zeta_i)\right)_k &= \max\left(\sum_{l=0}^{p-1} h_l \zeta_i^{k-l}\right) \\
\max\left(F(\zeta_i)\right)_k &= \sum_{l=0}^{p-1} h_l \max(\zeta_i^{k-l})
\end{aligned}
$$

(5.4.2)

The measurement error is bounded by Δ_{θ_i}.

$$
\max\left(F(\zeta_i)\right)_k = \sum_{l=0}^{p-1} h_l \Delta_{\theta_i}
$$

- For the terms which contain the derivative of the signal, we can write:

$$
\begin{aligned}
\max\left(F\left(A_i(\dot{\zeta}_i)\right)\right) &= F\left(\max\left(A_i(\dot{\zeta}_i)\right)\right) \\
&= A_i\left(\max\left(F(\dot{\zeta}_i)\right)\right)
\end{aligned}
$$

(5.4.3)

The derivative of the bounded random error of measurement can be calculated as follows:

$$
\dot{\zeta}_i^k \equiv \frac{\zeta_i^k - \zeta_i^{k-1}}{\Delta t}
$$

Applying the filter on this derivative:

$$F(\dot\zeta_i)_k = \frac{\sum\limits_{l=0}^{p-1} h_l\zeta_i^{k-l} - \sum\limits_{l=0}^{p-1} h_l\zeta_i^{k-l-1}}{\Delta t}$$

$$F(\dot\zeta_i)_k = \frac{(h_0\zeta_i^{k-0} + h_1\zeta_i^{k-1} + \cdots + h_{p-1}\zeta_i^{k-p-1})}{\Delta t}$$

$$- \frac{(h_0\zeta_i^{k-1} + h_1\zeta_i^{k-2} + \cdots + h_{p-1}\zeta_i^{k-p-2})}{\Delta t}$$

$$F(\dot\zeta_i)_k = \frac{h_0\zeta_i^{k-0} + (h_1 - h_0)\zeta_i^{k-1}}{\Delta t} + \cdots$$

$$+ \frac{(h_{p-1} - h_{p-2})\zeta_i^{k-p-1} - h_{p-1}\zeta_i^{k-p-2}}{\Delta t}$$

$$F(\dot\zeta_i)_k = \frac{h_0\zeta_i^{k-0} + \sum\limits_{l=0}^{p-2}(h_{l+1} - h_l)\zeta_i^{k-l-1} - h_{p-1}\zeta_i^{k-p-2}}{\Delta t}$$

$$\max\left(F(\dot\zeta_i)_k\right) = \frac{h_0\delta_{\theta_i} + \sum\limits_{l=0}^{p-2}(h_{l+1} - h_l)\delta_{\theta_i} + h_{p-1}\delta_{\theta_i}}{\Delta t}$$

The filter F can be applied on the residuals (r_1 and r_2) generated from the model of Figure 5.3.6, this filter F is a linear filter, so we can write the following relation using the two ARRs (ARR_1 and ARR_2):

$$\left|\begin{array}{l} ARR_1 : F(MSf - C_1\frac{dSSe}{dt} - SSf + \zeta_{SSf} + C_1\frac{\zeta_{SSe}}{dt}) = 0 \\ \Leftrightarrow F(MSf - C_1\frac{dSSe}{dt} - SSf) + F(\zeta_{SSf} + C_1\frac{\zeta_{SSe}}{dt}) = 0 \\ \Leftrightarrow F(MSf - C_1\frac{dSSe}{dt} - SSf) \leq \max(F(\zeta_{SSf} + C_1\frac{\zeta_{SSe}}{dt})) \end{array}\right.$$

$$\left|\begin{array}{l} ARR_2 : F(SSe - R_1SSf - Se + \zeta_{SSe} + R_1\zeta_{SSf}) = 0 \\ \Leftrightarrow F(SSe - R_1SSf - Se) + F(\zeta_{SSe} + R_1\zeta_{SSf}) = 0 \\ \Leftrightarrow F(SSe - R_1SSf - Se) \leq \max(F(\zeta_{SSe} + R_1\zeta_{SSf})) \end{array}\right.$$

5.4.2. Case of a mean filter. A mean filter named Simple Moving Average (SMA) is a moving average filter, where all coefficients are equal ($h_1 = h_2 = \cdots = h_{s-1}$). The equations 5.4.2 and 5.4.3 become as follows:

$$\max\left(F(A_i\zeta_i)_k\right) = A_i\sum\limits_{l=0}^{p-1} h_l\Delta_{\theta_i}$$

(5.4.4) $$\max\left(F\left(A_i(\dot\zeta_i)\right)\right)_k = A_i\left(\frac{h_0\Delta_{\theta_i} + h_{p-1}\Delta_{\theta_i}}{\Delta t}\right)$$

If the coefficients of the filter h_l ($l = 0, 1, 2..., p-1$) are equal to $\frac{1}{p}$, where ($p-1$) is the filter order. Equation 5.4.4 becomes as follows:

$$\max\left(F(A_i\zeta_i)_k\right) = A_i\Delta_{\theta_i}$$
$$\max\left(F\left(A_i(\dot\zeta_i)\right)\right)_k = A_i\left(\frac{2\Delta_{\theta_i}}{p\Delta t}\right)$$

The use of the filter on the residual signal can cause the non detection of certain faults, which appear with some frequencies that are filtered by the used filter.

A bank of filters can be used to keep the detectability of all faults, and to increase the sensitivity of the residuals to certain faults as shown in Figure 5.4.2.

Thresholds

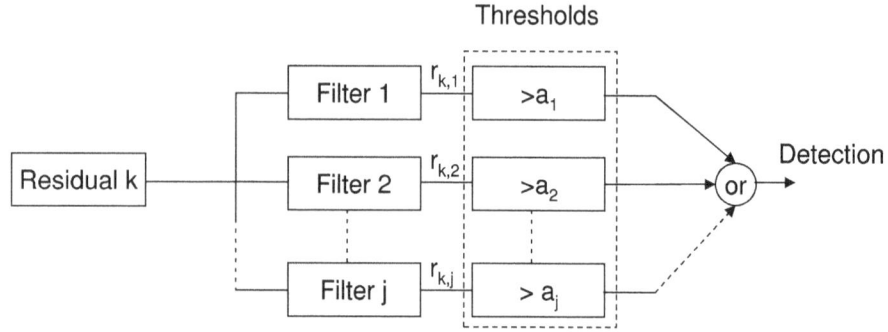

FIGURE 5.4.2. Residual evaluation with a bank of filters.

5.5. Application

In this Application, the presented procedure of robust ARRs generation with respect to output and input uncertainties is applied to an electromechanical subsystem of a robot named Robotino. The latter is composed of three electromechanical subsystems (three omni-directional wheels with three DC motors). Each Subsystem contains two detectors that measure the current and the angular speed of the DC motor, where the angles between the axes of the motors are equal to $120°$ (Figure 5.5.1).

5.5.1. Modeling. The considered subsystem is modeled by BG tool as shown in Figure 5.5.3. To obtain the model, the different physical components and phenomena are considered. The electrical part of the DC motor is modeled by R_a, L_a and U which represent the electrical resistance, inductance and the voltage, respectively. A gyrator (GY) represents the energy transfer between the DC motors electrical and the mechanical (mechanical resistance R_e and inertia J_e) parts. The reducer is modeled by a transformer TF and the inertia of the wheel is represented by J_s and the resistance with R_s.

In this application, the input uncertainties are modeled by replacing MSe and MSf by the sub model represented in Figure 5.5.4-(a) and Figure 5.5.4-(b).

From the model of Figure 5.5.5, the following ARRs can be obtained:

$$
\begin{aligned}
ARR_1 &= U - L_a \frac{di}{dt} - R_a i - m\dot{\theta}_m + a_1 \\
ARR_2 &= m\,i - J_e \frac{d\dot{\theta}_m}{dt} - R_e \dot{\theta}_m - \frac{J_s}{N^2}\frac{d\dot{\theta}_m}{dt} - \frac{R_s}{N^2}\dot{\theta}_m \\
&\quad + \Phi(F_x) + a_2 \\
a_1 &= -R_a \zeta_i - L_a \frac{d\zeta_i}{dt} - m\zeta_{\dot{\theta}_m} \\
a_2 &= -m\zeta_i - R_e \zeta_{\dot{\theta}_m} - J_e \frac{d\zeta_{\dot{\theta}_m}}{d} - \frac{J_s}{N^2}\frac{d\zeta_{\dot{\theta}_m}}{dt} \\
&\quad - \frac{R_s}{N^2}\zeta_{\dot{\theta}_m} + \frac{1}{N}\zeta_{\Phi(F_x)}.
\end{aligned}
$$

Where ζ_i and $\zeta_{\dot{\theta}_m}$ are the measurement errors on the current and velocity detectors respectively. $\zeta_{\Phi(F_x)}$ is the error on the input of the impact effort $\Phi(F_x)$. All the output and the input errors are considered bounded as follows:

$$
\begin{aligned}
|\zeta_i| &\leq \Delta_i = 0.2A \\
|\zeta_{\theta_m}| &\leq \Delta_{\theta_m} = \pi/500 rad \\
|\zeta_{\Phi(F_x)}| &\leq \Delta_{\Phi(F_x)} \leq 0.0016 \\
F_x &= (6 \pm 0.06)N
\end{aligned}
$$

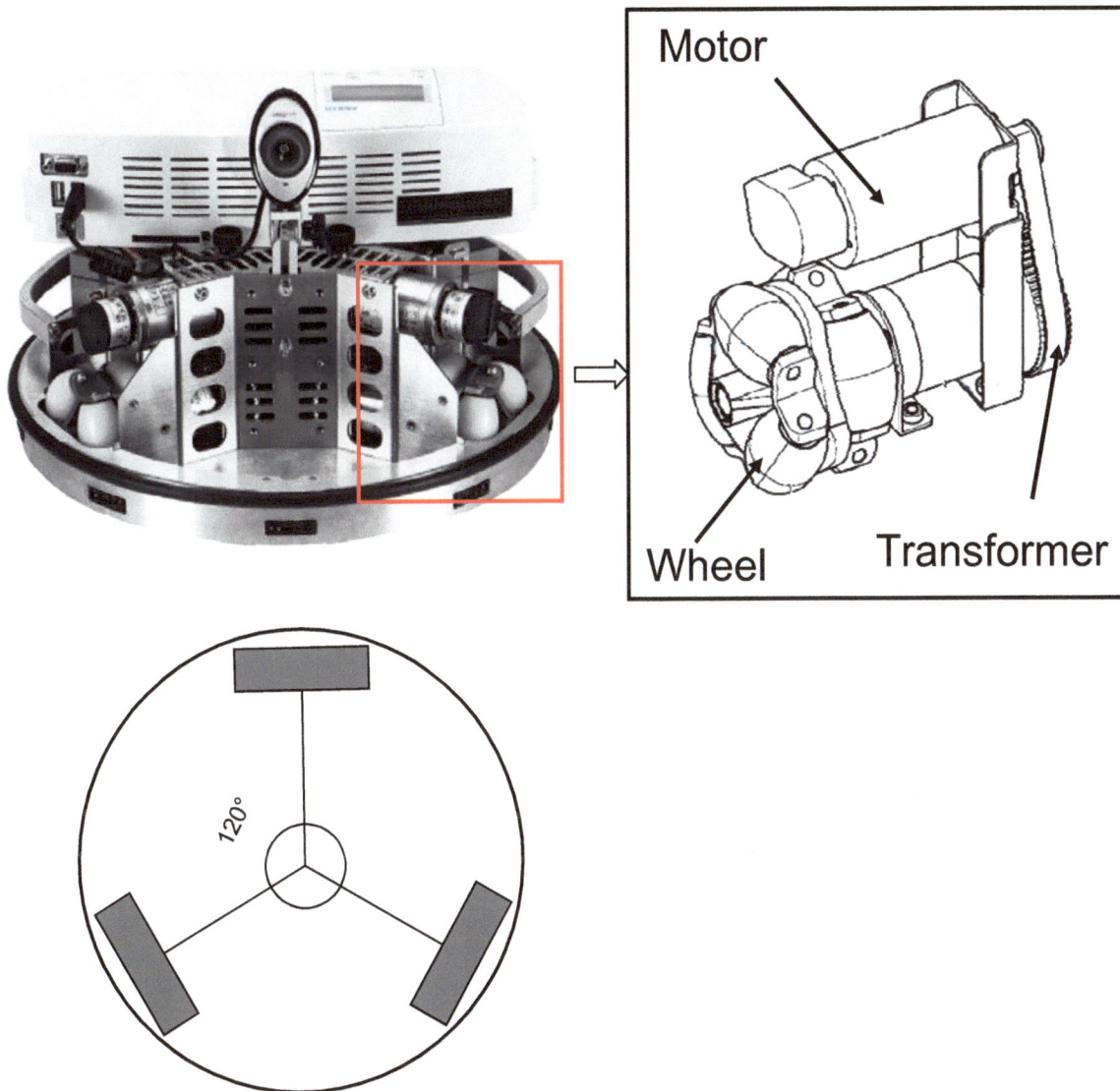

FIGURE 5.5.1. Electromechanical Robotino subsystem.

a_1 and a_2 are the uncertain parts of the two ARR, that can be used to calculate the thresholds $a_{th,1}$ and $a_{th,2}$ for residuals r_1 and r_2 respectively.

$$
\begin{aligned}
a_{th,1} &= \max\left(a_1\right); \\
&= -R_a \max\left(\zeta_i\right) - L_a \max\left(\frac{d\zeta_i}{dt}\right) - m \max\left(\zeta_{\dot{\theta}_m}\right); \\
a_{th,2} &= \max\left(a_2\right); \\
&= -m \max\left(\zeta_i\right) - R_e \max\left(\zeta_{\dot{\theta}_m}\right) - J_e \max\left(\frac{d\zeta_{\dot{\theta}_m}}{d}\right) \\
&\quad - \frac{J_a}{N^2} \max\left(\frac{d\zeta_{\dot{\theta}_m}}{dt}\right) n - \frac{R_a}{N^2} \max\left(\zeta_{\dot{\theta}_m}\right) + \frac{1}{N} \max\left(\zeta_{\Phi(F_x)}\right).
\end{aligned}
$$

5.5.2. Simulation and Experimental results. We have applied the developed approach of diagnosis on real experimental data of the dynamic electromechanical system. The input and the output signals are shown in Figure

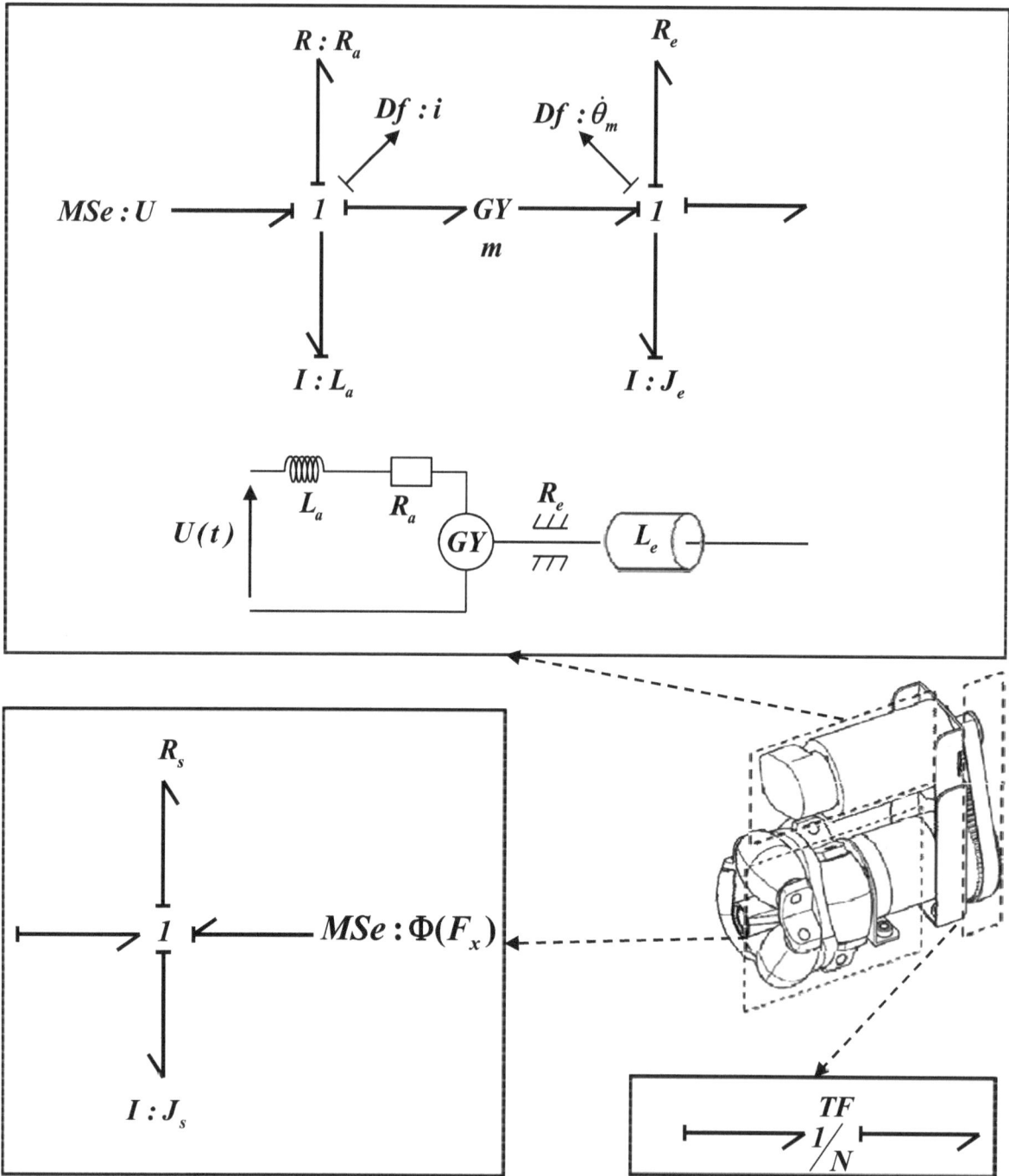

FIGURE 5.5.2. Different step of system modeling.

5.5.12. The input signal is not measured, but it is estimated from the desired velocity and the measured velocity. The contact force is not considered. The parameters of the model are the following (5.5.1).

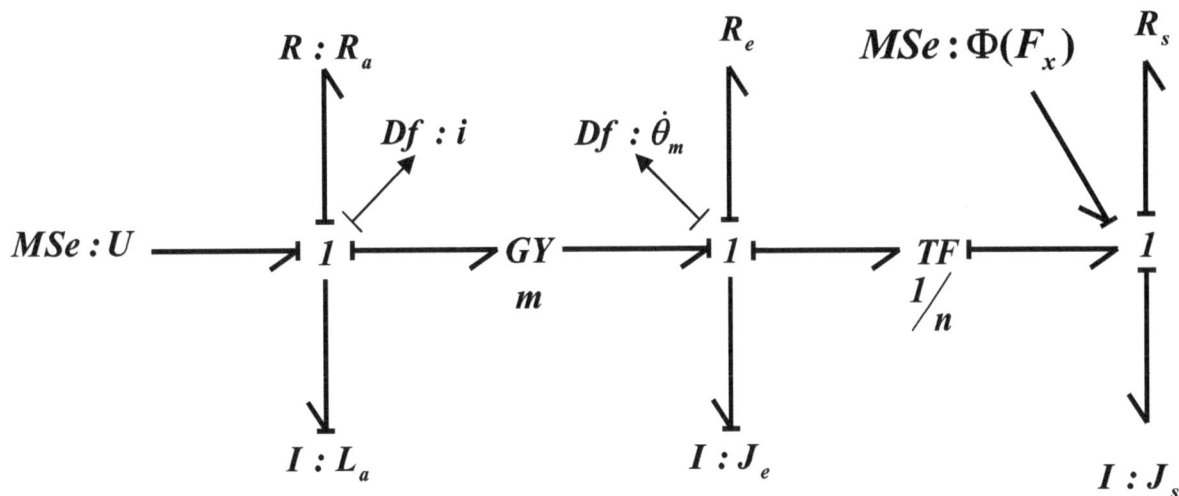

FIGURE 5.5.3. Bond graph model of the electromechanical subsystem.

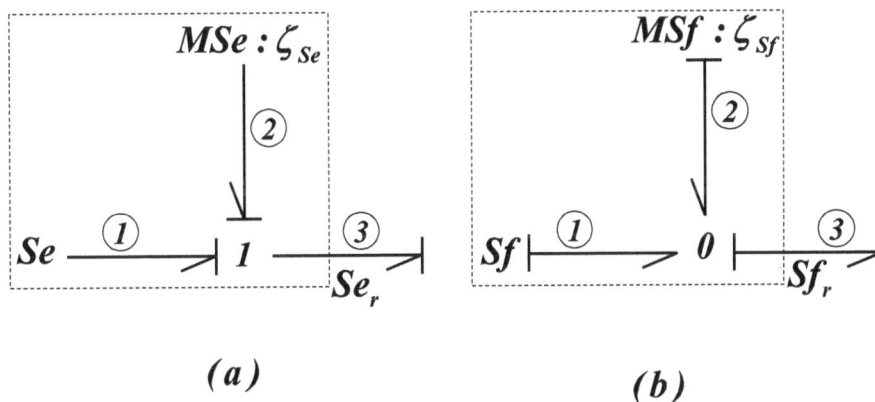

(a)   (b)

FIGURE 5.5.4. Input uncertainty modeling.

(5.5.1)

$$R_a = 8.13\Omega;$$
$$L_a = 0.0089H;$$
$$m = 0.04315V \cdot sec \cdot rad^{-1};$$
$$J_e = 0.00000795Kg \cdot m^2;$$
$$R_e = 0.000047Nm \cdot sec \cdot rad^{-1};$$
$$J_s = 0.00063Kg \cdot m^2;$$
$$R_s = 0.00002Nm \cdot sec \cdot rad^{-1};$$
$$N = 16;$$
$$r_w = 0.04m;$$

5.5.2.1. *Simulation results.* In Figure 5.5.6, the input $U(t)$ and the outputs of the system i and $\dot{\theta}_m$ in a normal situation are presented, and the faulty signal of the current sensor is shown in Figure 5.5.7. Either in the same figure the signal produced by the faulty current sensor is given. The residuals r_1 and r_2 in a normal situation are shown in Figure 5.5.9 and Figure 5.5.8, and in a faulty situation in Figure 5.5.10 and Figure 5.5.11. We notice that the fault is not detected by the residual r_2 without filtering, and clearly detected after applying the filters.

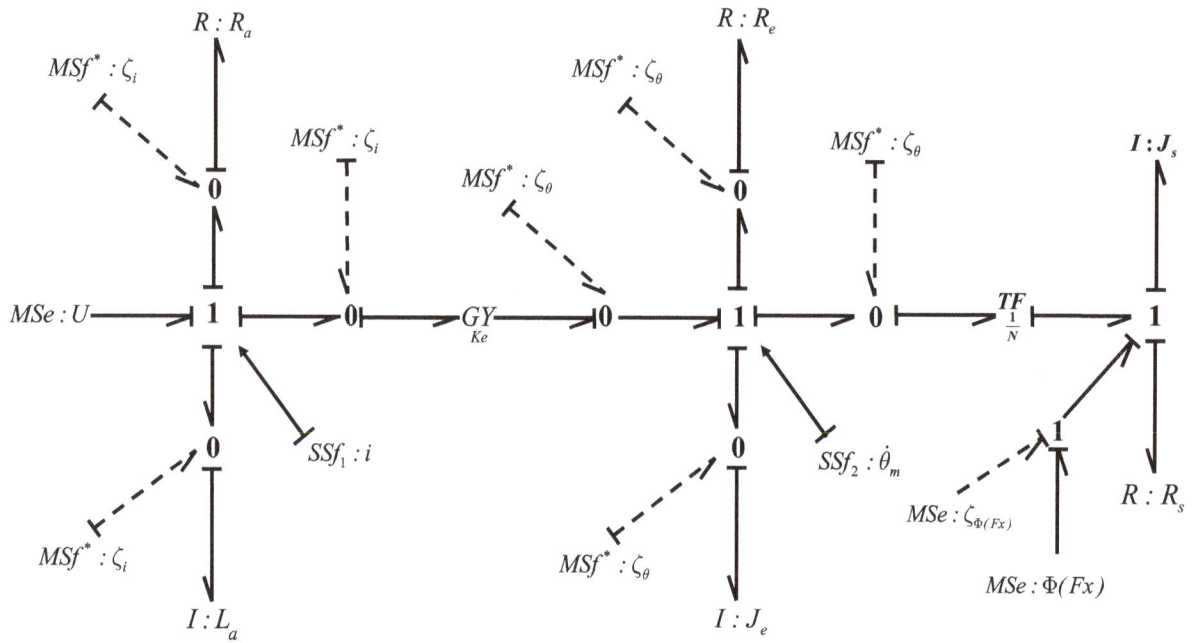

FIGURE 5.5.5. BG model of the electromechanical subsystem with input and output uncertainties.

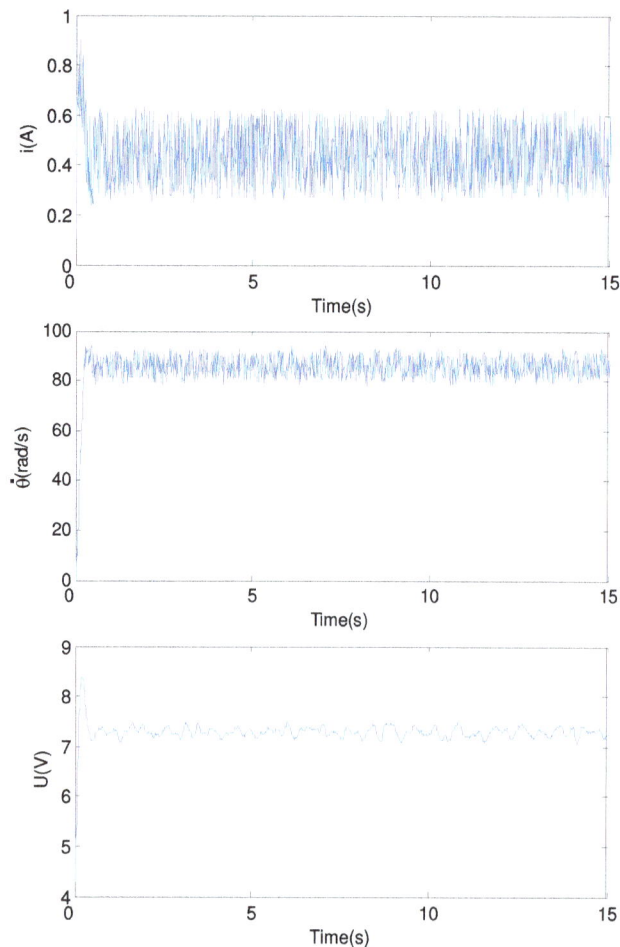

FIGURE 5.5.6. Simulation: Input and output signals of the system.

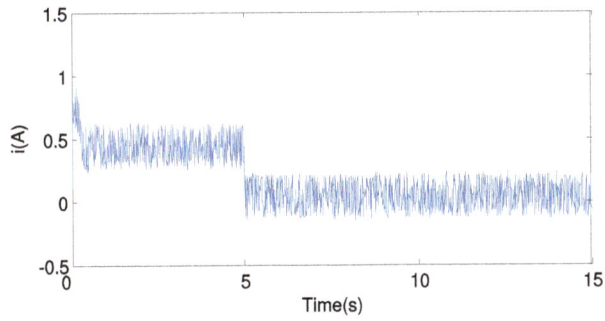

FIGURE 5.5.7. Simulation: Current sensor signal in a faulty situation.

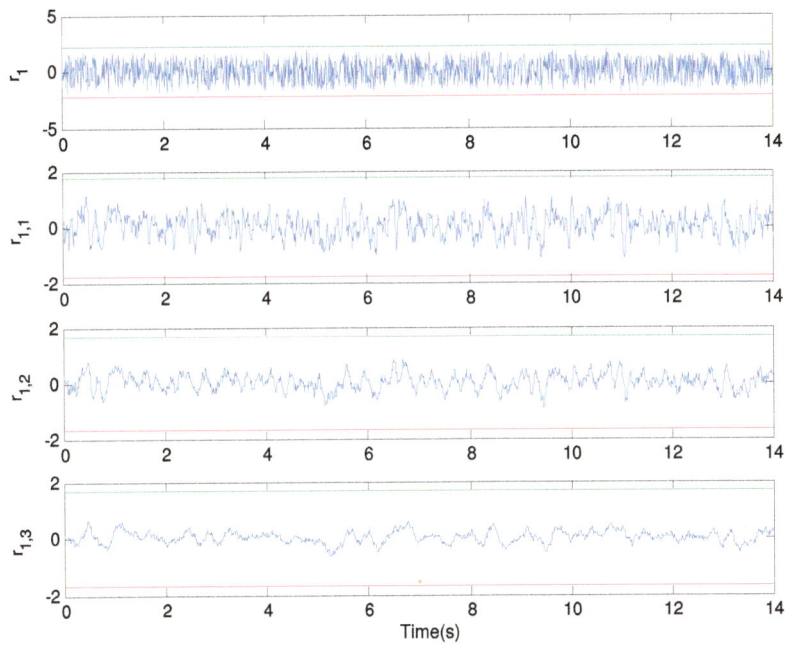

FIGURE 5.5.8. Simulation: r_1 in a normal situation.

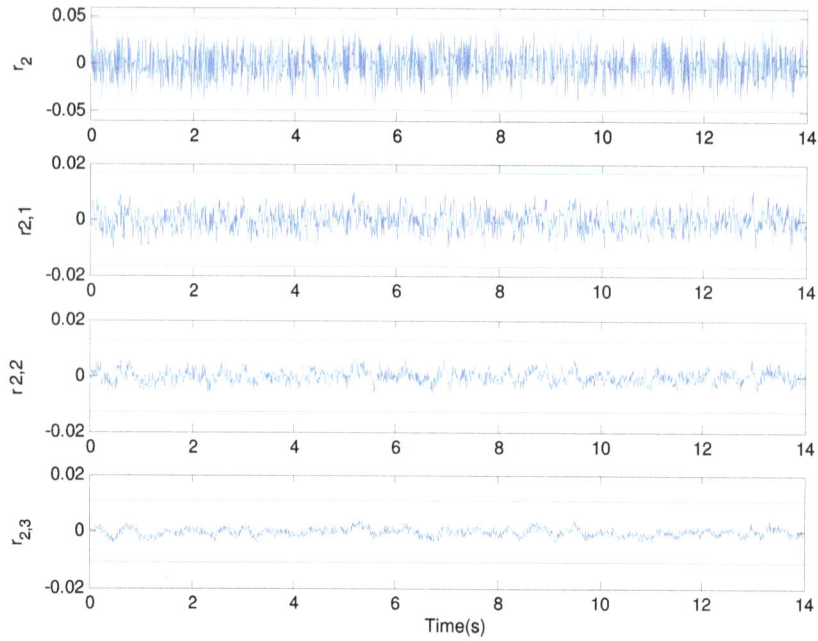

FIGURE 5.5.9. Simulation: r_2 in a normal situation.

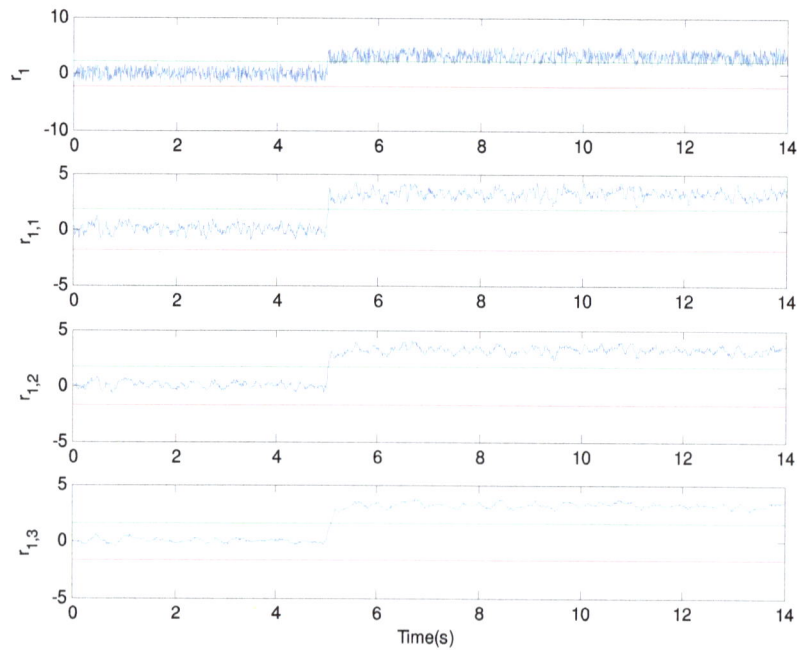

FIGURE 5.5.10. simulation: r_1 in a faulty situation.

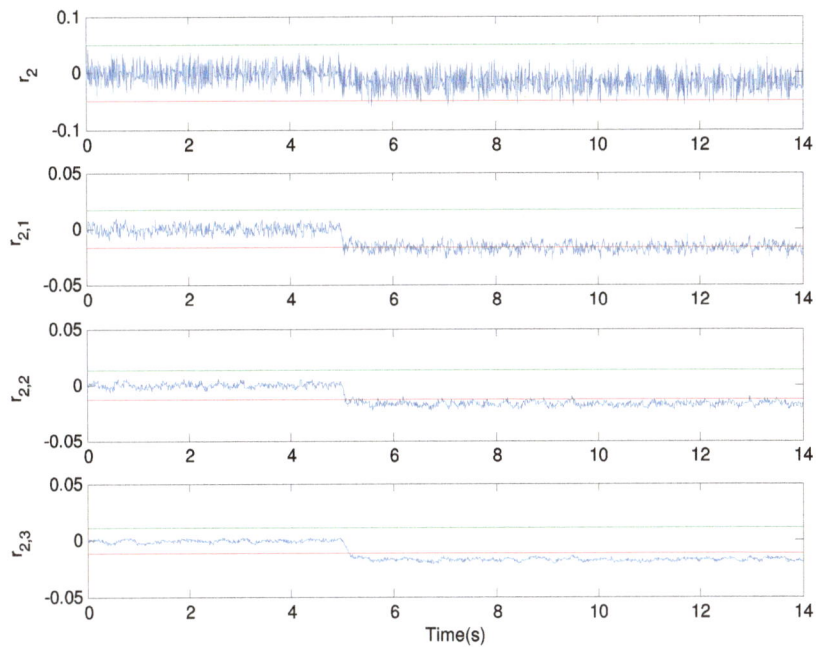

FIGURE 5.5.11. simulation: r_2 in a faulty situation.

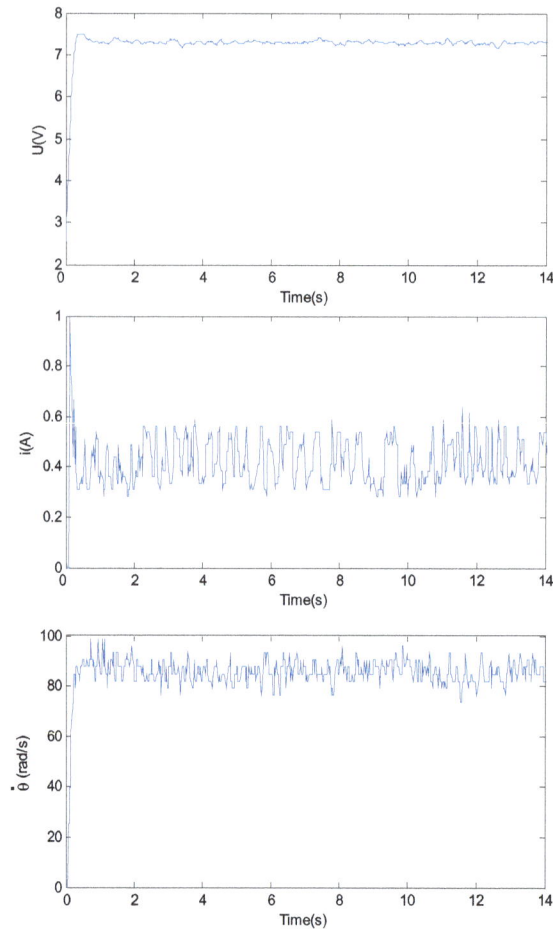

FIGURE 5.5.12. Experimental input and output signals.

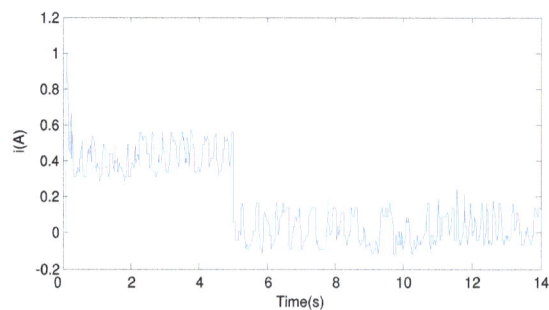

FIGURE 5.5.13. Faulty Signal of the current sensor.

5.5.2.2. *Experimental results.* The residuals r_1 and r_2 calculated without filtering in presence of measurement uncertainties in normal situation is shown in Figure 5.5.15 and Figure 5.5.14. We notice that the values of the residuals are greater-than the thresholds' values before $t = 1s$, and this may be due to the modeling errors of the contact effort.

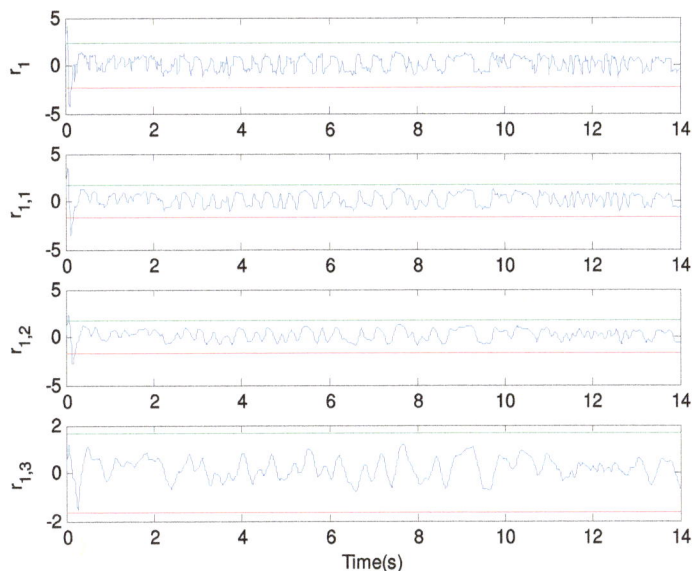

FIGURE 5.5.14. r_1 in a normal situation.

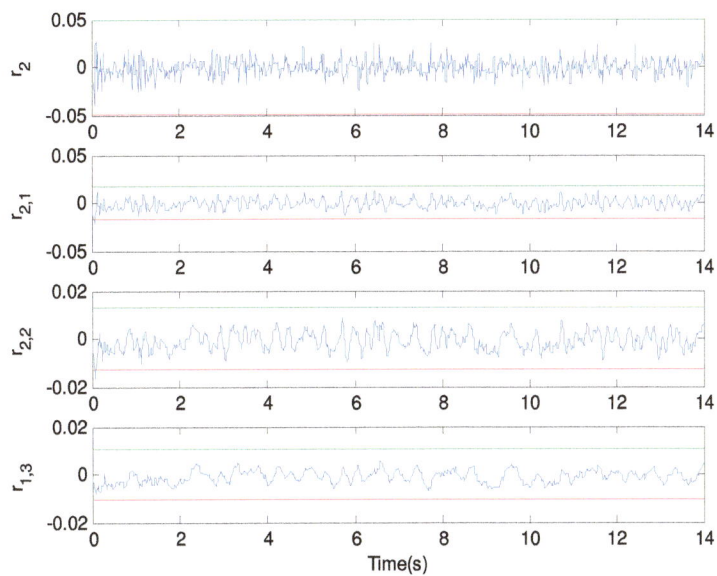

FIGURE 5.5.15. r_2 in a normal situation.

The residuals r_1 and r_2 are tested in a faulty situation by disturbing the signal of the current sensor. So, we add a constant signal to the measured signal to be disturbed. The faulty signal of the current sensor is shown in Figure 5.5.13.

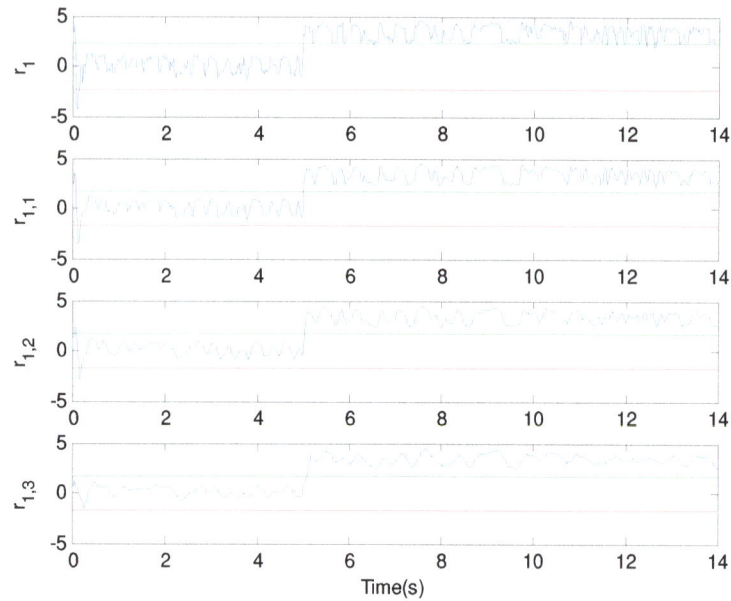

FIGURE 5.5.16. r_1 in a faulty situation.

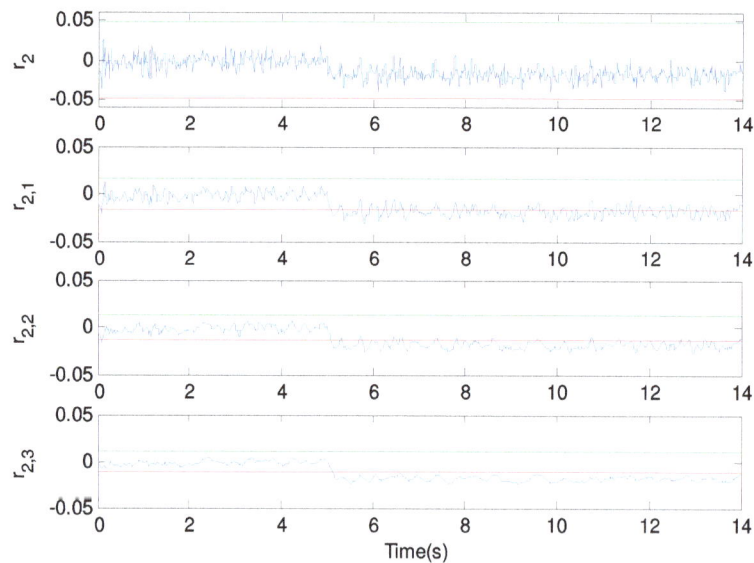

FIGURE 5.5.17. r_2 in a faulty situation.

The two residuals in the considered faulty situation are shown in Figure5.5.16 and Figure 5.5.17. The residual r_1 does not show a much sensibility to the fault on the current sensor. Consequently, the FDI system does not detect this fault. To solve this problem, we have applied three filters on the residual r_1 and we have calculated three new thresholds for the outputs of these filters.

5.6. Conclusions

In this chapter, the problem of robust fault detection and isolation has been studied in order to improve the detectability and the sensitivity of the residuals to certain faults which can not be detected by the basic residuals. This method is based on BG modeling and linear filtering. The BG model is used to model the measurement uncertainties and to generate the analytical redundancy relations. A specific form of linear filter is used to improve the detectability of the residuals and to recalculate the thresholds in order to avoid the problem of the over-estimation of the thresholds. The approach is applied on a mobile robot named Robotino and the results show the efficiency of this method.

Bibliography

[1] R.J. Patton and J. Chen, "observer-based fault detection and isolation: robustness and applications", Control Engineering Practice, 1997, vol. 5, pp. 671-682.

[2] R. J. Patton, P. M. Frank, and R. N. Clark. Fault Diagnosis in Dynamic Systems, Theory and Applications. Prentice-Hall, Englewood Cliff , NJ, 1989.

[3] Youcef touati, Rochedi Merzouki, Belkacem Ould Bouamama. Fault detection and isolation in presence of input and output uncertainties using bond graph approach. IMAACA 2011,pp. 221-227, 2011.

[4] A. Medvedev. "Fault detection and isolation by a continuous parity space method", Auromatica, Vol. 31, No. 7, pp. 1039-1044, 1995.

[5] Arun K. Samantaray and Belkacem Ould Bouamama, "Model-based Process Supervision: A Bond Graph Approach",1st Edition. 2008, 473 p. Springer.

[6] Gautam Biswas, Gyula Simon, Nagabhushan Mahadevan, Sriram Narasimhan, John Ramirez, and Gabor Karsai. A robust method for hybrid diagnosis of complex systems. In 5th IFAC Symposium on Fault Detection, Supervision and Safety of Technical Processes (SAFEPROCESS), pages 1125-1130, Washington, D.C, 2003.

[7] M.A. Djeziri, R. Merzouki, B. Ould-Bouamama, and G. Dauphin-Tanguy. Robust fault diagnosis using bond graph approach. Int. Journal of IEEE/ASME Transaction on Mechatronics, 12 (6):599-611, 2007.

[8] Z. Han ,W. Li , S. L. Shah. (2002). Fault detection and isolation in the presence of process ncertainties. 15th IFAC Wold Congress. pp.1887-1892.

[9] H. M. Paynter, Bond graphs and diakoptics, Matrix Tensor Q., Vol. 19, No. 3, pp. 104-107, 1969.

[10] Soumitro Banerjee. Dynamics for Engineers, john wily & Sons, Ltd ,(2005).

CHAPTER 6

Bond Graph Model-Based Fault Detection and Isolation : Application to Intelligent Autonomous Vehicles

Samir Benmoussa[1],
Polytech-Lille, LAGIS, UMR CNRS 8219
Avenue Paul Langevin, 59655 Villeneuve D'Ascq, France
samir.benmoussa@polytech-lille.fr

Rochdi Merzouki
Polytech-Lille, LAGIS, UMR CNRS 82193
Avenue Paul Langevin, 59655 Villeneuve D'Ascq, France
rochdi.merzouki@polytech-lille.fr

Belkacem Ould Bouamama
Polytech-Lille, LAGIS, UMR CNRS 8219
Avenue Paul Langevin, 59655 Villeneuve D'Ascq, France
belkacem.ouldbouamama@polytech-lille.fr

ABSTRACT. In this chapter, a methodology for plant fault detection and isolation using Bond Graph modeling tool is presented, the motivation using bond graph is that the dynamic model of the system and fault detection and isolation are performed using the same tool. Based on bond graph properties, the conditions of diagnosability are carried out structurally on the graph representing the system. The considered plant fault is a break of the transmission axle on an electromechanical system representing a quarter of an autonomous vehicle, named Robucar. Co-simulation using a vehicle simulator CALLAS/Simulink is given to show the efficiency of the proposed methodology.

Keywords: Fault Detection and Isolation, Bond Graph, Algebraic approach, Modeling, Intelligent systems.

6.1. Introduction

Intelligent Autonomous Vehicles (*IAVs*) have known a remarkable development in recent years. This is due to various reasons such as: economics (where they are cheaper to operate), reduction in acoustic noise, and less CO_2 pollution than conventional vehicles. These vehicles are designed to be operated autonomously and to be able to adapt to their surrounding environment. For that propose, it is important to improve the supervision of such autonomous systems in normal and faulty situations, in order to avoid hazards and increase safety. To reach this objective, it is necessary to add a monitoring system embedded in these vehicles making possible a real time fault detection and isolation (FDI).

In the literature, FDI methods can be divided in two classes. Methods based on the model of the system (model-based FDI), and the ones based on the process data (data-based FDI). Based on the model of the system, several methods have been developed to solve the problem of residual generation: parity equations [9], observer model-based [15], analytical redundancy relations (ARRs) [17], and the algebraic approach [7].

---

[1]Corresponding author

Among the model-based approaches, there are the ones whose objectives are to create a graph describing the model of the system. One of the approaches is *Digraph* $G(S, A)$ [4], where nodes S represent states, inputs, and measurement output variables. A is the edge between the nodes. The Bond Graph (*BG*) is also a graphical representation of the physical system since the nodes S represent input, output, and physical components of the system. A represents bonds and, it describes the power transfer between the passive and active components of multi-physical systems. The BG is the interface between the physical and the mathematical model of the system.

Several problems have been solved structurally using this graphical approach, amongst them: observability and controllability [18], system inversion [12], and FDI [6, 16]. BG model-based FDI uses analytical redundancy relations (*ARRs*) which are generated in a systematic way from the BG model of the system using its causal and structural properties. From *ARRs*, a fault signature matrix (*FSM*) is defined allowing the study of fault monitorability and isolability.

This chapter presents a methodology for plant fault detection and isolation. Combining the algebraic approach definition and BG properties, simple conditions to detect and isolate plant faults are proposed. These conditions are performed on the BG model using the notion of bicausality and disjoint causal path analysis. BG model-based FDI has several advantages:

- Both modeling and monitoring are performed using the same tool
- The detectability and isolability of plant fault is performed structurally on the graph;
- The plant fault can be considered as an additional input energy source(effort or flow);
- The concept of the fault is more physical than mathematical.

The presented methodology is applied to detect and isolate a break in the transmission axle of the electromechanical system on a quarter of of autonomous electric vehicle, named Robucar.

This chapter is structured as follows, in Section 2 some definitions and problem formulation are expressed. Model-based FDI using algebraic approach and BG are given in the section 3. The way it is applied to an electrical vehicle is given in Section 4. Co-simulation results are presented in section 5. The chapter ends with general synthesis and remarks.

6.2. Definitions and problem formulation

To guarantee systems safety, a subsystem called *monitoring system* is integrated into the former. This system consists of algorithms programed in electronic chips, which have the ability to detect and isolate various kinds of faults.

Definition A fault is an unpermitted deviation of at least one characteristic property (feature) of the system from the acceptable, usual, standard condition [10]

There is a difference between a fault and a failure. When a fault occurs, the system functions degrade, but continue their mission. In the case of a failure, the system cannot achieve its purpose. Nevertheless, when a mathematical model of the system is used, faults and failures have the same model representation. In the literature, one can find two kinds of faults: additive faults and multiplicative faults. According to [10], additive and multiplicative faults are defined by :

Definition An additive fault is a change of the signal $y(t)$ by an addition of $f(t)$.

Definition A multiplicative fault is a change in the parameter, because another variable $u(t)$ is multiplied by $f(t)$.

Figure 6.2.1 shows a graphical representation of the given definitions.

As one can notice, additive faults may represent faults on the sensor or actuator. Component (plant) faults can be considered as multiplicative ones. The first step in FDI model-based approaches is the faulty system modeling, based on the nature of the fault (Additive or multiplicative), the faulty system model takes different representation. A general representations of the faulty system when all possible (sensor, actuator, component) faults may occur has the following form [3]:

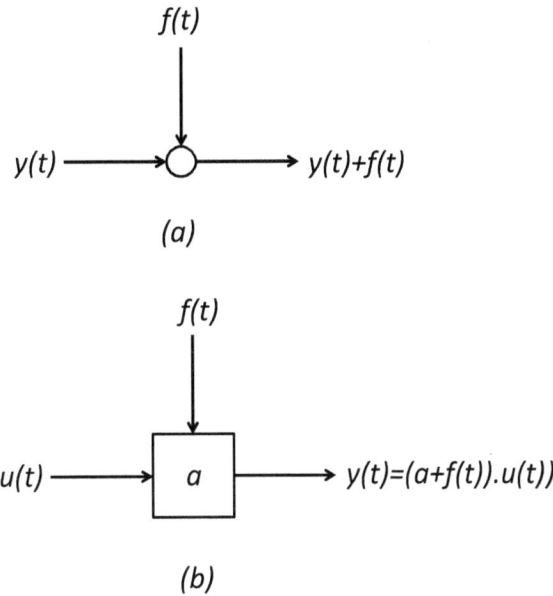

$$f(t)$$

$$y(t) \longrightarrow \bigcirc \longrightarrow y(t)+f(t)$$

(a)

$$f(t)$$

$$u(t) \longrightarrow \boxed{a} \longrightarrow y(t)=(a+f(t)).u(t))$$

(b)

FIGURE 6.2.1. (a): Additive fault for an output signal modeling; (b): Multiplicative fault modeling

(6.2.1)
$$\begin{cases} \dot{x} = Ax + Bu + H_1 f \\ y = Cx + Du + H_2 f \end{cases}$$

Where $x \in R^n$ is the state vector, $u \in R^m$ is the input vector, $y \in R^p$ is the output vector. $f \in R^q$ is the fault vector, and each element f_i corresponds to a specific fault. A, B, C are system matrices of appropriate dimensions. H_1, and H_2 are the entry fault matrices, which represent the effect of the faults on the system.

On this chapter, only component faults, which have the specificity that could be expressed as a change in the system parameters, are considered. For example, a fault on the ith row and j^{th} column of the element of matrix A can be modeled by:

(6.2.2)
$$\begin{cases} \dot{x} = Ax + Bu + I_i \Delta a_{ij} x_j \\ y = Cx \end{cases}$$

where x_j is the jth element of the state vector x, and I_i is a n- dimensional vector filed by zeros except in the ith elements. For the sake of simplicity, the plant faults are considered as unknown inputs to the system. So, if q system's components have to be monitored, the mathematical model of the system in nominal situation is augmented by q unknown inputs, and it takes the following form:

(6.2.3)
$$\begin{cases} \dot{x} = Ax + Bu + Mf \\ y = Cx \end{cases}$$

Where $M \in R^{n \times q}$ is a known matrix, and F is an unknown vector that needs to be monitored.

As a remark, the number of faults is equal or less than to the number of sensors ($q \leq p$).

6.3. Model-fased fault detection and isolation

6.3.1. Algebraic approach for FDI. In Fliess's theoretical approach [7], a linear system (Λ^{pert}) is regarded as a finitely generated free $k[s]$−module , where $k[s]$ is a commutative principal ideal domain of linear differential operators of the form $\sum_{finite} c_v s^v$, $c_v \in k$: s is the usual symbol of derivation, and k is the field of real or complex numbers.

In Λ^{pert}, two finite subsets are distinguished: the fault variable F and the perturbation variable π, which do not "interact", i.e., $span_{k[s]}(F) \cap span_{k[s]}(\pi) = \{0\}$ [7]. The nominal system is defined by $\Lambda = \Lambda^{pert}/span_{k[s]}(\pi)$.

Definition An *input output system* is a linear system Λ^{pert}, equipped with an input u and an output y [7], such that :

- The *input* of the linear system Λ^{pert} is a finite sequence $u = (u_i)_{1 \leq i \leq m}$ of elements of Λ^{pert} such that $\Lambda^{pert}/[u]_{k[s]}$ is torsion, the input u is assumed independent.
- The *output* of the linear system Λ^{pert} is a finite sequence $y = (y_i)_{1 \leq i \leq p}$ of elements of Λ^{pert}.

Assumptions : the following properties are assumed to be satisfied:

- $span_{k[s]}(u) \cap span_{k[s]}(\pi) = \{0\}$
- $span_{k[s]}(u) \cap span_{k[s]}(F) = \{0\}$

Which means that the control variable u does not interact with the perturbation and the fault variables.

Fault detection and isolation is done in an algebraic framework using the following theorem and definitions:

Theorem 2. *The system Λ^{pert} is observable (in the sense that the state is observable with respect to u and y), then it is diagnosable if, and only if, F is observable with respect to u, y, and x [5].*

Definition (Algebraic Detectability) : a fault F is said to be *detectable* if, it is observable over u and y [7].

Definition (Algebraic Isolability) : any fault variable in F is said to be isolable if, and only if, there exists a system of parity equation [7]

(6.3.1)
$$M \begin{pmatrix} F_1 \\ \vdots \\ F_q \end{pmatrix} = Q \begin{pmatrix} u_1 \\ \vdots \\ u_m \end{pmatrix} + S \begin{pmatrix} y_1 \\ \vdots \\ y_p \end{pmatrix}$$

where $M \in k[s]^{q \times q}$, $Q \in k[s]^{q \times m}$, $S \in k[s]^{q \times p}$, $\det M \neq 0$.

In other words, it is required that :

- The system must be observable: the states of the system can be expressed as a function of outputs and their derivatives,
- Each fault variable has to be written under a polynomial equation format F_i and finitely many time derivatives of u and y with coefficients in $k[s]$.

(6.3.2)
$$\varphi(F_i, u, \dot{u}, ..., y, \dot{y}, ...) = 0$$

In the next section, an alternative to algebraic fault detection and isolation will be presented. The proposed method is based on the BG tool, since the latter is not only used for system modeling, but also for structural properties analysis such as: observability and controllability. The interpretation of the fault detection and isolation conditions proposed in the algebraic approach are given on BG by [2]. The latter is done using the bicausality notion and disjoint causal paths analysis.

6.3.2. Bond Grap-H Model Based FDI. A linear bond graph model is described by the following state space equation

(6.3.3)
$$\dot{x} = Ax + Bu$$

where x is the state variable associated to the storage element in integral causality on the BG model in the preferred integral causality, and u is the input variables associated with source (effort or flux) elements. A and B are constants matrices of appropriate dimensions expressing the junction structure interconnection.

By setting $\dot{x} = \delta x$, the Eq.6.3.3 can be rewritten as :

$$(6.3.4) \qquad \delta x = A\delta^{-1}\delta x + Bu$$

Definition A ring BG over a ring k and a k-module Ω means a BG described by a set of equations of the form:

$$(6.3.5) \qquad x_i = \sum_{j=1}^{n} a_{i,j}x_j + \sum_{k=1}^{m} b_{i,k}u_k \qquad 0 \le i \le n$$

where $a_{i,j}$, $b_{i,j}$ are in k, and x_i, x_j and u_k are in Ω, [**1**].

Using this definition, A BG model is seen as k-module Ω, thus the BG approach can absorb definitions proposed in the algebraic approach. Before announcing the said conditions, some definitions and notions will be recalled.

6.3.2.1. *Plant fault modeling on BG.* On the BG approach, the system's component are modeled by R, C, and I elements. So if a plant fault occurs in the system, one of these elements is affected. To show how a plant (multiplicative) fault is modeled on the BG model, let us consider a resistive (R) element in resistance causality with the following characteristic equation:

$$(6.3.6) \qquad e = R.f$$

If the element R is faulty, then an additional value R_f is added to the nominal value R_n, so equation (6.3.6) can be rewritten as

$$(6.3.7) \qquad e = (R_n + R_f).f = R_n.f + R_f.f = e_n + e_d$$

wh

O
BG by
faulty

epresented on
unction of the

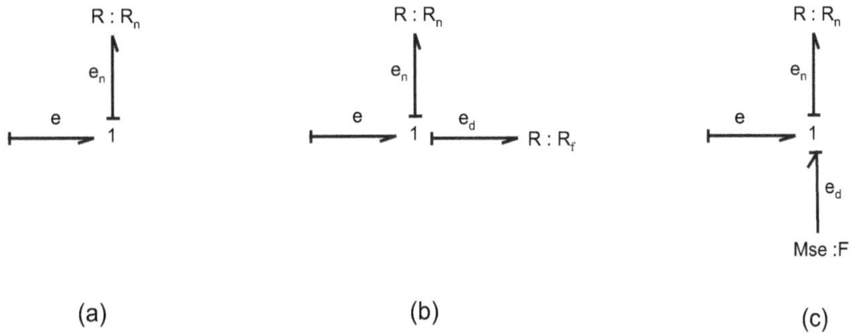

FIGURE 6.3.1. The BG model of the (a):healthy resistance; (b): faulty resistance; (c) using a modulated energy source.

Let's use this modeling technique on a RC circuit example given in the Figure 6.3.2. The RC circuit is composed of energy source U, a resistance R_e, and a capacitor C_e. These elements are modeled on BG by : a source of effort Se, a passive element R, and a storage element C as given in Figure 6.3.2-(b). If the R-element is faulty, then using the technique of plant fault modeling described above, a modulated source of effort $MSe : F$ is added to the 1-junction as

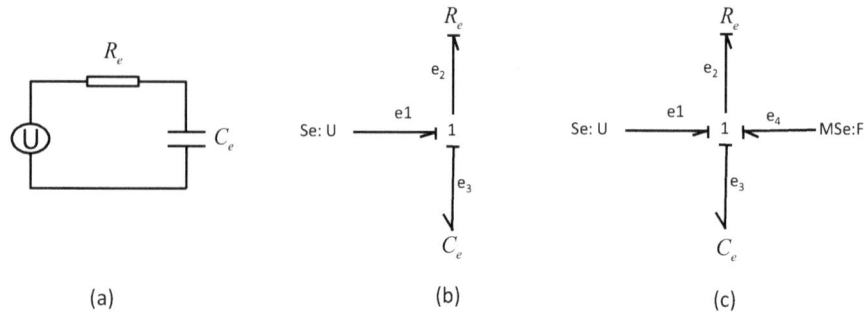

FIGURE 6.3.2. (a): A RC circuit, (b): BG model of the RC circuit, (c): BG model of the faulty RC circuit.

given on BG model of Figure 6.3.2-(c). From the same graphical tool, a state space form of the circuit can be derived systematically by exploiting the causal and structural properties of the former as follows:

from the 1-junction, the following structural and behavioral equations can be obtained :

$$(6.3.8) \qquad f = f_1 = f_2 = f_3 = f_4; e_1 - e_2 - e_3 + e_4 = 0$$

$$(6.3.9) \qquad e_1(t) = U(t), \ e_2(t) = R_e.f(t), \ e_3 = \frac{1}{C_e} \int f, e_4(t) = F(t)$$

As result :

$$(6.3.10) \qquad U(t) - R_e f(t) - \frac{1}{C_e} \int f(t) + F(t) = 0$$

Knowing that the energy variables *generalized moment* $p(t)$ and *generalized displacement* $q(t)$ are associated with state variables. They are obtained by integration of the power variable with respect to time:

$$(6.3.11) \qquad \begin{aligned} p(t) &= \int_{-\infty}^{t} e(\tau)d\tau \\ q(t) &= \int_{-\infty}^{t} f(\tau)d\tau \end{aligned}$$

Using these variables, The Eq.6.3.10 can be rewritten as :

$$(6.3.12) \qquad U(t) - R_e \dot{q}(t) - \frac{1}{C_e} q(t) + F(t) = 0$$

So, the RC circuit can be described under state space form as :

$$(6.3.13) \qquad \dot{x}(t) = -\frac{1}{R_e C_e} x(t) + \frac{1}{R_e}(U(t) + F(t))$$

where $x = \{q(t)\}$, $u = \{U(t), F(t)\}$, of the form $\dot{x} = Ax + Bu$.

Using this modeling technique, the fault variable is decoupled from the state one. The algebraic detectability and isolability can be applied since the system is on a linear state space form and the fault variable F is seen as a disturbance input added to the system.

6.3.2.2. *Plant fault detection and isolation based on BG.* From BG model in integral causality of a healthy system (without faults), one can get directly from the graph the number of inputs, outputs, and states without passing by the mathematical model of the system. So, the number of inputs: $m = card\{Se, Sf\}$, the output: $p = card\{Df, De\}$, and the number of the states: $n = card\{I, C\}$ in integral causality. The storage elements in derivative causality are not associated to the state variables. If q faults are considered, the BG model of the system is completed by a q modulated input $\{Mse, Msf\}$ element. Note that the number of faults q must be equal or inferior to the number of outputs, for the theorems quoted below to be valid.

After adding the unknown inputs representing the effort (or flow) brought by the faults, the observability of the former is done on the bicausal BG model of the system. The latter is obtained by applying the *Sequential Causality Assignment Procedure* (SCAP) for FDI as described in [**2**].

6.3.2.3. *Bicausal Bond graph.* The bicausal BG is introduced to study control problems such as system inversion, state estimation, and unknown parameter estimation [**8**]. It overcomes the assignment statements that can not be derived from the constraint equations of a so-called *unicausal BG* model, Figure 6.3.3.-(a), which implies two *assignment statements* :

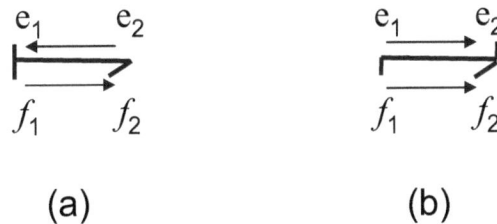

$$(a) \qquad\qquad (b)$$

FIGURE 6.3.3. (a) : a bond with causality, (b) : a bicausal bond

Bicausal introduces also some additional BG elements [**12**], among which SS (Source-Sensor), AE (Amplifier of effort), and AF (Amplifier of flow). Figure 6.3.4 gives causality assignment for Source-Sensor element.

In the context of fault detection and isolation, the output variable can be imposed on the system without modifying the energy structure (or constraint equations). This can be performed by using a SS element having a flow source/effort source causality (Figure 6.3.4). Then the output to be imposed plays the role of an input variable of that SS element while its conjugate is set to a null value leading to a null power propagation on that bond. Similarly, the fault variable to be isolated, will be observed on another SS element with a flow sensor/effort sensor causality. Our purpose is to present in the next subsection the fault isolability performed by the bicausal BG.

6.3.2.4. *Causality assignment algorithm for FDI.* The same *Sequential Causality Assignment Procedure for Inversion* (SCAPI) algorithm as for system inversion [**13**] is used to study fault detection and isolation, the difference is that the input variable is replaced by the fault variable, and a derivative causality is assigned instead of an integral one. So the algorithm for FDI is as follows:

- Step 1 : Determine the shortest path between the fault input variable and the output variable on the BG model in integral causality

'unicausal stroke'	SS element type	'Bicausal stroke'	SS element type
 ⟶⊣ SS	Flow source/ Effort sensor	 ⌐↗ SS	Flow source/ Effort source
 ⊢⟶ SS	Effort source/ Flow sensor	 ⌐⟶⌐ SS	Flow sensor/ Effort sensor

FIGURE 6.3.4. Source / sensor causality assignment

- Step 2 : Replace the source associated with the fault input variable by an SS element and connect an SS element to the output variable.
- Step 3 : Assign the flow source/effort source causality to the output SS element, and propagate the causal information toward the input SS element along the shortest path determined in step 1. Extend the causal information.
- Step 4 : Choose any energy storage element (C or I) without causality, and assign a preferred derivative causality to it. Propagate the causal information as far as possible. Repeat this step until all C or I elements are causalled.
- Step 5 : Choose any unassigned R element and assign to it an arbitrary causality. Propagate the causal information as far as possible. Repeat this step until the bond graph is causalled.

The shortest path between the input and the output is chosen using the following definition:

Definition The length of the causal path is equal to the number of storage elements C and I that belongs to that path, on the BG model of the system in integral causality

6.3.2.5. Fault detection and isolation based on bicausal Bond Graph. The conditions of fault detection and isolation are done directly on the BG model of the system using the following theorems:

Theorem 3. *(fault detectability)*
If all storage elements (I, C) take the derivative causality, and there exists a causal path from the output to each storage element (which represents the state variable), so the fault variables modeled by modulated inputs are observable (detectable) with respect to the input $\{Se, Sf\}$ and the output $\{De, Df\}$ variables.

The fault isolability is done on the bicausal BG model by analyzing the causal path from the SS element associated to the output variable to the SS element associated to the fault variable through the storage elements.

Definition Two causal paths are said to be disjoint if, and only if, they do not share a common variable [12].

Theorem 4. *(fault isolability)*
q faults are structurally isolable if, and only if, there are q disjoint causal paths linking the sensor to the fault through all storage elements C and I, that exist in the path.

To summarize, the algorithm for structural fault detection and isolation is given as follows:
- Model the system using the BG approach
- Add modulated inputs of the number of considered faults to the BG on integral causality,
- Apply SCAP for FDI algorithm to obtain a bicausal BG model of the system,
- Verify the fault/system observability,
- Verify the fault isolability

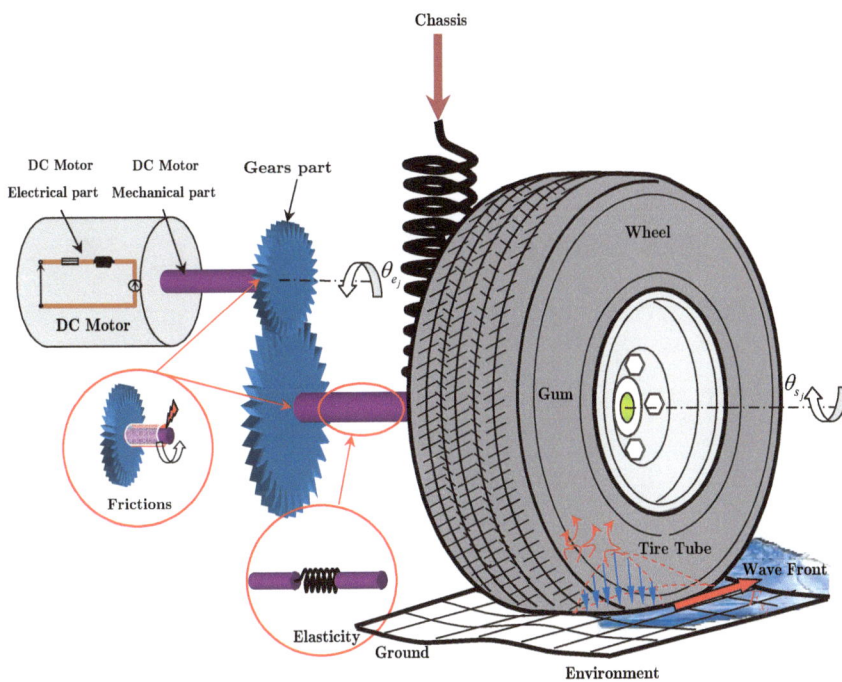

FIGURE 6.4.1. The electromechanical system

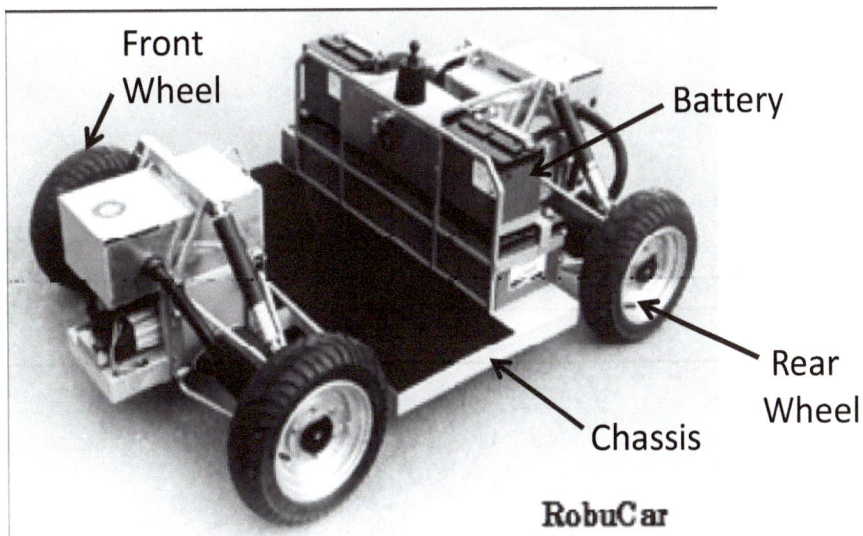

FIGURE 6.4.2. RobuCar

6.4.1. BG model of the system. The BG model of the electromechanical system is given in in Figure 6.4.3, a detailed modeling of this system can be found in [**11**]. The former can be divided on three principle parts: the DC motor part, the gear part, and the wheel part.

- The DC motor part is a combination of an electrical part and a mechanical part. The electrical part corresponds to the RL circuit. It is composed of an input voltage source U_0, an electrical resistance R_e, an inductance L, and a electromotive force feedback EMF (with a constant k_e), which is linear to the angular velocity of the rotor. The mechanical part is characterized by the rotor inertia J_e, a viscous friction parameter f_e, and a transmission axle rigidity K.
- The gear part concerns the mechanical gear which links the mechanical part to the load part with a reduction constant N, this part is represented on the BG model with a transformer element (TF) between the velocities ~~of the motor and the wheel.~~
- ~~...~~ its inertia J_s, effort source,

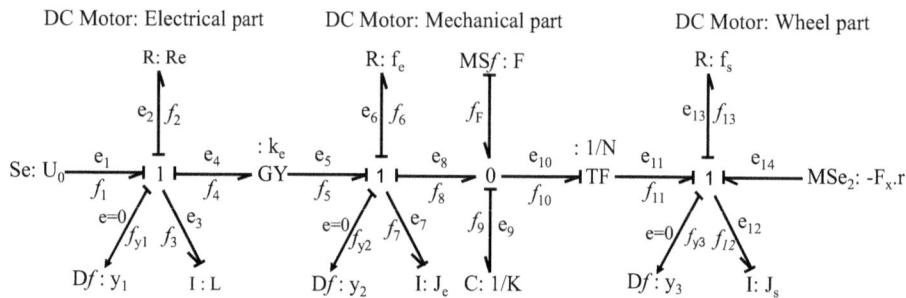

FIGURE 6.4.3. BG model of the faulty electromechanical system in integral causality.

The available sensors are the current, the angular velocity of the rotor, and the angular velocity of the wheel. They are represented in BG model by y_1, y_2, and y_3 "D_f" element respectively.

The plant fault considered in this paper is a fault on the transmission axle rigidity K, represented on the model by a storage element C. The characteristic equation of the C element is written in integral causality as:

$$(6.4.1) \qquad e = Fc \int f dt \; ; \; Fc = K$$

As the C element is connected to 0 junction, so the known variable is the effort e not the flow f, thus equation (6.4.1) is rewritten as :

$$(6.4.2) \qquad f = \frac{1}{Fc} \frac{d}{dt} e$$

If the C element is composed of the nominal value $\frac{1}{K}$ and an additive fault F, equation (6.4.2) can be rewritten as :

$$(6.4.3) \qquad f_n + f_F = \frac{1}{Fc} \frac{d}{dt} e + F_K \frac{d}{dt} e$$

The BG representation of the additional part f_F, which represents the fault variable, is a modulated flow source MSf, placed at the same junction of the storage element C, as is shown in Figure 6.4.3.

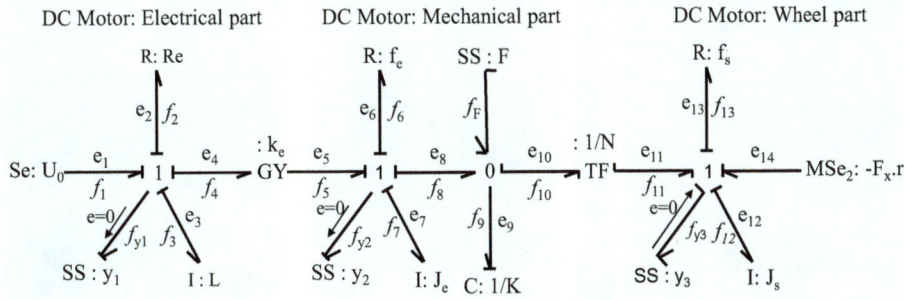

FIGURE 6.4.4. The bicausal BG model of the faulty Electromechanical system.

6.4.2. BG-based fault detectability and isolability. FDI is done on BG model of the system in bicausal causality, the latter given in Figure 6.4.4 is obtained by applying the causality assignment algorithm for FDI on the BG model of the system in integral causality as described in [**2**].

The electromechanical system modeled using BG and given by Figure 6.4.3 is diagnosable since all storage elements take a derivative causality in the bicausal BG given in Figure 6.4.4, and there are causal paths from outputs to each state (storage elements), which means that the system is observable in presence of the fault, and the fault variable F is observable with respect to the input u and the output y.

The isolability is performed by analyzing the disjoint causal path from outputs y to faults on the bicausal BG model of the system. From Figure 6.4.4, we can find a causal path from $SS : y_3$ to $SS : F$ passing through all the storage elements existing in this path

$$(6.4.4) \qquad f_{y3} \rightarrow f_{12} \rightarrow e_{12} \rightarrow e_{11} \rightarrow e_{10} \rightarrow e_9 \rightarrow f_9 \rightarrow f_F$$

So, we conclude that the fault F satisfies theorem 2, thus, it is isolable.

Remark For further details, the reader is referred to consult [**2**], where different cases of fault isolability are given.

6.5. Results of the co-simulations

The simulation results are obtained by co-simulating the dynamic behavior of one electromechanical traction system of RobuCar with the behavior of the whole vehicle using two separate platforms, Matlab and Callas/SCANeR Studio [**14**] softwares. The latter is a vehicle driving simulator dedicated for research and development. The co-simulation scheme is done according to Figure 6.5.1. The specification of the electromechanical system is given in Table 1.

TABLE 1. Specification of the electromechanical system

R_e	1.32	(Ω)
L	2.30	(H)
k_e	0.0655	Nm/A
J_e	0.002	(Nms^2/rad)
f_e	0.003	(Nms/rad)
J_s	0.14	$(dNms^2/°)$
f_s	0.036	$(dNms/°)$
K	1	$(dNm/°)$
N	13	-
r	0.21	m

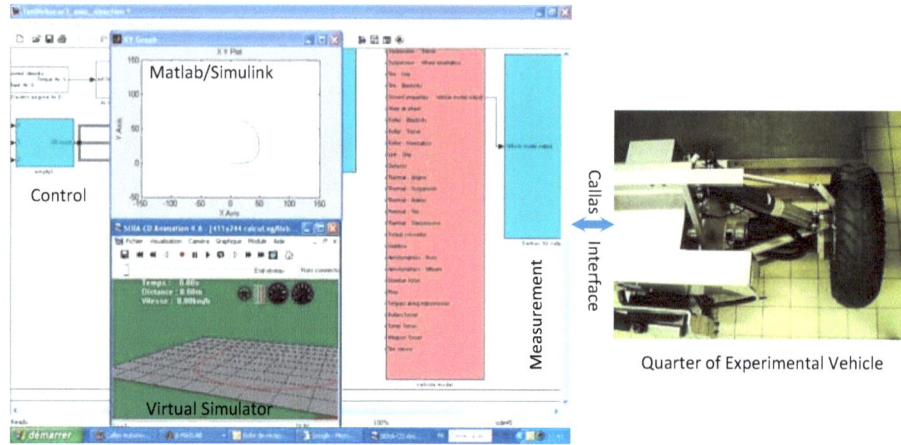

FIGURE 6.5.1. Co-simulation scheme : Callas / a quarter of a vehicle

6.5.1. Fault indicator generation. Fault indicators are obtained from the causal relations between BG variables (constraints and characteristics equations). Thus from the 0-junction:

$$(6.5.1) \qquad f_F = -f_8 + f_9 + f_{10}$$

The flows f_8, f_9, and f_{10} are expressed in term of the inputs ($Se = U_0, MSe_2 = -F_x.r$), the outputs f_{y1}, f_{y2}, f_{y3}, and their derivatives by following the causal path, between the flow and the given output.

The flow f_8 is linked directly to f_{y2} by (6.5.2), it is given by $f_8 = f_{y2}$

$$(6.5.2) \qquad f_{y2} \to f_8$$

The flow f_{10} is linked to the output f_{y3} by passing through the TF element using the following causal path

$$(6.5.3) \qquad f_{y3} \to f_{11} \to TF : 1/N \to f_{10}$$

Thus, $f_{10} = N f_{y3}$.

The flow f_9 is linked to the SS output element f_{y3} by

$$(6.5.4) \qquad f_{y3} \to f_{12} \to e_{11} \to e_{10} \to e_9 \to f_9$$

$$(6.5.5) \qquad f_9 = \frac{J_s}{NK}\ddot{f}_{y3} + \frac{f_s}{NK}\dot{f}_{y3} - \frac{1}{NK}\dot{M}Se_2$$

So, the fault indicator f_F is written as a function of the input, the outputs, and their derivatives by:

$$(6.5.6) \qquad F = -y_2 + \frac{J_s}{NK}\ddot{y}_3 + \frac{f_s}{NK}\dot{y}_3 + Ny_3 - \frac{1}{NK}\dot{u}_2$$

The variables f_F, Se, MSe_2, f_{y1}, f_{y2}, and f_{y3} are replaced by F, u_1, u_2, y_1, y_2, and y_3 respectively.

The residual r is obtained by deriving twice the fault indicator F on the Laplace domain and multiplying its both sides by s^{-2} as following:

The fault indicator is written in Laplace domain by:

$$(6.5.7) \qquad \begin{aligned} F = -y_2 + \left(s^2\frac{J_s}{NK} + s\frac{f_s}{NK} + N\right)y_3 - y_3(0) - \dot{y}_3(0) \\ -s\frac{1}{NK}u_2 - \dot{u}_2(0) \end{aligned}$$

F is derived twice in order to eliminate the initial conditions which may be unknown:

$$
\begin{aligned}
\frac{d^2}{ds^2} F = {} & s^2 \frac{J_s}{NK} \frac{d^2}{ds^2} y_3 \\
& + s \left(\frac{4J_s}{NK} \frac{d}{ds} y_3 + \frac{f_s}{NK} \frac{d^2}{ds^2} y_3 - \frac{1}{NK} \frac{d^2}{ds^2} u_2 \right) \\
& + \left(-\frac{d^2}{ds^2} y_2 + \frac{2J_s}{NK} y_3 + \frac{2f_s}{NK} \frac{d}{ds} y_3 + N \frac{d^2}{ds^2} y_3 \right. \\
& \left. - \frac{2}{NK} \frac{d}{ds} u_2 \right)
\end{aligned}
$$

(6.5.8)

After multiplying equation (6.5.8) by s^{-2}, and returning to the time domain, the obtained residual r is the following:

(6.5.9)
$$
\begin{aligned}
r = {} & \frac{J_s}{NK} t^2 y_3 \\
& + \int_0^t \left(-4 \frac{J_s}{NK} \lambda y_3 + \frac{f_s}{NK} \lambda^2 y_3 - \frac{1}{NK} \lambda^2 u_2 \right) d\lambda \\
& + \int_0^t \int_0^\alpha \left(-\alpha^2 y_2 + \frac{2J_s}{NK} y_3 - \frac{2f_s}{NK_s} \alpha y_3 + N\alpha^2 y_3 \right. \\
& \left. + \frac{2}{NK} \alpha u_2 \right) d\alpha d\lambda
\end{aligned}
$$

6.5.2. Simulation results. Figures 6.5.2, 6.5.3, 6.5.4, and 6.5.5 present the behavior of the electromechanical system in a nominal case, and in the fault one (the break on the axle occurred at time $t = 10s$). Figure 6.5.6 depicts the res   curred, that th   own in Figure

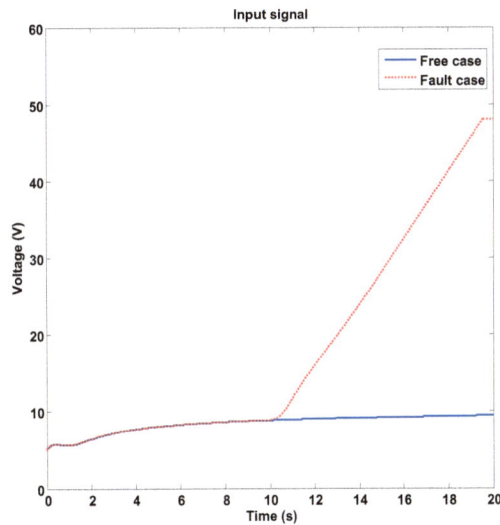

FIGURE 6.5.2. The input signal : $Se = U_0$

6.6. Conclusions

In this chapter, a methodology for plant diagnosis was applied to a quarter of an autonomous vehicle to detect and isolate a fault on the axle transmission. The former is based on the bond graph approach since it can be used for both systems modeling and monitoring. Based on the tool properties, fault detection and isolation are performed structurally directly from the graph. Simulation results show the effectiveness of the proposed methodology.

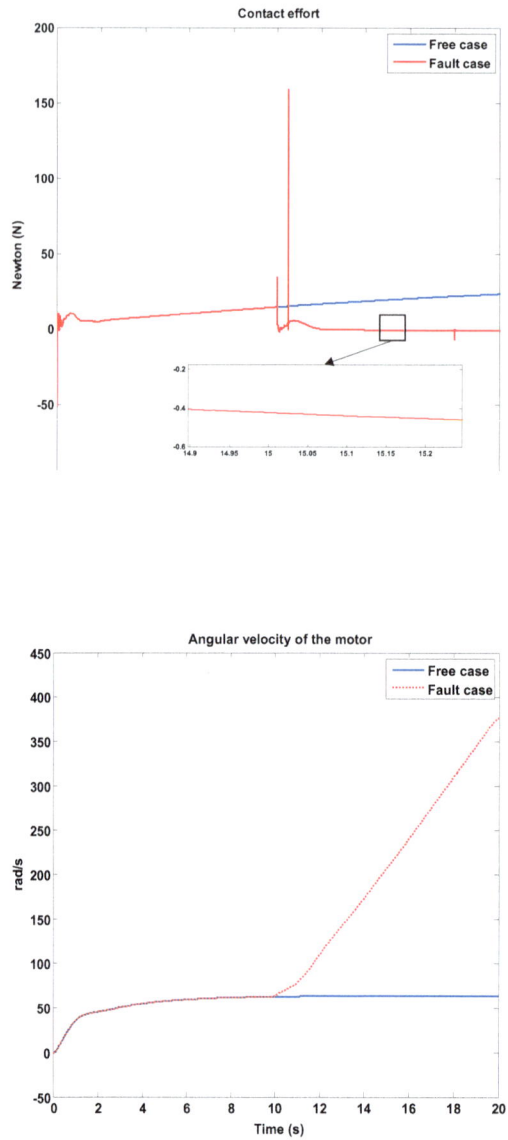

FIGURE 6.5.4. Angular velocity of the rotor in a fault-free case and in a faulty case.

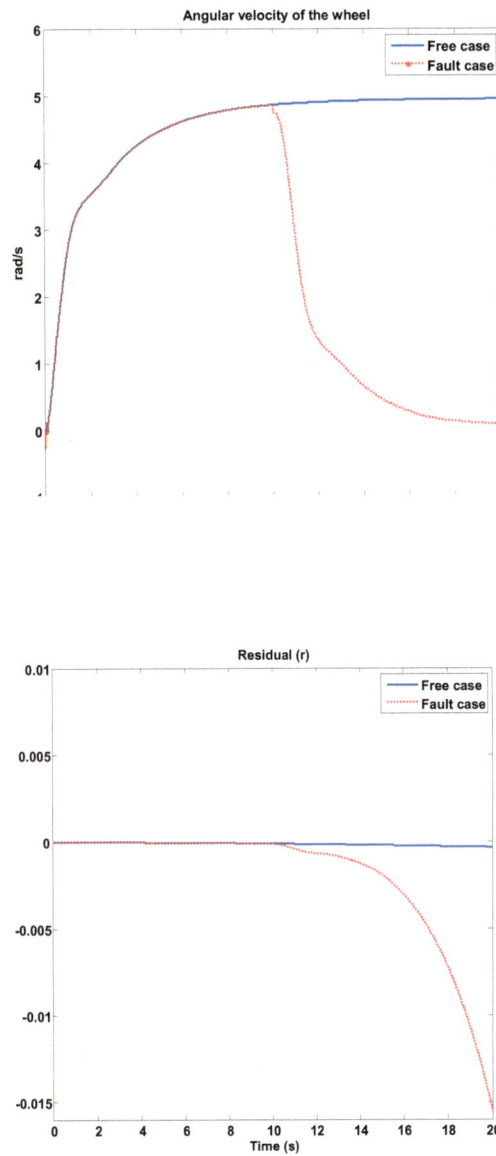

FIGURE 6.5.6. Residual in a fault-free case and in a faulty case.

Bibliography

[1] A. Achir, C. Sueur, and G. Dauphin-Tanguy. Ring bond graphs over non commutative rings: application to varitional bond graphs. In *I3M-IMMACA*, 2004.

[2] S. Benmoussa, B. Ould-Bouamama, and R. Merzouki. Component fault detection and isolation comparaison between bond graph and algebraic approach. In *IMAACA*, pages 214–220, 2011.

[3] J. Chen and R.J. Patton. *Robust Model-Based Fault Diagnosis in dynamic systems*. Kluwer academic publishers, 1999.

[4] J. M. Dion, C. Commault, and J. van der Woude. Generic properties and control of linear structured systems: a survey. *Automatica*, 39:1125–1144, 2003.

[5] S. Diop and R. Martinez-Guerra. On an algebraic and differential approach of nonlinear systems diagnosis. In *proceedings of the 40th IEEE : Conference On Decision and Control*, pages 585–589, Orlando, FL, USA, 2001.

[6] W. El-Osta, B. Ould Bouamama, and C. Sueur. Monitorability indexes and bond graphs for fault tolerance analysis. In *Safeprocess IFAC*, Bejing, Chine, 29-1 Sept 2010 2006.

[7] M. Fliess and C. Join. An algebraic approach to fault diagnosis for linear systems. In *CESA*, July 2003.

[8] P.J. Gawthrop. Estimating physical parameters of nonlinear systems using bond graph models. In *SYSID*, 2000.

[9] J. Gertler. Fault detection and isolation using parity relations. *Control Engineering Practice*, 5:653–661, 1997.

[10] Rolf Isermann. *Fault-Diagnosis Systems : An Introduction from Fault Detection to Fault Tolerance*. Springer, 2006.

[11] R. Merzouki, B.Ould-Bouamama, M.A. Djeziri, and M. Bouteldja. Modelling and estimation for tire-road system using bond graph approach. In *International Conference on Intelligent Robots and Systems*, pages 3785–3790, October 9-15 2006.

[12] R.F. Ngwompo and P.J. Gawthrop. Bond graph-based simulation of non-linear inverse systems using physical performance specifications. *Journal of the Franklin Institute*, 336:1225–1247, 1999.

[13] R.F. Ngwompo, S. Scavarda, and D. Thomasset. Inversion of linear time-invariant siso systems modelled by bond graph. *Journal of the Franklin Institute*, 336:157–174, 1996.

[14] OKTAL. Scaner driving simulation engine, 2011.

[15] R.J. Patton and J. Chen. observer-based fault detection and isolation : robustness and applications. *Control Engineering Practice*, 5:671–682, 1997.

[16] A.K. Samantary and S.K. Ghoshal. Bicausal bond graph for supervision : from fault detection and isolation to fault accommodation. *Journal of the Franklin Institute*, 345:1–28, 2008.

[17] M. Staroswiecki and G. Comtet-Varga. Analytical redundancy relations for fault detection ans isolation in algebraic dynamic systems. *automatica*, 37:687–699, 2001.

[18] C. Sueur and G. Dauphin-Tanguy. Bond-graph approach for structural analysis of mimo linear systems. *Journal of the Franklin Institute*, 328:55–70, 1991.

CHAPTER 7

Structural Reconfigurability Analysis for an Over-Actuated Electric Vehicle

Rui Loureiro[1],
Polytech-Lille, LAGIS, UMR CNRS 8219
Avenue Paul Langevin, 59655 Villeneuve D'Ascq, France
rui-jose.loureiro@polytech-lille.fr

Rochdi Merzouki,
Polytech-Lille, LAGIS, UMR CNRS 8219
Avenue Paul Langevin, 59655 Villeneuve D'Ascq, France
rochdi.merzouki@polytech-lille.fr

Belkacem Ould Bouamama,
Polytech-Lille, LAGIS, UMR CNRS 8219
Avenue Paul Langevin, 59655 Villeneuve D'Ascq, France
belkacem.ouldbouamama@polytech-lille.fr

ABSTRACT. In this chapter the problem of structural reconfigurability analysis is introduced and applied on an over-actuated electric vehicle. This task is performed through the use of the bond graph tool, which is an adequate tool for dynamic modeling of complex systems, and for fault detection and isolation. The latter is performed by exploiting the structural and causal properties of the bond graph model to generate analytical redundancy relations in a systematic manner. Then, the actual system diagnosis information is exploited to study the different possibilities of system structural reconfigurability conditions. Finally, an algorithm is proposed, so that this study can be performed in a systematic way.

Keywords: Fault reconfigurability, Modeling, Fault diagnosability, Electric vehicles.

7.1. Introduction

In the last 30 years, automatic industrial systems have made large improvements in order to boost their performance. The constraints related to performance lead to the development of systems with higher complexity, which increases the risk of abnormal behaviors. Hence, process systems safety has become an important issue in the field of automatic control systems. In order to avoid faults provoking a system shut down, or even accidents, there is the necessity of implementing algorithms that are able to study the presence of faults, together with control laws that work even if a fault is present in the system. When these algorithms are implemented, the system is defined as a fault tolerant system (FTs) .

FTs can be achieved through the application of two different approaches of Fault Tolerant Control (FTC). In [21] FTC strategies were divided into passive (PFTC) and active (AFTC) approaches. The former makes used of robust control techniques, where the controller synthesis is done by considering the occurrence of possible faults. However, often a limited number of possible faults is considered. The active approach requires an algorithm to study the faults (Fault Detection and Diagnosis (FDD)), and then the control system is modified based on the FDI information so that

---

[1]Corresponding author
Part of information included in this chapter has been previously published in IEEE Transactions on Vehicular Technology Volume 61, Number 3, March 2012, Pages 986-997.

the FTs objectives remain achievable even with less performance. AFTC strategies are essentially based on redundancy analysis. Redundancy is available in a system when a task can be performed by alternative means and it can be made available in the form of material and/or analytical redundancy. As referred in [31], the former exists when identical hardware/software components are arranged in parallel, and it can be presented in the form of duplex or, triplex redundancy, ... etc. Analytical redundancy has to do with the ability to use static/dynamic relations between variables associated to the system components. Moreover, AFTC strategies can be divided into fault accommodation and system reconfiguration [2]. The former uses estimations of the fault provided by the FDD algorithm to change the control system in such a way that closed-loop stability and acceptable performance is guaranteed. The latter, requires the isolation of the fault from the Fault Detection and Isolation (FDI) algorithm, and then the control is reconfigured simply by using the healthy part of the system.

As referred in [32], countless works regarding fault accommodation and system reconfiguration can be found in the literature. However, most of FDD and FDI techniques are developed as diagnostic or monitoring tools, instead of an integral part of FTC systems [32]. The same can be stated about most of FTC approaches, they assume that the fault information provided by the FDD or FDI algorithms is perfect. The combination of these methods in a simple way (computationally tractable), and taking into account the limitations of both algorithms is still a debatable issue in the automatic control community. To achieve a system tolerant to faults, it is of primary importance that the FTC method interprets correctly the information provided by the FDD or FDI algorithm. For instance, there are works devoted to study the level of systems FT [11, 27, 31, 9, 1], however they assume that the FDI step is performed, instead of considering the diagnosis capabilities of the system. These works use analytical models under state space format. Moreover, their main idea is to verify systems input or output redundancy, and if the energy required to perform reconfiguration strategies is acceptable. However, in these referred works, only actuator and sensor faults are considered. Also, the mathematical model of the system is supposed to be under a linear or a class of non-linear systems format. Furthermore, analytical approaches require accurate models, and numerical values of the parameters, which are not always available in real systems. This is why graphical approaches based on the system model can be an alternative to study fault diagnosability and fault reconfigurability possibilities, before industrial design.

The existing graphical approaches (digraphs [7], signed digraphs [15], bipartite graph [2], and bond graphs [25]), allow conclusions to be made about structural properties of the system without knowing their numerical values, where performing complex calculations is not required. Most graphical approaches are concerned with control analysis (observability, controllability [29]), and diagnosability [2, 5, 30, 12, 4, 3, 16]. Consequently, fewer works are devoted to control re-design analysis e.g., [24]. Recently in [6], control reconfiguration has been evaluated by using a functional modeling technique named Multilevel Flow Modeling (*MFM*) introduced by *Lind* in [13]. In this work, faults are assumed identified, located in the control system, and the reconfigurability analysis is performed offline.

Among the graphical approaches, the bond graph (BG), which makes the modeling systematic by following the energy flow in the system under study, contains some interesting properties such as: behavioral, structural, and causal that can be exploited not only for modeling but also for analysis and synthesis. BG has proven to be a well-adapted tool for modeling multi domain mechatronic systems [10] and for diagnosis[25]. In the following sections we will use the BG tool not only for dynamic modeling and FDI, but also to study the structural capabilities of the system to be reconfigured by taking into account the FDI algorithm effectiveness. For a detailed explanation of the proposed technique, an over-actuated electric vehicle, named RobuCar is considered throughout this chapter.

7.2. Graph and Bond graph

A graphical approach is used to study and analyze structured systems that are independent of the system parameter values. This approach requires a low computational burden, which allows dealing with large scale systems. The existing contributions related to the graph analysis proved that observability, controllability, input-output decoupling, ... etc., can be simply deduced from the structural properties of the graph.

There are different graphical methods used in the literature: Digraph [7], signed digraph [15], bipartite graph [2], and bond graph [25].

Definition The digraph, denoted $G(S, A)$, is deduced from state space equations. It is composed by a set of vertices (S), $S = \{U \cup Y \cup X\}$ which correspond to the system inputs, outputs, and states. The interactions between these nodes are represented by directed edges (A). The graph is called signed (SDG) if the edge considers the qualitative influence between variables.

Definition A graph is bipartite, if its vertices can be partitioned into two disjoint subsets Z (set of variables that defines the dynamic behavior of the system), and C (set of equations that defines the relations among the variable set), $S = \{C \cup Z\}$. The relations between these two subsets are represented by edges (A).

Definition A bond graph which is also a graph, $G(S, A)$ is a unified graphical language for multi-physic domains. Unlike the other graphs mentioned above, S represents physical components, subsystems, and other basic elements called junctions. While the edges A, called power bonds represent the power exchanged between nodes. This power is labeled by two conjugated power variables, named effort (e), and flow (f)

7.3. Structural system reconfigurability analysis using BG model

AFTC strategies intend to ensure the continuity of the process operation in the presence of faults [25]. This strategy requires two steps. Firstly, diagnostic tools have to be implemented to detect, localize and estimate the fault. Then, the system can remain in process either by performing system reconfiguration , or fault accommodation . Only the former is considered in this chapter. Moreover, we do not treat the problem of creating an AFTC but we tackle the problem of structural analysis on fault reconfigurability conditions, making a decision on the possible synthesis of an adequate control. To tackle this problem it is necessary to use the information obtained from the diagnosis procedure which is also obtained by using the BG model.

Let us define a control problem as $< \Sigma_o, C(\theta), U >$ ([2, 28]), where Σ_o and U are respectively, the system (Σ) objectives and the set of admissible control laws. Moreover $C(\theta)$ are the constraints C that represent the Σ behavior, while θ is the parameter that C depends on. When a fault occurs in a Σ, the control problem changes, and as a consequence, these changes must be studied to keep achieving the Σ_o.

(Objectives) : system objectives (Σ_o) are a set of specifications, which the Σ should respect, where $\Sigma_o = \{o_1, o_2, \ldots, o_k\}$, where k is equal to the total number of Σ_o.

(System reconfiguration) : System reconfiguration is a strategy in which the faulty Σ is modified so that only the healthy part of the Σ (Σ') is controlled in order to achieve the desired Σ_o (solves the control problem $< \Sigma_o, C'_n(\theta'_n), U'_n >$).

The faults may be located in the Σ inputs (actuators), outputs (sensor), and in the plant. In a BG model, Σ_o are associated to a flow, an effort, or the both (power). Basically these objectives represent what the Σ is expected to achieve.

A general structure of an AFTC through system reconfiguration is illustrated by the scheme of Figure 7.3.1 (white blocks).

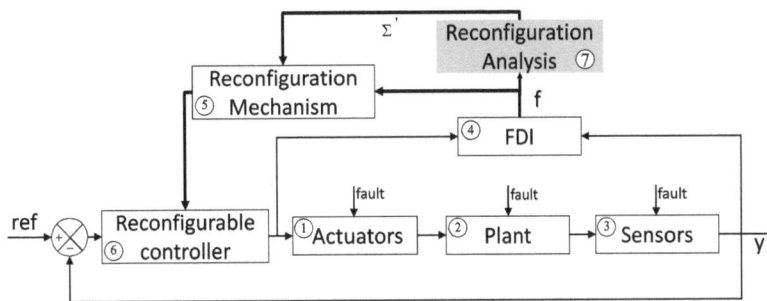

FIGURE 7.3.1. General structure of a system reconfiguration strategy

The first block represents the controlled inputs of the system. If any of the actuators is not working properly, there is an actuator fault in the system. The second block represents the Input/Output dynamics of the system. Any fault that changes this dynamics is defined as a plant fault. Finally, the third block symbolizes the measurement architecture of the system. A fault in this block is a sensor fault. Moreover, the complete structure of an AFTC strategy requires a FDI algorithm, as illustrated in the fourth block, a reconfiguration mechanism of the controller (block five), and a reconfigurable controller (block six).

In this chapter, the introduction of block seven (see Figure 7.3.1-grey block) is proposed. In this block the possible synthesis of system reconfiguration based on the fault information is analyzed, and when possible Σ' is provided to the reconfiguration mechanism block.

To execute the proposed analysis, the structural and causal properties of the BG tool will be exploited. The interest of using this tool is that the BG can be used for dynamic modeling, for fault diagnosis, and to study structurally the possible synthesis of an adequate control ensuring system reconfiguration.

7.3.1. Modeling of an over-actuated electric vehicle using the BG tool. For a detailed explanation of the proposed analysis an over-actuated multi-input multi-output autonomous electric vehicle named, RobuCar (see Fig.6.4.2) is considered throughout this chapter. It is composed by four actuated wheels that can be piloted manually or automatically inside confined and safety areas. Traction torque, computed by decentralized controllers, is supplied to each wheel by a DC-motor. This vehicle example is structured as a graph of four independent quarters of vehicle (Fig.6.4.1), presented in Section 6.4, where each jth subsystem $j \in [1, 4]$ represents an electromechanical system in interaction with the ground (DC-actuator and wheel). Moreover, the measurement scheme of the RobuCar is composed of one inertial sensor measuring longitudinal, lateral and vertical accelerations and angular velocities of the yaw ($\dot{\psi}$), the roll and the pitch. Longitudinal ($\dot{u}$) and lateral ($\dot{v}$) velocities are estimated directly from the acceleration measurements.

The considered dynamics, while modeling the vehicle, are the electromechanical systems for traction (Fig.6.4.1), together with the longitudinal, lateral and yaw dynamics of the chassis. The maximal applied velocity to the vehicle is *18km/h* therefore, some dynamics have neglected effects on the whole vehicle motion during the vehicle motion such as the pitch and the roll ones. Furthermore, the road surface is assumed uniform hence, the suspension dynamics are modeled. In conclusion, the modeled dynamics for this case study are the electromechanical model of the traction system and longitudinal, lateral and yaw dynamics on the center of gravity (CoG). The Σ_o of the RobuCar are defined as driving it at a desired longitudinal ($\dot{u}^d$), lateral ($\dot{v}^d$), and yaw ($\dot{\psi}^d$) velocities ($\Sigma_o = \{\dot{u}^d, \dot{v}^d, \dot{\psi}^d\}$).

There numerical parameters of the electromechanical system were presented in Table 1 presented in Section 6.4. Note that in this section the under-script j is added to the elements and $L_{ej} = L$, $R = r$, and $Fl_j = F_x$.

Dynamic equations in state space format can be systematically derived from a BG model. The state equations of the jth electromechanical system appear below as equation (7.3.3), where p_{Lj}, p_{Jj}, q_{Kj} and p_{Sj} are the energy variables of L_{ej}, J_{ej}, K_j, and J_{sj} elements, respectively. Development of the dynamic equations from the BG model is presented in (7.3.2). Equations are developed for the first two $1-$junctions, and the same methodology is generated for the remaining ones. Then, by following the same procedure the complete dynamic equations in state space format of the jth electromechanical system (7.3.3) are deduced.

A : Structural equations and system states :

$$1 - junction \Rightarrow f_1 = f_2 = f_3 = f_4 = i_{mj}; \quad e_1 - e_2 - e_3 - e_4 = 0;$$

$$1 - junction \Rightarrow f_5 = f_6 = f_7 = f_8 = \dot{\theta}_{ej}; \quad e_5 - e_6 - e_7 - e_8 = 0;$$

(7.3.1)

$$x_1 = p_{Lj} = \int e_3; \quad x_2 = p_{Jj} = \int e_7.$$

B : Behavioral equations :

$$e_1 = U_{0j}; \quad e_2 = R_{ej}.f_2; \quad f_3 = \frac{1}{L_{ej}} \int e_3; \quad e_4 = k_{ej}.f_5;$$

$$e_5 = k_{ej}.f_4; \quad e_6 = f_{ej}.f_6; \quad f_7 = \frac{1}{J_{ej}} \int e_7.$$

C : Dynamic equations :

$$\dot{p}_{Lj} = e_3 = -\frac{R_{ej}}{L_{ej}} p_{Lj} - \frac{k_{ej}}{J_{ej}} p_{Jj} + U_{0j};$$

(7.3.2)

$$\dot{p}_{Jj} = e_7 = \frac{k_{ej}}{L_{ej}} p_{Lj} - \frac{f_{ej}}{J_{ej}} p_{Jj} - e_8.$$

D : State Space representation of dynamic equations of a quarter of RobuCar :

$$
\overbrace{\begin{bmatrix} \dot{p}_{Lj} \\ \dot{p}_{Jj} \\ \dot{q}_{Kj} \\ \dot{p}_{Sj} \end{bmatrix}}^{\dot{X}} = \overbrace{\begin{bmatrix} -\frac{R_{ej}}{L_{ej}} & -\frac{k_{ej}}{J_{ej}} & 0 & 0 \\ \frac{k_{ej}}{L_{ej}} & -\frac{f_{ej}}{J_{ej}} & -K_j & 0 \\ 0 & \frac{1}{J_{ej}} & 0 & -\frac{N_j}{J_{sj}} \\ 0 & 0 & N_j K_j & -\frac{f_{sj}}{J_{sj}} \end{bmatrix}}^{A} \overbrace{\begin{bmatrix} p_{Lj} \\ p_{Jj} \\ q_{Kj} \\ p_{Sj} \end{bmatrix}}^{X} + \overbrace{\begin{bmatrix} 1 & 0 \\ 0 & 0 \\ 0 & 0 \\ 0 & -1 \end{bmatrix}}^{B} \overbrace{\begin{bmatrix} U_{0j} \\ Fl_j.r \end{bmatrix}}^{U},
$$

(7.3.3)

$$
\overbrace{\begin{bmatrix} i_{mj} \\ \dot{\theta}_{ej} \\ \dot{\theta}_{sj} \end{bmatrix}}^{Y} = \overbrace{\begin{bmatrix} \frac{1}{L_{ej}} & 0 & 0 & 0 \\ 0 & \frac{1}{J_{ej}} & 0 & 0 \\ 0 & 0 & 0 & \frac{1}{J_{sj}} \end{bmatrix}}^{C} \overbrace{\begin{bmatrix} p_{Lj} \\ p_{Jj} \\ q_{Kj} \\ p_{Sj} \end{bmatrix}}^{X} .
$$

Fl_j is the longitudinal contact effort generated by the interaction between the wheel and the ground. The latter is generated in terms of the longitudinal velocity ($\dot{u}$), and of the angular one ($\dot{\theta}_{sj}$), as illustrated by the canonical curve estimated from the nonlinear tire model [20], presented in (7.3.4).

(7.3.4)
$$
Fl_j = [\delta_0 - \delta_1 e^{-\beta|(\dot{u}-r\dot{\theta}_{sj})|} - \delta_2(\dot{u} - r\dot{\theta}_{sj})].sign(\dot{u} - r\dot{\theta}_{sj}),
$$

where, δ_0 and δ_1 are the dry and stiction forces. δ_2 is the viscous friction coefficient, respectively. β is the stiction coefficient. Knowing that the slip velocity ($\dot{u} - R\dot{\theta}_{sj}$) is small at a maximum operated velocity CoG ($18Km/s$), the stiction and viscous behaviors of Fl_j are neglected. Thus 7.3.4 is simplified to coulomb friction behavior [22] as follows:

(7.3.5)
$$
Fl_j = [\delta_0].sign(\dot{u} - r\dot{\theta}_{sj}),
$$

The complete model of the RobuCar is based on a simplified vehicle dynamic model as illustrated in Figure 7.3.2.

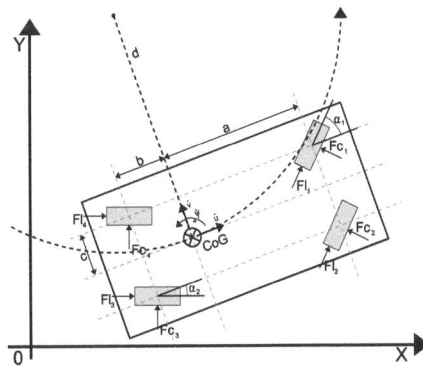

FIGURE 7.3.2. RobuCar simplified dynamic model.

Fc_j is the cornering force transmitted to the wheel, and it is calculated by the following equation [26]:

(7.3.6)
$$
Fc_j = \frac{m(\dot{\theta}_{sj}.r)^2}{d},
$$

where d is the radius of the bend. α_1 and α_2 are respectively the front and rear wheel angles. a and b are the distance from the vehicle center of gravity to the front and rear axles, respectively, and c is the distance of the left and right wheels from the longitudinal vehicle axis. Finally, m is the mass of the vehicle, which is about $310Kg$.

The complete BG of RobuCar with the considered dynamics is illustrated in Figure 7.3.3. In this model it is possible to see four electromechanical systems in the upper and lowers part of the model. In the center of the model, the longitudinal dynamics (junction associated to $Df : \dot{u}$), the lateral dynamics (junction associated to $Df : \dot{v}$), and

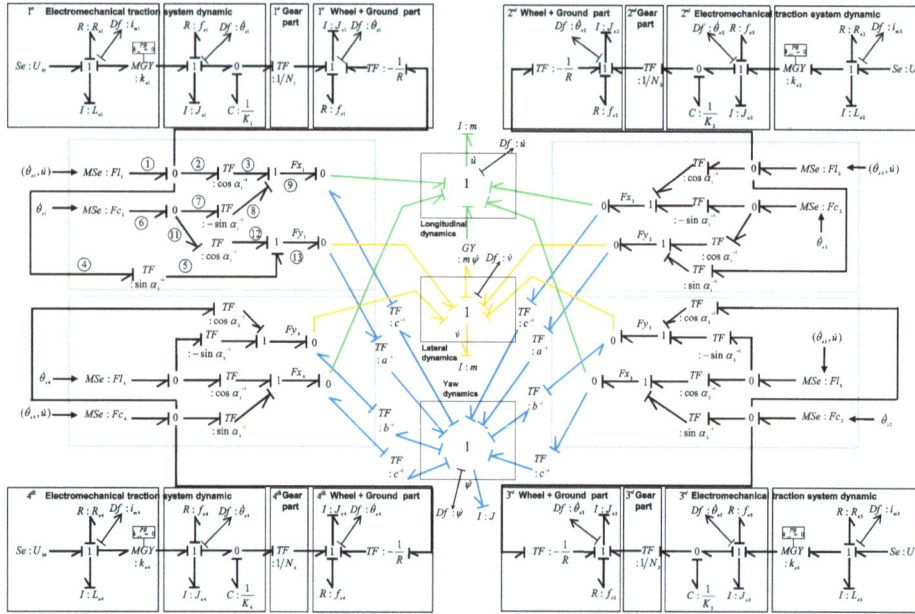

FIGURE 7.3.3. BG model of the RobuCar.

the yaw dynamics (junction associated to $Df : \dot{\psi}$) are modeled. Fx_1, and Fy_1 are the components of the tire forces along the longitudinal and lateral vehicle axes, respectively. These forces are computed by following the paths from bonds ① and ②, to bonds ⑨ and ⑬ of Figure 7.3.3.

$$e_1 = e_2 = e_4 = Fl_1, \qquad e_3 = Fl_1 cos(\alpha_1),$$
$$e_6 = e_7 = e_{11} = Fc_1, \qquad e_8 = -Fc_1 sin(\alpha_1),$$
$$e_5 = Fl_1 sin(\alpha_1), \qquad e_{12} = Fc_1 cos(\alpha_1),$$
$$Fx_1 = e_9 = e_3 + e_8, \qquad Fy_1 = e_{13} = e_5 + e_{12},$$
$$Fx_1 = Fl_1 cos(\alpha_1) - Fc_1 sin(\alpha_1), \qquad Fy_1 = Fl_1 sin(\alpha_1) + Fc_1 cos(\alpha_1).$$

The same can be obtained for the remaining forces. In addition, from the part of the bond graph model representing the longitudinal dynamics in integral causality (junction where detector ($Df : \dot{u}$) is associated), the mathematical expression of this dynamic is presented in (7.3.7), the lateral one is presented in (7.3.8), and the yaw in (7.3.9).

$$(7.3.7) \qquad m\ddot{u} = Fx_1 + Fx_2 + Fx_3 + Fx_4 + m\dot{\psi}\dot{v},$$

$$(7.3.8) \qquad m\ddot{v} = Fy_1 + Fy_2 + Fy_3 + Fy_4 - m\dot{\psi}\dot{u},$$

$$(7.3.9) \qquad J\ddot{\psi} = [-Fx_1 + Fx_2 + Fx_3 - Fx_4]c + [Fy_1 + Fy_2]a - [Fy_3 + Fy_4]b$$

7.3.2. Bond graph for control analysis. In contrast to the algebraic approach which allows analyzing the Σ controllability and observability, by generating and calculating the rank of the controllability and observability matrix, from a bond graph model the same can be structurally concluded without the use of any calculations and parameters assignment, as proposed in [**29**].

Theorem 5. *The system is structurally controllable/observable if and only if two conditions are verified:*

- *There is a causal path connecting a source/detector to each I, and C element in integral causality;*
- *All I, and C elements accept a derivative causality. If the latter is not completely respected, a dualization of the source/detector is required to put all the I and C elements in derivative causality* [**29**].

From a BG model, the order of the system is equal to the number of integrally causaled storage elements. The BG rank is equal to the number of storage elements that accept a derivative causality when it is assigned to the model [29]. The number of structural null modes is the number of storage elements that cannot accept a derivative causality, when a preferred derivative causality is assigned to the model [23]. Finally observers (actuators) are superfluous when the system remains observable (controllable) if these observers (actuators) were not included in the system. These properties have been automated in dedicated software such as [17].

7.3.3. Bond graph for FDI analysis. As previously mentioned, the causal and structural properties of a BG model can be exploited to perform fault diagnosis. It requires an accurate model of the Σ, to compare its actual behavior with the expected one, and then its results (residuals) are evaluated. FDI with BG approach makes use of analytical redundancy relations (ARRs) . Classically, an ARR is a constraint derived from an over-constrained subsystem and expressed in terms of known variables of the process. In a BG sense, an $ARR = f(SSe, SSf, Se, Sf, MSe, MSf, \theta)$, where θ is the parameter's vector, and f is a constraint function. The causal properties of a BG model can be exploited for an automatic generation of ARRs through the procedure presented in [19]. This approach performs a dualization of the process measurements, to transform it into sources of information, and assigns a preferred derivative causality to the BG model. Then, ARRs can be computed by following the causal paths to eliminate the unknown variables [14]. These ARRs are then used to compute a Boolean Fault Signature Matrix (FSM). Ideally, in a non-faulty situation, the residuals are expected to approach zero. On the other hand, when a residual exhibits a significant value, a fault is detected in the Σ and then this information is used for fault isolation. However in the presence of measurement, parameters, and model uncertainties the diagnosis information may be corrupted, and some residuals can be triggered in the presence of some faults even if they are not sensitive to them.

In this section a methodology of FSM generation is presented by using the BG model, without ARRs generation. The latter is obtained by exploiting the concept of covering path (oriented graph). For a clear explanation, the BG can be represented as a graph, as depicted in Figure 7.3.4:

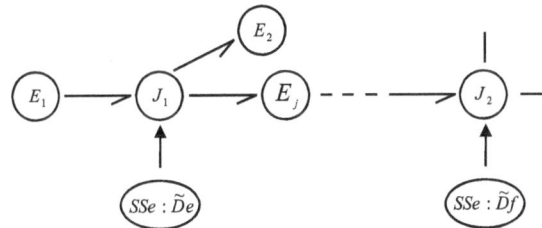

FIGURE 7.3.4. Bond graph represented as a graph

Some helpful explanations are now given.

- Let *(m)* be the number of components *(E)* to be monitored in a BG $(m = Card\{E\})$, where $E = \{R \cup C \cup I \cup TF \cup GY \cup Se \cup Sf \cup De \cup Df\}$.
- *(r)* is the number of redundant sensors, i.e., more than one sensor in the same junction.
- The number of ARRs *(n)* are equal to the number of junctions linked to at least one sensor, plus the number of redundant sensors *(r)* [25].

Recalling that each ARR candidate is based on covering path method, which utilizes sequences of linkages that have identical causal direction called causal path. The ARR candidate is based on the constitutive relation of the junction linked to at least one sensor. The sensor *(De, or Df)* is dualized to the source of signal $(SSe : \tilde{D}e,$ or $SSf : \tilde{D}f)$, where f_p, or e_p are unknown variables (for J_1 or J_2). The dualized detector becomes a signal source of information to the physical Σ, but it does not exchange any power with it.

The ARR is then generated by eliminating the unknown variables (e_p, or f_p), by covering the path from known $(SSe, SSf, Se,$ or $Sf)$ to the unknown ones.

Let us denote the path by a set of BG elements (vertices) it consists of, i.e.,

(7.3.10) $$\{E^{(i)}\} = \{E_0^{(i)}, E_1^{(i)}, \ldots, E_{q-1}^{(i)}, E_q^{(i)}\},$$

where $i = 1, \ldots, n$.

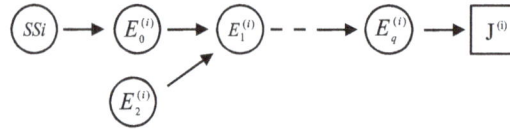

FIGURE 7.3.5. Oriented graph obtained to compute the ith ARR

TABLE 1. Fault signature matrix (FSM)

	r_1	r_2	$\cdots$	r_n	M_b	I_b
E_1	s_{11}	s_{12}	$\cdots$	s_{1n}	mb_1	ib_1
E_2	s_{21}	s_{22}	$\cdots$	s_{2n}	mb_2	ib_2
$\vdots$	$\vdots$	$\vdots$	$\cdots$	$\vdots$	$\vdots$	$\vdots$
E_m	s_{m1}	s_{m2}	$\cdots$	s_{mn}	mb_m	ib_m

Hence, $\{E^{(i)}\}$ is the set of component faults which can be detected by the ith ARR. Thus, the causal path leads to an oriented graph $G_{ARR}^{(i)}(S,A)$ (Figure 7.3.5):

$$(7.3.11) \qquad ARR_i = f(SSi, E_0^{(i)}, \ldots, E_q^{(i)}),$$

The length of a path is equal to the number of covered BG elements to calculate the unknown variable, and q is the length of the ith path. The same procedure is used for all ARRs.

Material redundant sensor can also be used to generate an ARR, however in this case, ARRs are equal to the difference between the measures of the two redundant sensors. As described in [**25**], material sensor redundancy is presented in a BG model by a sensor which can not be dualized without violating the causality assignment rules in the associated junction, and there are direct causal paths from one or more sensors in inverted causality to the non-inverted one without passing through any passive or two-port element (see Figure 7.3.6).

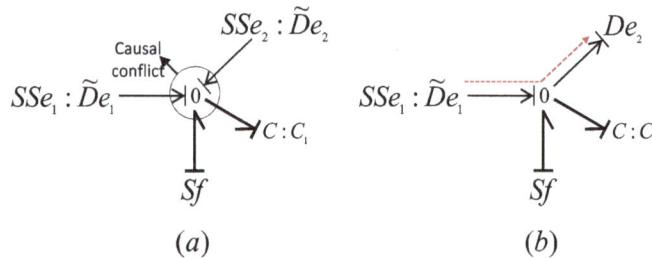

FIGURE 7.3.6. Direct sensor redundancy in a BG model for diagnosis in preferred derivative causality: *(a)* Violation of the causality assignment rules if both $SSe : \tilde{D}e_1$, and $SSe : \tilde{D}e_2$ are dualized, *(b)* only $SSe : \tilde{D}e_1$ is dualized (no violation)

Finally, the FSM can then be obtained from the oriented graphs $G_{ARR}^{(i)}(S,A)$ without any calculation as presented in Table 1.

Where mb_j and ib_j are equal to one if the fault that may affect the jth component is detectable and isolable, respectively. Moreover, the values of s_{ji} are assigned as follows:

$$(7.3.12) \qquad s_{ji} = \begin{cases} 1 & \text{if the vertex } E_j^{(i)} \in G_{ARR}^{(i)}, \\ 0 & \text{otherwise.} \end{cases}$$

Where, $E_j^{(i)}$ is the jth component ($j = 1, \ldots, m$) which can be detected by the ith ARR. The signature vector of each component fault E_j is given by the row vector: $V_{E_j}{}_{j=1,\ldots,m} = [s_{j1}, s_{j2}, \ldots, s_{jn}]$.

A component fault E_j is detectable (M_b (mb_j)) if at least one s_{ji} of its signature vector V_{E_j} is different from zero ($\exists i_{(i=1,...,n)} : s_{ji} \in V_{E_j} \neq 0$). A component fault E_j can be isolated (Ib (ib_j)) if its signature vector V_{E_j} is different from all others,

$$(7.3.13) \qquad ib_j = \begin{cases} 1 & \text{if} \quad \forall l_{(l=1,...,m)} : \quad V_{E_j} \neq V_{E_l} (j \neq l), \\ 0 & \text{otherwise.} \end{cases}$$

Thus, to obtain the oriented graph, a derivative causality is assigned to the BG model of the jth electromechanical as illustrated in Figure 7.3.7.

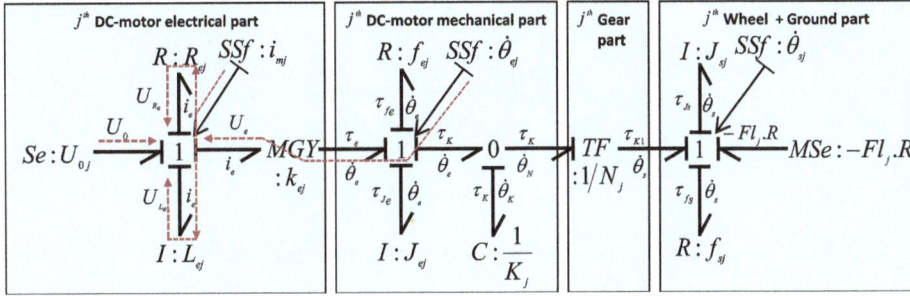

FIGURE 7.3.7. BG model of the jth electromechanical system for diagnosis in derivative causality.

In this model, the C-element remains in integral causality, otherwise a causal conflict would occur at its associated 0-junction. In the case of unknown initial conditions, the ARRs that require the effort (τ_k) imposed at this junction cannot be generated because the initial conditions associated with the C-element are required. However, in this case the initial conditions related to the initial angular positions of the actuator and the wheel are known ($\theta_{ej} = \theta_{sj} = 0$), hence the ARRs can be computed [8].

Example : This approach can be applied to the jth electromechanical system. To represent the procedure to obtain the oriented graph, the propagation from known variables to unknown is represented by the causal paths (dashed line) illustrated in Figure 7.3.7. Then, after applying the explained procedure to the jth electromechanical system, the oriented graph illustrated in Figure 7.3.8 is obtained. To compute the oriented graph obtained from the sensor associated to the 1−junctions of the electrical part (C_{J1E}), U_0, U_{R_e}, U_{L_e}, and U_e have to be known. To obtain U_{R_e}, the measured current i_m propagates through the constraint of measure (C_{m1}), and then it flows by the characteristic equation constraint of the R_e element (C_{Re}). The same is done to compute U_{L_e} but before passing through the characteristic equation constraint of the L_e element (C_{L_e}), it has to pass by the derivative one (C_{d1}). Moreover, U_0 is a known input and U_e is obtained from the measured $\dot{\theta}_e$ which propagates by the gyrator constraint (C_{GY_1}). To obtain the complete oriented graph, this procedure is applied on the two remaining junctions associated to a sensor, where, C_{J1M}, C_{J1W} are respectively, the constitutive relations associated to the 1−junctions of the mechanical and load parts. C_{J0} is the constitutive relations associated to the 0−junction. The rest of the constraints $C_{J_e}, C_{f_e}, C_{GY_2}, C_{TF_1}, C_{TF_2}, C_K, C_{J_s}, C_{f_s}$) represent the characteristic equations of the BG model elements. Moreover, from the oriented graph, the first three collumns of the FSM presented in Table 2 are obtained. From this FSM, one can conclude that all of the Σ components are monitorable. However, only sensor $Df : \dot{\theta}_{ej}$ is isolable if it is subject to a fault.

For simulation purposes, the structural analysis is not enough, so $ARRs$ have to be generated. In this case, the structural constraint of the 1−junction associated to $Df : i_{mj}$, and the characteristic equation constraint of each element is represented mathematically in (7.3.14), (7.3.15), respectively. The obtained ARR (ARR_1) is depicted in (7.3.16). The procedure used for $ARRs$ generation from the BG model was presented in Section 5.2

$$(7.3.14) \qquad U_0 - U_{R_e} - U_{L_e} - U_e = 0,$$

$$(7.3.15) \qquad U_{R_e} = R_{ej}.i_{mj}, \quad U_{L_{ej}} = L_{ej}.\frac{di_{mj}}{dt}, \quad U_e = k_{ej}.\dot{\theta}_{ej},$$

$$(7.3.16) \qquad ARR_{1j} : U_{0j} - R_{ej}.i_{mj} - L_{ej}.\frac{di_{mj}}{dt} - k_{ej}.\dot{\theta}_{ej} = 0.$$

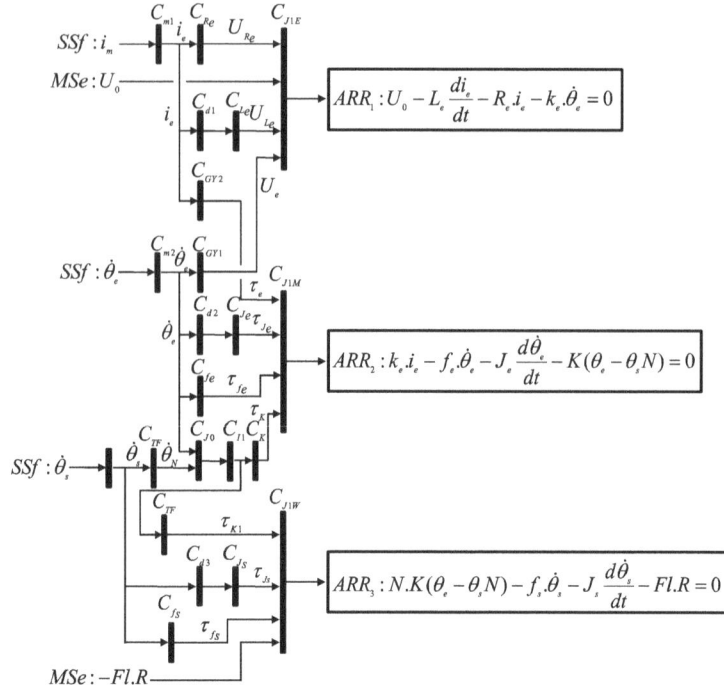

FIGURE 7.3.8. Oriented graph obtained from the BG model for diagnosis in derivative causality of the jth electromechanical system, with its associated ARRs.

The same procedure is applied to the remaining measured junctions, and the following ARRs are obtained (7.3.17), (7.3.18):

$$(7.3.17) \qquad ARR_{2j} : k_{ej}.i_{mj} - f_{ej}.\dot{\theta}_{ej} - J_{ej}.\frac{d\dot{\theta}_{ej}}{dt} - K_j.(\theta_{ej} - \theta_{sj}.N_j) = 0,$$

$$(7.3.18) \qquad ARR_{3j} : N_j.K_j(\theta_{ej} - N_j.\theta_{sj}) - f_{sj}.\dot{\theta}_{sj} - J_{sj}.\frac{d\dot{\theta}_{sj}}{dt} - Fl_j.R = 0.$$

From the inertial sensor presented in the vehicle body the computed residuals (Eq. 7.3.19, 7.3.20, and 7.3.21) are the following:

$$(7.3.19) \qquad ARR_l : Fx_1 + Fx_2 + Fx_3 + Fx_4 + m.\dot{v}.\dot{\psi} - m.\frac{d\dot{u}}{dt} = 0,$$

$$(7.3.20) \qquad ARR_c : Fy_1 + Fy_2 + Fy_3 + Fy_4 - m.\dot{u}.\dot{\psi} - m.\frac{d\dot{v}}{dt} = 0,$$

$$(7.3.21) \qquad ARR_\psi : [-Fx_1 + Fx_2 + Fx_3 - Fx_4]c + [Fy_1 + Fy_2]a - [Fy_3 + Fy_4]b - J.\frac{d\dot{\psi}}{dt} = 0.$$

If all the sensors of the system are used to generate ARRs, three faults can now be isolable ($\dot{\theta}_{ej}, \dot{\theta}_{sj}, J$).

7.3.4. Bond graph for reconfigurability analysis. After performing the diagnosis, the following step is to use this information in order to conclude structurally for which faults the system can be reconfigured. Three different methodologies are considered. One for sensor faults, one for actuators, and one for plant or non-isolable faults. Note that since fault accommodation is not considered, plant and non-isolable faults require the same methodology.

Sensor faults: For this type of faults, it is required that the fault is isolable. The latter can be verified by the FSM (Table 2) of the Σ, as described in subsection 7.3.3. Additionally, the faulty sensor is switched off (meaning that from a graphical point of view its associated outer vertex is removed, thus causing a change on the graphical architecture). For such faults Σ' can be held in operation if one of the two following conditions is fulfilled:

TABLE 2. Fault Signature Matrix (FSM)

Part	Comp.	j^{th} Elect. system			Chassis			M_b	I_b
-	-	r_{1j}	r_{2j}	r_{3j}	r_l	r_c	r_ψ	–	–
Electrical	$Se:U_{0j}$	1	0	0	0	0	0	1	0
	L_{ej}	1	0	0	0	0	0	1	0
	R_{ej}	1	0	0	0	0	0	1	0
	$Df:i_{mj}$	1	1	0	0	0	0	1	0
	k_{ej}	1	1	0	0	0	0	1	0
Mechanical	$Df:\dot{\theta}_{ej}$	1	1	1	0	0	0	1	1
	f_{ej}	0	1	0	0	0	0	1	0
	J_{ej}	0	1	0	0	0	0	1	0
	K_j	0	1	1	0	0	0	1	0
	N_j	0	1	1	0	0	0	1	0
Wheel	$Df:\dot{\theta}_{sj}$	0	1	1	1	1	1	1	1
	f_{sj}	0	0	1	0	0	0	1	0
	J_{sj}	0	0	1	0	0	0	1	0
	R	0	0	1	1	1	1	1	0
Chassis	$Df:\dot{u}$	0	0	1	1	1	1	1	0
	m	0	0	0	1	1	1	1	0
	$Df:\dot{v}$	0	0	0	1	1	1	1	0
	J	0	0	0	0	0	1	1	1
	$Df:\dot{\psi}$	0	0	0	1	1	1	1	0

Condition 1: Material redundancy is available, in this case the same control law can still be used;

Condition 2: Σ' remains structurally observable simply by using the remaining healthy sensors (analytical measurement redundancy is presented in Σ and it can be verified by Theorem 5). In this case, it is structurally concluded that system reconfiguration is required.

Furthermore, in order to detect future faults it would also be interesting to see if Σ' remains monitorable. Thus, the algorithm provided in section 7.3.3 should be re-worked without considering the faulty sensor.

Example : consider the jth electromechanical system illustrated in Figure 6.4.1, and let us assume the presence of a fault in sensor $Df:\dot{\theta}_{ej}$, that can be isolated (see Table 2). As exposed in Figure 6.4.1 there is no material measurement redundancy hence condition 2 presented above has to be verified. The latter can be concluded by applying Theorem 5 to Σ' (illustrated in Figure 7.3.9), which does not consider the faulty sensor. It can then be concluded that the system remains structurally observable and that system reconfiguration can be performed.

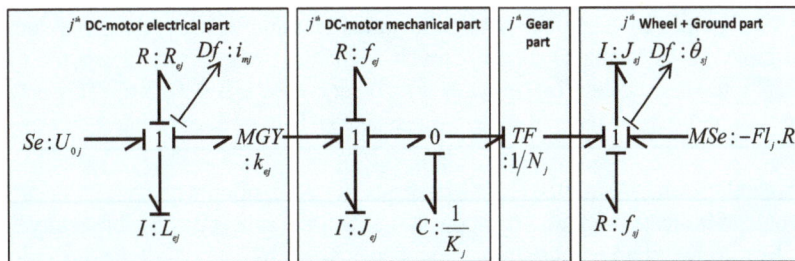

FIGURE 7.3.9. BG model of the jth electromechanical system when sensor $Df:\dot{\theta}_{ej}$ is faulty

Actuator fault: Because of the duality between observability and controllability, actuator and sensor faults are treated in a similar way. Nevertheless, in this case it is related to Σ controllability. Again, the FDI procedure should be able to isolate the fault. For such faults Σ' can be held in operation if one of the two following conditions is respected:

Condition 1: Material redundancy is available, in this case, the same control law can still be used;

Condition 2: Σ' remains structurally controllable simply by using the remaining healthy sensors (analytical measurement redundancy is presented in Σ and it can be verified by Theorem 5). In this case, it is structurally concluded that system reconfiguration is required.

Moreover, the faulty actuator is not used for control purposes, (meaning from a graphical point of view that its associated outer vertex is removed, thus causing a change on the graphical architecture).

Hypothesis 1. *The dynamics of the sources can always be turned off so that they do not provide flow or effort to the system.*

Material input redundancy is presented in Σ if the shortest direct causal paths linking two or more sources to a detector meet the same passive and two-port elements. The shortest causal path from a source (Se/Sf) to a detector (De/Df) is the one involving the minimal number of passive elements when following the path from Se/Sf to De/Df. In Figure 7.3.10 an example of material input redundancy is given. Its shortest causal paths between each source to a detector are described in (7.3.22) for sources of flow (Sf), and (7.3.23) for sources of effort (Se), corresponding to part *(a)*, and *(b)* of Figure 7.3.10, respectively.

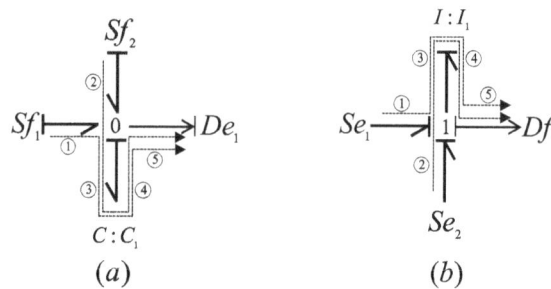

FIGURE 7.3.10. Material input redundancy: Direct causal path from flow sources (S_f), part *(a)* (effort sources (S_e), part *(b)*) to sensor involving the same elements

(7.3.22)
$$(a) \begin{cases} Sf_1 & \rightarrow & C:C_1 & \rightarrow & De_1, \\ Sf_2 & \rightarrow & C:C_1 & \rightarrow & De_1. \end{cases}$$

(7.3.23)
$$(b) \begin{cases} Se_1 & \rightarrow & I:I_1 & \rightarrow & Df_1, \\ Se_2 & \rightarrow & I:I_1 & \rightarrow & Df_1. \end{cases}$$

Both paths in Figure 7.3.22 involve the same elements, meaning that both inputs affect the output exactly in the same way (existence of material input redundancy). Hence, Sf_1, and Sf_2 fall in this type of redundancy. In a similar way, the same can also be concluded about the sources Se_1, and Se_2.

Example : consider the jth electromechanical system illustrated in Figure 6.4.1, and let us assume the presence of an input fault U_{0j}. From Table 2 it is possible to conclude that its signature vector is the same as the signature vector of the electrical inductance (L_j) and resistance (R_ej), $V_{U_{0j}} = V_{L_{ej}} = V_{R_{ej}} = [1, 0, 0, 0, 0, 0]$. This fault cannot be isolated. Therefore, the necessary condition of isolability is not respected. It is then concluded that system reconfiguration cannot be performed.

Plant and non-isolable faults: As noticed in the previous example, often when a fault occurs in a process Σ, the FDI algorithm is not able to indicate the faulty component, i.e., a finite subset of possible faulty components have the same fault signature. However, system reconfiguration may be possible to apply if all the dynamics of the finite subset of possible faulty components can be removed from Σ_o. To remove the faulty dynamics, a cut of the power transfer between the faulty subsystems to the healthy one is necessary. This is called *Path Breaking* (PB).

Definition Physically a PB means the cut of power propagation between a subsystem to Σ_o. In other words, the dynamics of a subsystem will not affect the dynamics of the overall Σ.

From the physical structure of a BG model, it is possible to study the causal path propagation, between components to Σ_o. Bonds in a BG model represent the power propagation between system components or subsystems. Thus, in a BG sense a PB occurs when the propagation of power ($P = effort.flow$) through the bond connecting two subsystems is equal to zero. This may be caused by using some controlled elements, such as: *Modulated R, I, C, MGY, MTF, Se, or Sf*. Note that due to physical constraints of these controlled elements, the locations of PBs have to be indicated by human experts during the system design stage, and a label is added to it. In this label the floating value (z) is included, where z is chosen so that $P_1 = e_1.f_1$, and $P_2 = e_2.f_2$ are equal to zero ($\exists z \in \Re^+ : P_1 = 0 \wedge P_2 = 0$), where $e_{1,z} = \Phi_z(f_{1,z})$, thus causing a PB. P_1 is the power that propagates to the controlled element that causes the PB, and P_2 is the power that flows from the faulty subsystem to the healthy one. To clearly understand the concept of PB consider Figure 7.3.11-(*a,b,c*), where three controlled elements that can cause a PB are illustrated. A_f and B are a faulty and a healthy subsystem, respectively.

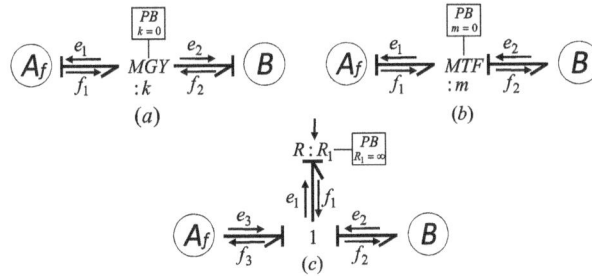

FIGURE 7.3.11. Some controlled BG elements that can cause a PB: part (*a*) Controlled gyrator ($MGY : k$), (*b*) Controlled transformer ($MTF : m$), and (*c*) Controlled resistive element ($R : R_1$).

In part (*a*), the PB is provoked if $k = 0$. The latter can be easily concluded from the behavioral equations (7.3.24), and from its development presented in (7.3.25).

(7.3.24)
$$
\begin{aligned}
e_1 &= k.f_2, \quad e_2 = k.f_1, \\
if \quad & z = k = 0, \\
e_1 &= 0.f_2 = 0, \quad e_2 = 0.f_1 = 0,
\end{aligned}
$$
(7.3.25)
$$P_1 = e_1.f_1 = 0, \quad P_2 = e_2.f_2 = 0.$$

As concluded from the behavioral equations (7.3.26), and from its development presented in (7.3.27), a PB is caused in part (*b*) if $m = 0$.

(7.3.26)
$$
\begin{aligned}
e_1 &= m.e_2, \quad f_2 = m.f_1, \\
if \quad & z = m = 0, \\
e_1 &= 0.e_2 = 0, \quad f_2 = 0.f_1 = 0,
\end{aligned}
$$
(7.3.27)
$$P_1 = e_1.f_1 = 0, \quad P_2 = e_2.f_2 = 0.$$

Finally, in part (*c*) the PB can be concluded from equations 7.3.28:

$$
\begin{aligned}
e_1 &= R_1.f_1, \quad if \quad z = R_1 = \infty, \\
f_1 &= 0.e_1 = 0, \quad f_2 = f_1 = f_3 = 0,
\end{aligned}
$$
(7.3.28)
$$P_1 = e_1.f_1 = 0, \quad P_2 = e_2.f_2 = 0.$$

Hypothesis 2. *The elements used to provoke a PB can always be set to their necessary values (z), even if this element is subject to a fault.*

Taking into consideration the information contained in FSM, let us define $F_{set} = \{F^b_{sig} | b \in (1 \ldots B)\}$, where F^b_{sig} represents the set of components (E) with the same fault signature and B the total number of different signatures. Based on the triggered fault signature (b), the controlled BG components that need to be used to cause the PB can be automatically deduced as follows:

PB conditions:

Condition 1: $(Modulated \quad R, \quad I \quad or \quad C, \; MGY, \text{ or } MTF)$ in which all the causal paths from the components (E) belonging to F_{sig}^b $(E \in F_{sig}^b)$, must pass by the component that causes the PB or by its associated junction before it achieves the Σ_o.

Condition 2: All the causal paths from the sources $(Se, \text{ or } Sf)$, that pass by the faulty component or by their associated junction before achieving the Σ_o must be stopped.

From a graphical point of view, the outer vertices that belong to A_f are removed and Σ' is obtained. In this case, the dimension of the Σ states, sources, and detectors may decrease so controllability/observability must be verified for Σ'. Finally, the FDI algorithm has to be performed to Σ' in order to verify if the monitorability conditions remain respected.

Moreover, a possible PB is presented in each jth electromechanical system. The converter of electrical energy to mechanical torque may provoke this situation, because (k_{ej}) is controlled. Then, it is possible to deduce that by setting $k_{ej} = 0$, the dynamics of the state p_L do not affect the ones of state p_J, and vice-versa. This can be mathematically verified by (7.3.3).

In the BG model, this situation is represented by the component $MGY : k_{ej}$, when $k_{ej} = 0$. This was previously explained in (7.3.24) and (7.3.25). Thus a label is added to this component as illustrated in Figure 7.3.12.

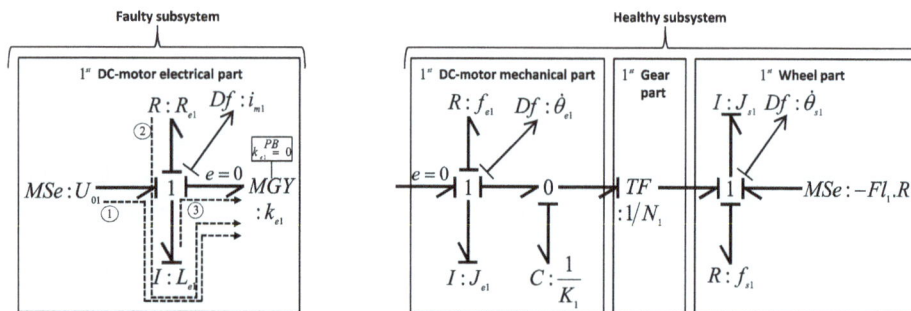

FIGURE 7.3.12. BG of the jth electromechanical system with a PB label in the $MGY : k$ element.

Example : consider the 1st electromechanical system illustrated in Figure 6.4.1, and let us assume that there is a fault in the Σ inductance (L_{e1}), which triggers r_{11} (see Table 2). In this case, the applied FDI algorithm is not able to isolate the faulty component because $Se : U_{01}$, R_{e1}, and L_{e1} have the same fault signature (see Table 2). Noting that by setting $k_{e1} = 0$, the causal path ①, ②, and ③, are removed from the healthy part of the system, hence the fault does not propagate until the Σ_o. This is represented in Figure 7.3.12 by separating the faulty subsystem (DC-motor electrical part) from the healthy one(DC-motor mechanical part). Hence, bearing in mind conditions (1), and (2) presented above, to cause a PB U_{01} and k_{e1} should be set to zero. Logically, after the PB, the 1st electromechanical system does not have a directed actuated input. Nevertheless, the contact effort (Fl_1) is an undirected controlled source. Due to the remaining healthy electromechanical subsystems, the overall vehicle remains controllable, as it can be concluded by Theorem 5. Moreover it is easy to conclude that the system also remains observable and monitorable. It is then structurally concluded that system reconfiguration can be performed to cope with this fault.

7.3.5. Algorithm for verifications of structural reconfigurability and case study. To represent the described procedure for structural system reconfigurability analysis in a systematic manner, the algorithm illustrated in Figure 7.3.13 is proposed. When a fault is detected in the system, the algorithm is activated and its FDI information is exploited, in order to conclude structurally that the system reconfiguration can be performed. This algorithm is detailed in the next section.

7.4. RobuCar study structural reconfigurability conditions

To explain the evolution of the algorithm of Figure 7.3.13, let us consider the overall BG model of the RobuCar that is based on a simplified vehicle dynamic model as illustrated in Figure 7.3.2.

To detail the procedure of the proposed algorithm (Figure 7.3.13), three single fault scenarios were introduced in RobuCar at 15 sec. An input $(Se : U_{01} (-2 \text{ V}))$, a sensor $(Df : \dot{\theta}_{s1} (+10 \text{ \%}))$, and a plant $(R_{e1} (+2 \text{ } \Omega))$ fault. The results are performed on a vehicle dynamic simulator software $(CALLAS/SCANeR \text{ } Studio)$ [18] using real data

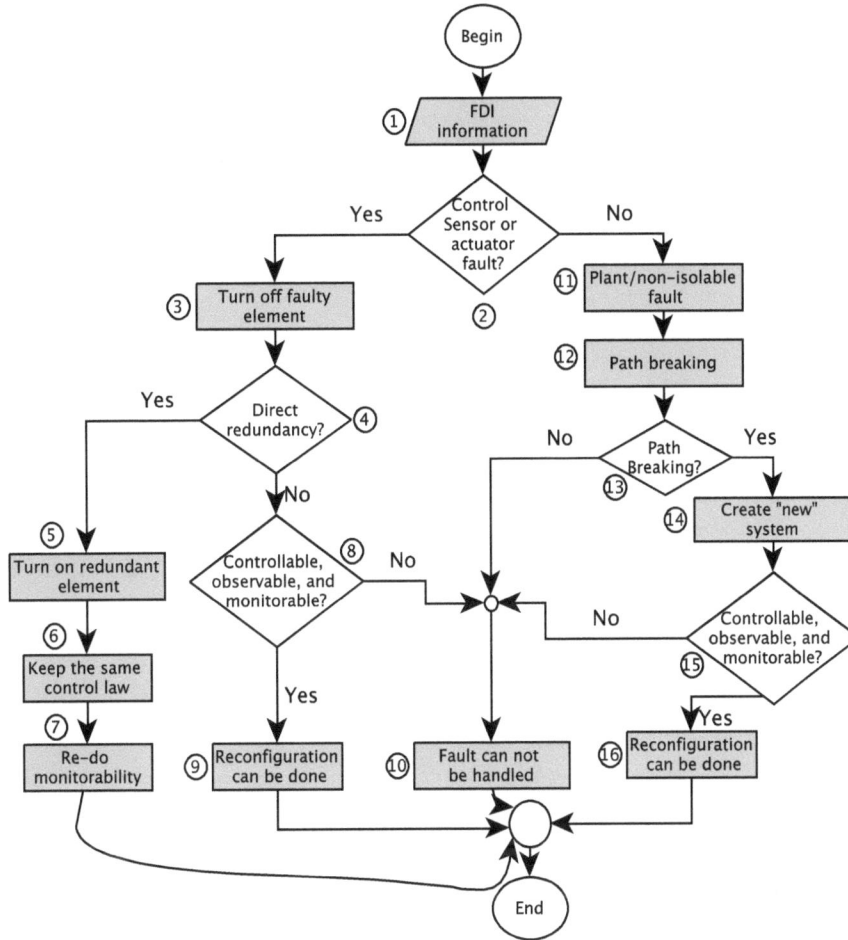

FIGURE 7.3.13. Algorithm of structural fault reconfigurability analysis

of the system. The latter is an automotive driving simulator, dedicated to engineering and research. In the following co-simulation, results are produced using two external programs under independent supports, where data is exchanged between the two programs. In this case, we co-simulate between a program implemented under *Matlab/Simulink* environment and *CALLAS/SCANeR Studio* simulator. For the following results, the whole dynamics of RobuCar are validated from experiments on *CALLAS/SCANeR Studio*. So we use this platform to co-simulate with the dynamics of the electromechanical traction system developed on *Matlab/Simulink* (Figure 6.5.1).

In nominal conditions, RobuCar, follows the trajectory illustrated in Figure 7.4.1 (solid line). And, as expected, none of the residuals are triggered while the vehicle is operating in nominal conditions because they remain bounded between the defined thresholds (Figure 7.4.2-7.4.4).

Example: Consider the input fault $Se : U_{01}$, which triggers ARR_{11} (see Table 2, and Figure 7.4.2-7.4.4 (dashed line)). From Figure 7.4.1 it can be seen that the RobuCar trajectory with a faulty input deviates from the one in fault-free mode.

To verify the structural ability of the system to cope with this fault signature, the algorithm (Figure 7.3.13) begins and then progresses to step ①. It is concluded from Table 2, that the FDI algorithm is not able to isolate the fault because the components $Se : U_{01}$, R_{e1}, and L_{e1} have the same fault signature. Thus, the answer from step ② is "No" because we are not able to isolate either an actuator or a sensor fault. Then, the algorithm follows to ⑪ (the fault is in the group of plant or non-isolable fault). So the algorithm looks for a PB ⑫. Bearing in mind PB conditions (1) and (2) presented above, a PB is found, which allows to conclude that the unknown faulty component dynamics can be removed from Σ_o of RobuCar (see Figure 7.3.12). Consequently, the answer of step ⑬ is "Yes", and Σ' is created

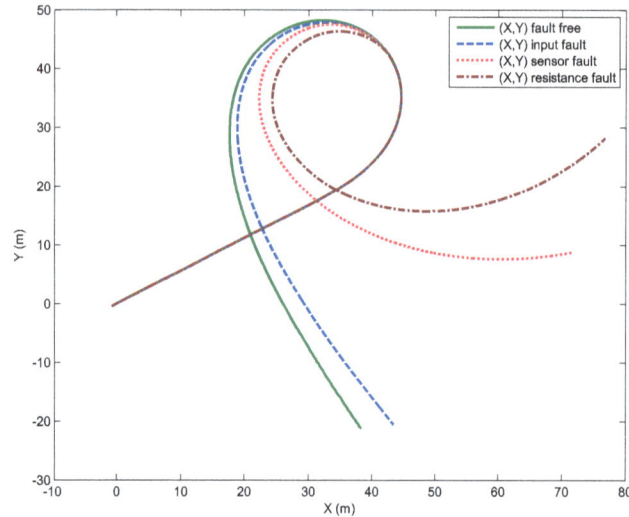

FIGURE 7.4.1. Tracked trajectory of RobuCar when subjected to healthy and faulty conditions

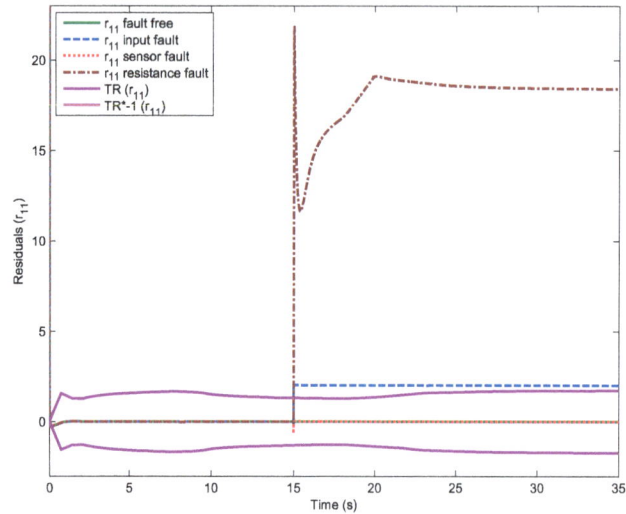

FIGURE 7.4.2. Residual 1 of the 1st electromechanical system of RobuCar under healthy and faulty conditions

⑭. The 1st electromechanical system does not have a directed actuated input. Nevertheless, the contact effort (Fl_1) is an undirected controlled source, hence the answer of step ⑮ is "Yes." It is then concluded, from a structural point of view that the existing fault can be handled by applying a reconfiguration technique.

Example: Consider the sensor fault $(Df : \dot{\theta}_{s1})$, as it can be seen from Table 3 and Figure 7.4.2-7.4.4 (dotted line), the r_{21}, r_{31}, r_l, r_c, r_ψ are the ones sensitive to this fault. Note that results of the last three residuals are not illustrated in this chapter. The deviation of the desired trajectory is depicted in Figure 7.4.1. Again, the algorithm is initialized, and in step ① it is obtained that the fault has a unique signature. Hence, the answer of step ② is "Yes", and $Df : \dot{\theta}_{s1}$ is turned off ③. Therefore, the algorithm evolves to step ④ where the answer is "No". Finally, the answer of step ⑧ is "Yes" so it is concluded that, reconfiguration can be performed.

Example: Consider the fault in the electrical resistance (R_{e1}). In this case, the procedure to conclude structurally if this fault can be dealt with, is exactly the same as for the input fault, because they have the same fault signature (Table 2). So it is structurally obtained that the system can be reconfigured. From Figure 7.4.4 (dashed-dotted line),

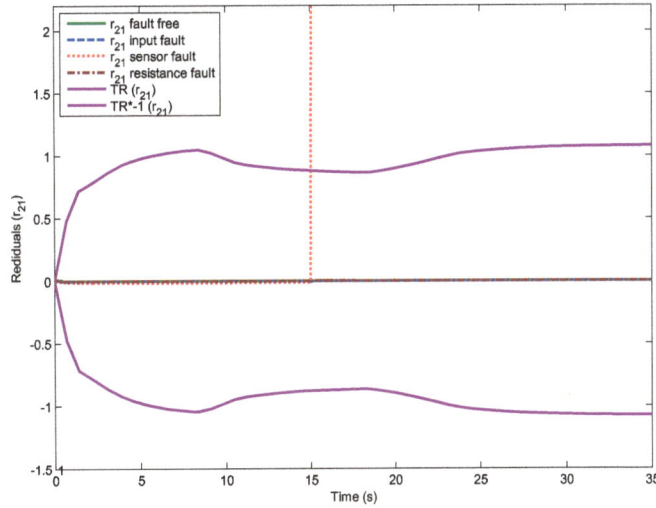

FIGURE 7.4.3. Residual 2 of the 1st electromechanical system of RobuCar under healthy and faulty conditions

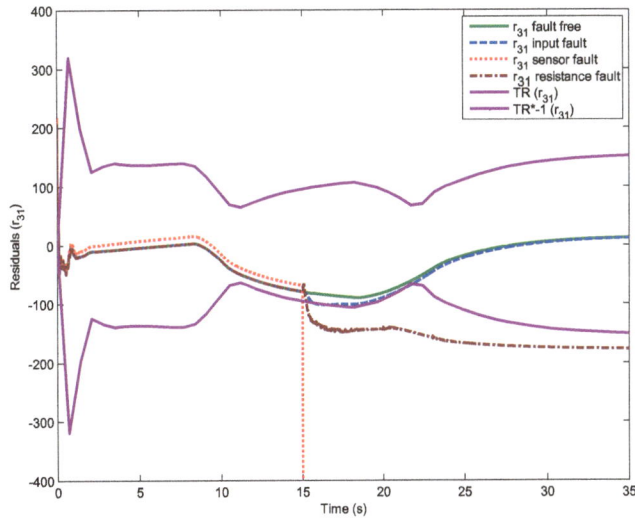

FIGURE 7.4.4. Residual 3 of the 1st electromechanical system of RobuCar under healthy and faulty conditions

one can notice that the fault in R_{e1} also triggers r_{31} when it should not. The latter may be justified by some unmodeled dynamics.

Moreover, because the diagnosis information is taken into consideration, this algorithm is not applied for each possible fault but for each F_{sig}^{b}. Hence, by performing the latter, the following binary Fault Signature and Reconfigurability Matrix (FSRM) is obtained:

Where, SR stands for structural reconfigurability, it is a binary column that is filled as follows:

$$(7.4.1) \qquad SR_j = \begin{cases} 1 & \text{if when the fault signature } F_{sig}^{b} \text{ is triggered } \Sigma \text{ can be reconfigured,} \\ 0 & \text{otherwise.} \end{cases}$$

From Table 3, one can define the set of critical fault signatures ($CF_{sig} = \{F_{sig}^{b} : SR_b = 0\}$) and the set of critical faulty elements ($CF_{ele} = \{E_j\} \in \{CF_{sig}\}$).

TABLE 3. Fault Signature and Recoverability Matrix (FSRM)

Part	Comp.	j^{th} Elect. system			Chassis			M_b	I_b	SR
-	-	r_{1j}	r_{2j}	r_{3j}	r_l	r_c	r_ψ	—	—	—
Electrical	$Se : U_{0j}$	1	0	0	0	0	0	1	0	1
	L_{ej}	1	0	0	0	0	0	1	0	1
	R_{ej}	1	0	0	0	0	0	1	0	1
	$Df : i_{mj}$	1	1	0	0	0	0	1	0	1
	k_{ej}	1	1	0	0	0	0	1	0	1
Mechanical	$Df : \dot{\theta}_{ej}$	1	1	1	0	0	0	1	1	1
	f_{ej}	0	1	0	0	0	0	1	0	0
	J_{ej}	0	1	0	0	0	0	1	0	0
	K_j	0	1	1	0	0	0	1	0	0
	N_j	0	1	1	0	0	0	1	0	0
Wheel	$Df : \dot{\theta}_{sj}$	0	1	1	1	1	1	1	1	1
	f_{sj}	0	0	1	0	0	0	1	0	0
	J_{sj}	0	0	1	0	0	0	1	0	0
	R	0	0	1	1	1	1	1	0	0
Chassis	$Df : \dot{u}$	0	0	1	1	1	1	1	0	0
	m	0	0	0	1	1	1	1	0	0
	$Df : \dot{v}$	0	0	0	1	1	1	1	0	0
	J	0	0	0	0	0	1	1	1	0
	$Df : \dot{\psi}$	0	0	0	1	1	1	1	0	0

7.5. Conclusion

In this chapter the same graphical tool is used for dynamic modeling, diagnosability analysis and structural verification of fault reconfigurability conditions of an over-actuated electric vehicle named RobuCar. An algorithm that concludes structurally for which faults the system remains able to achieve its objectives without complex calculations, was presented. It is concluded that in some situations, component fault isolability is not a necessary condition for system reconfigurability. These situations can be determined from the BG model of the system in study. Therefore, the number of faults that a system can tolerate when proper control actions are applied may increase. As a limitation of this approach, the energy required to control and observe the system is not known because of the use of a structural approach that is independent of the system's numerical values. Moreover, due to the latter, closed-loop stability and system performance cannot be guaranteed at this stage.

Bibliography

[1] A. Aitouche and B. Ould-Bouamama. Fault-tolerant control with respect to actuator failures application to steam generator process. In *Proceedings European Symposium on Computer Aided Process Engineering - 15*, pages 1471–1476, 2005.

[2] Mogens Blanke, Michel Kinnaert, Jan Lunze, and Marcel Staroswiecki. *Diagnosis and Fault-Tolerant Control*. Springer, 2003.

[3] Anibal Bregon, Gautam Biswas, and Belarmino Pulido. Compilation techniques for fault detection and isolation: A comparison of three methods. In *Proceeding 19th International Workshop on Principles of Diagnosis Dx*, Blue Mountains, Australia, 2008.

[4] C. Commault, J. M. Dion, O. Sename, and R. Motyeian. Observer-based fault detection and isolation for structured systems. *IEEE transactions on automatic control*, 47(12):2074–2079, 2002.

[5] M. Cordier, P. Dague, M. Dumas, F. Levy, J. Montmain, M. Staroswiecki, and L. Travé-Massuyès. A comparative analysis of ai and control theory approaches to model-based diagnosis. In *14th Eur. Conf. Artificial Intelligence (ECAI 2000), Berlin, Germany*, 2000.

[6] José Luis de la Mata and Manuel Rodríguez. Accident prevention by control system reconfiguration. *Computers and Chemical Engineering*, 34:846–855, 2010.

[7] Jean-Michel Dion, Christian Commault, and Jacob van der Woude. Generic properties and control of linear structured systems: a survey. *Automatica*, 39:1125–1144, 2003.

[8] M.A. Djeziri, R. Merzouki, and B. Ould-Bouamama. Robust monitoring of electric vehicle with structured and unstructured uncertainties. *IEEE Transactions on Vehicular Technology*, page In Press, 2009.

[9] B.M. Gonzaléz-Contreras, D. Theilliol, and D. Sauter. On-line reconfigurability evaluation for actuator faults using input/output data. In *7th IFAC Symposium on Fault Detection, Supervision and safety of Technical Processes, Safeproces'09*, Barcelona, Spain, 2009.

[10] D.C. Karnopp, D. Margolis, and R. Rosenberg. *Systems Dynamics: A Unified approach*. John Wiley, New York, second edition, 1990.

[11] A. Khelassi, D. Theilliol, and P. Weber. Reconfigurability analysis for reliable fault-tolerant control design. In *7th Workshop on Advanced Control and Diagnosis, ACD'2009*, Zielona Gora, Poland, 2009.

[12] Mattias Krysander, Jan Aslund, and Mattias Nyberg. An efficient algorithm for finding minimal overconstrained subsystems for model-based diagnosis. *IEEE Transactions on Systems, Man, and Cybernetics - Part A: Systems and Humans*, 38(1):197–206, 2008.

[13] Morten Lind. Representing goals and functions of complex systems - an introduction to Multilevel Flow Modeling. Technical report, Institute of Automatic Control System, Technical University of Denmark, Lyngby, 1990.

[14] Chang. Boon. Low, Danwei. Wang, Shai. Arogeti, and Jing. Bing. Zhang. Monitoring ability analysis and qualitative fault diagnosis using hybrid bond graph. In *Proceedings of the 17th World Congress The International Federation of Automatic Control Seoul, Korea, July 6-11*, pages 10516–10521, 2008.

[15] M. R. Maurya, R. Rengaswamy, and V. Venkatasubramanian. Application of signed digraphs-based analysis for fault diagnosis of chemical process flowsheets. *Engineering Applications of Artificial Intelligence*, 17(5):501–518, 2004.

[16] Pieter J. Mosterman and Gautam Biswas. Diagnosis of continuous valued systems in transient operating regions. *IEEE Trans. on Systems, Man and Cybernetics*, 29(6):554–565, 1999.

[17] A. Mukherjee and A. K. Samantaray. System modelling through bond graph objects on SYMBOLS 2000. In *Int. Conf. Bond Graph Modeling and Simulation*, volume 33, pages 164–170, Phoenix, Arizona, 2001. SCS.

[18] OKTAL. Scaner driving simulation engine, July 2011.

[19] B. Ould-Bouamama, A.K. Samantaray, M. Staroswiecki, and G. Dauphin-Tanguy. Derivation of constraint relations from bond graph models for fault detection and isolation. In *International Conference on Bond Graph Modeling and Simulation*, pages 104–109, 2003.

[20] H. B. Pacejka and R. S. Sharp. Shear force developments by pneumatic tires in steady state conditions, a review of modelling aspects. *Vehicle systems dynamics*, 20:121–176, 1991.

[21] Ron J Patton. Fault-tolerant control systems. the 1997 situation. In *Proceedings of the 3rd IFAC symposium on fault detection, supervision and safety for technical processes*, pages 1033–1055, August 1997.

[22] M.A. Djeziri R. Merzouki, B. Ould-Bouamama and M. Bouteldja. Modelling and estimation for tire-road system using bond graph approach. *MECHATRONICS*, 17:93–108, 2007.

[23] A. Rahmani, C. Sueur, and G. Dauphin-Tanguy. Pole assignement for systems modelled by bond graph. *Journal of the Franklin Institute*, 331B(3):299–314, 1994.

[24] A.K. Samantaray and S.K. Ghoshal. Bicausal bond graphs for supervision: From fault detection and isolation to fault accommodation. *Franklin Institute*, 345(1):1–28, 2008.

[25] A.K. Samantaray and B. Ould-Bouamama. *Model-based Process Supervision. A Bond Graph Approach*. Springer Verlag, 2008.

[26] Société de technologie Michelin. *The tyre Grip*, 2001.

[27] Marcel Staroswiecki. On reconfigurability with respect to actuator failures. In *Proceeding of 15th Triennial World Congress*, Barcelona, Spain, 2002.

[28] Marcel Staroswiecki and Anne-Lise Gehin. From control to supervision. *Annual Reviews in Control*, 25:1–11, 2001.

[29] C. Sueur and G. Dauphin-Tanguy. Bond graph approach for structural analysis of MIMO linear systems. *Journal of the Franklin Institute*, 328(1):55–70, 1991.

[30] L. Trave-Massuyes, T. Escobet, and X. Olive. Diagnosability analysis based on component supported analytical redundancy relations. *IEEE Transactions*, 36 (6):1146 –1160, 2006.

[31] N. Eva Wu, Kemin Zhou, and Gregory Salomon. Control reconfigurability of linear time-invariant systems. *Automatica*, 36:1767–1771, 2000.

[32] Youmin Zhang and Jin Jiang. Bibliographical review on reconfigurable fault-tolerant control systems. *Annual Reviews in Control*, 32:229–252, 2008.

Mechatronic & Innovative Applications, 2012, 147-167

CHAPTER 8

Robust Fault Decision: Application to an Omni Directional Mobile Robot

Nizar Chatti[1],
Polytech-Lille, LAGIS, UMR CNRS 8219
Avenue Paul Langevin, 59655 Villeneuve D'Ascq, France
nizar.chatti@polytech-lille.fr

Anne Lise Gehin
Polytech-Lille, LAGIS, UMR CNRS 8219
Avenue Paul Langevin, 59655 Villeneuve D'Ascq, France
anne-lise.gehin@polytech-lille.fr

Belkacem Ould-Bouamama
Polytech-Lille, LAGIS, UMR CNRS 8219
Avenue Paul Langevin, 59655 Villeneuve D'Ascq, France
Belkacem.Ouldbouamama@polytech-lille.fr

ABSTRACT. Fault diagnosis is crucial for ensuring the safe operation of complex engineering systems and avoiding the execution of an unsafe behavior. This chapter deals with Robust Decision Making(RDM) for fault detection of electromechanical systems by combining the advantages of Bond Graph(BG) modeling and Fuzzy logic reasoning. A fault diagnosis method implemented in two stages is proposed. In the first stage, the residuals are deduced from the BG model allowing the building of a Fault Signature Matrix(FSM) according to the sensitivity of residuals to different parameters. In the second stage, the result of FSM and the robust residual thresholds are used by the fuzzy reasoning mechanism in order to evaluate a degree of detectability for each set of components. Finally, in order to make robust decision according to the detected fault component, an analysis is done between the output variables of the fuzzy system and components having the same signature in the FSM. The performance of the proposed fault diagnosis methodology is demonstrated through experimental data of an omni directional robot.

Keywords: Bond graphs; Diagnosis; Robust decision making; Modeling; Graphical approach; Analytical Redundancy Relations; Supervision; Fuzzy logic; Monitoring; Dynamic behavior; Fault indicators; Model-Based diagnosis; Fault detection and isolation; Parameter uncertainties; Mobile robotics.

8.1. Introduction

Nowadays, the growing demand for safety and reliability of modern engineering systems motivates the development of new fault diagnosis algorithms for the decision support system. It is worth noting, that a wide variety of concepts, methods and tools have been developed to address decision challenges that confront a large degree of uncertainty for evaluating residuals including both the fault detection and the isolation capabilities. Fault detection and isolation (FDI) procedures consist of comparison between the real and the reference process behaviors. In the literature, different approaches for the FDI have been developed based on quantitative and qualitative models. A good account of these methods is given in [1], [13], [3]. Among robust diagnosis quantitative methods, [4] deals with the generation of fault indicators and residual thresholds in the presence of parameter uncertainties by using a bond graph representation.

Models allowing the production of fault indicators are very important in the design of a safe system. Nevertheless, to determine the exact mathematical relationships between the physical values characteristic of the system is not

---

[1]Corresponding author

always easy, especially if the system makes appear phenomena of different natures as in an electromechanical system. Bond Graph is a multidisciplinary and unified graphical modeling language which has proved its adequacy to represent energy exchanges in mixed systems [5]. Bond Graphs have first been used as modeling tool, and their causal and structural properties (observability, controllability, monitorability) have been subsequently used to generate fault indicators in a systematic and generic way [6], [7]. Moreover, Bond Graph modeling has also been used in the past for different FDI approaches [8], [9].

In [10], residual expressions are generated from Bond Graph model in derivative causality and detection is based on residual fixed thresholds around a working point and adaptive thresholds that vary according to the set point. One of the drawbacks of this classical approach is the uncertainty and instability of the decision especially in relation to time when the residual values are close to the thresholds. In order to remedy this situation, [11] proposes a linear fractional transformation (LFT) uncertain model with filtering methods for residuals' generation and evaluation. A robust FDI with respect to parameter uncertainties using Bond Graph modeling approach in the LFT configuration is also proposed in [12]. Nevertheless, even if some applications of these approaches exist, fault detection and isolation procedure is not trivial. Basically, when the residual is weakly sensitive to the fault or to a value very near to the threshold boundaries, the fault alarms still exist. To overcome these limitations, we present in this chapter a robust decision making approach (RDMA). It represents a particular set of methods and tools designed to support decision making under conditions of deep uncertainty. This RDMA is based on Fuzzy logic methodology and aims to facilitate the identification of vulnerabilities by identifying the residuals' change through time. The proposed RDMA is implemented in two stages. In the first stage, the residuals are deduced from the Bond Graph model and then a Fault Signature Matrix (FSM) is built according to the sensitivity of residuals to different parameters. In the second stage, the result of FSM and the robust chosen thresholds are used by the fuzzy reasoning mechanism to evaluate a degree of detectability for each set of parameters having the same signature in the FSM.

The membership functions are applied in order to generate simple descriptions of regions in the space of uncertain input parameters (namely the set of residuals) that best describe the cases where the classical approaches are unsuccessful. Then, the rule base of the Fuzzy system is tuned to optimize both predictability and fault detection by decision making.

This chapter will be organized as follows. In the second part, a short introduction recalls the principles of Bond Graph (BG) methodology and the generation of Analytical Redundancy Relations (ARRs). Section 3 is devoted to the proposed Fuzzy logic methodology. In section 4, we illustrate the proposed approach to a traction system of an omni directional mobile robot named Robotino and finally section 5 concludes the chapter by highlighting the strengths of the proposed approach.

8.2. Bond Graph methodology

The Bond Graph Model (BGM) is a graphical description of dynamic behavior of physical systems. This means that systems from different domains (electrical, mechanical, hydraulic, thermodynamic ...) are described by physical values of the same type, effort and flow (see Table 1). Indeed, a BGM consists of subsystems linked together by a set of half arrows representing power bonds and direction of power (see Figure8.2.1). Each process is then described by a pair of variables (effort e and flow f) and their product is the power [31], [32]. Furthermore, Bond Graph modeling is a powerful tool for modeling engineering systems, especially when different physical domains are involved.

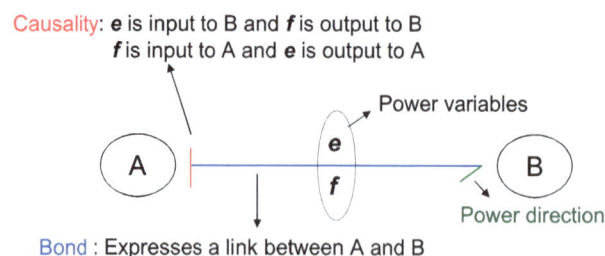

FIGURE 8.2.1. Principle of BG representation

System	Effort e	Flow f
Translational Mechanical	Force	Velocity
Rotational Mechanical	Torque	Angular velocity
Electrical	Voltage	Current
Hydraulic	Pressure	Volume flow rate
Thermal	Temperature	Entropy change rate
Chemical	chemical potential	Mass flow rate

TABLE 1. Effort and flow variables in some physical domains

8.2.1. Basic Elements of Bond Graph. Since details of the bond graph theory can be easily found in the literature [31], [32], this section briefly recalls the elementary notions of bond graph tool for modeling and explains how BG structural and causal properties can be used for Fault Detection and Isolation (FDI).

Definition The key of bond graph modeling is the representation (by a bond) of exchanged power as the product of two generic power variables (named efforts e and flows f) with elements acting between these variables and junction structures to reproduce the global system as interconnected subsystems. A BG can be considered as a graph $G(S, A)$ whose vertices S denote subsystems, basic elements (junction) or components while the edges A (called power bonds) represent the instantaneous mutual power transfer between nodes.

As shown in Figure 8.2.2a, the exchanged power between two systems ΣA and ΣB is represented by a bond labeled by two conjugated variables named effort (e) and flow (f). Effort is the intensive variable (*e.g.* pressure, electrical potential, temperature, chemical potential, force, torque, etc.) and flow is the derivative of extensive variable (*e.g.* volume flow, current, entropy flow, velocity, molar flow, etc.). Effort variable is labeled above the bond and flow variable is labeled below the bond. The positive direction of power flow (a generalized coordinate system used in bond graph models) is represented by the half-arrow on the bond.

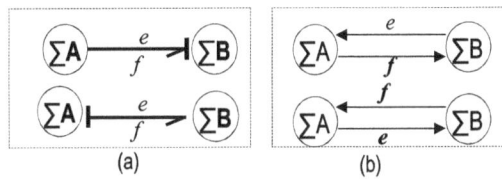

FIGURE 8.2.2. Bond graph definition and causality concept

From a supervision (control and diagnosis analysis) point of view, one important and powerful property of the bond graph is the concept of causality. Each hardware component is described by its constraints (called in bond graph constitutive equations). The constraint restricts the trajectory of the vector formed by the component variables to the space defined by the physical law associated to the component. A constraint is naturally acausal. But, if we need to simulate the physical phenomena (the model), we have to decide in which order the variables (effort and flow) will be computed. Consequently, we need to make a series of cause and effect decisions: this is the concept of causality. Indeed, the determination of causes and effects (by a covering causal path) in the system is directly deduced from the graphical representation. Causality is denoted by a cross-stroke (see Figure 8.2.2a). By convention, the causal stroke is placed near (respectively far from) the bond graph element (system $\sum$) for which the effort (respectively flow) is known. As in the example of Figure 8.2.2a, assigned causalities mean that system ΣA imposes effort (or flow) to system ΣB. The corresponding block diagrams are given by Figure 8.2.2b. The causal properties of the BG have been exploited for control analysis (structural observability and controllability) [14], and identification and model inversion using bicausality analysis [15]. Elements in a bond graph model are classified into three groups:

- **active elements** which supply power to the process (source of effort and flow noted Se, Sf)
- **passive elements** which transform the received power into a dissipated power ($R - element$), stored potential energy ($C - element$) or kinetics ($I - element$)

• **conservative multiport elements** used to reproduce the constraint architecture of the global system to be modeled. They consist of *junction structures, TransFormers* and *GYrators*. Junction structures are used to interconnect bond graph elements (R, C and I) by a "0" junction when they have a common effort and different flows, and by a "1" junction when they have a common flow and different efforts. TransFormer noted TF and GYrator (GY) are used to represent energy transformation from one domain to another. Sensors are modeled by effort (De) and flow (Df) detectors.

8.2.2. Structural Analysis based on bond graph.

The structural and causal properties of the bond graph can be used for FDI purposes. Before developing this part, the concept of structural analysis is reminded. Structural analysis rests on the concept of matching on a bipartite graph. A bipartite graph is a graph $G(C, A, Z)$ which represents the model structure of a system by two sets of vertices: the constraints $C = \{c_1, c_2, ..., c_m\}$ and the variables $Z = \{z_1, z_2, ..., z_n\}$, and the set of edges A which link the constraints to the variables. The set of edges A is defined as follows: $(c_i, z_j) \in A$ if the variable appears in the constraints c_i. The variables which can be quantitative, qualitative, fuzzy,... are partitioned into:

$$(8.2.1) \qquad\qquad Z = \{X\} \cup \{K\}$$

where $\{X\}$ is the subset of the unknown variables and $\{K\}$ is the subset of the known variables composed of control variables U and measured variables Y. In classical approaches [16], the bipartite graph is deduced from a set of equations obtained from the behavioral model of the system. In our case, it is directly deduced from the physical bond graph model. Indeed, the set of constraints C, from the bond graph theory includes information about the structure C_s, the behavior C_b, the measurement C_m and the control system C_c:

$$(8.2.2) \qquad\qquad C = \{C_s\} \cup \{C_b\} \cup \{C_m\} \cup \{C_c\}$$

Structural equations $\{C_s\}$ represent the conservation laws and are deduced from the junction equations, transformers and gyrators. The constraints $\{C_b\}$ are associated with the behavior model (expressing how the energy is transformed). In the BGM, they describe the physical phenomena which are represented in lumped-parameter bond graph elements (R, C, and I). Measurement constraints $\{C_m\}$ express the way in which the sensors transform some state variables of the process into output signals which can be used for FDI and control purposes. The unknown variables $\{X\}$ are the pair of power variables that label the bonds:

$$X = \{e_i, f_i\} \ (i = 1, n) \, , n \text{ is the number of passive bond graph elements.}$$

In the bond graph sense, the set of known variables $\{K\}$ are represented by: the flow (Df) and the effort (De) sensors, the flow (Sf) and the effort (Se) sources, the modulated flow (MSf) and effort (MSe) sources and the control inputs (u).

$$(8.2.3) \qquad K = \{Df\} \cup \{De\} \cup \{Sf\} \cup \{MSf\} \cup \{Se\} \cup \{MSe\} \cup \{u\}$$

Extra constraints can be added while the model is used in derivative causality for model based diagnosis. They correspond to analytical relations between the system's variables. Indeed, based on the structural analysis theory, it was shown in [16] that any system can be uniquely decomposed into an over-constrained (observable and redundant), a just-constrained (observable but not redundant) and an under-constrained (not observable and not redundant) subsystems. Only the over-constrained subsystem is monitorable. Compared to other graphical models such as Signed Directed Graph (SDG) [17], digraphs, bipartite graphs ...) where vertices correspond to variables and the edges represent mutual influence between those variables, BG is also a graph, but the vertices are BG element (physical components and junctions) labeled by power variables, and the edges represent the exchanged power.

Since its structural, behavioral and causal properties, BG can be used not only for dynamic modeling where it has proved its convenient but also for control analysis (observability, controllability, generation of transfer function, state equations) for deterministic [14] and uncertain models [18]. The next subsection shows how the BG is also an appropriate tool for the FDI system design.

8.2.3. Fault Indicators generation from bond graph. An interesting analogy exists between structural diagnosability based on canonical decomposition of bipartite graph methodology and bond graph based structural analysis. A BGM with correct causality means that the corresponding system of equations is solvable and then the set of unknown variables can be calculated.

Recall that FDI procedure consists of two steps: the detection and the isolation. In the detection step, the theoretical behavior of the system, described by the system constraints is compared to the actual behavior obtained from the known variables (the measured outputs and the control inputs). If a discrepancy exists, a fault is detected. In the isolation step, the nature of the violated constraints is exploited to isolate the faulty component. This detection and isolation can be performed only if some redundancy exists in the system. Analytical Redundancy Relations (ARRs) are the constraints that express this redundancy. They are derived from the over-constrained subsystem and are expressed in terms of known variables of the process. They have the form $f(K) = 0$. Numerical evaluation of an ARR yields a residual: $r = Eval\,[f(K)]$. K represents a set of bond graph elements.

Residuals are never equal to theoretical zero in any online application involving real measurements. Indeed, due to the sensor noises and parameter uncertainties, residual values contain small variances.

For illustration, consider the simple RL electrical example given in Figure 8.2.3-a. To avoid the problem of the initial conditions which are not known in real processes, ARRs are directly generated from the BG model given in derivative causality. Indeed, models are given in integral causality (Figure 8.2.3-b) when they are for physical simulation purpose while they are in derivative causality (Figure 8.2.3-c) when they are used for ARRs generation.

Here, the effort (or flow) detectors are transformed into signal sources $SSe = \widetilde{De}$ (or $SSf = \widetilde{Df}$) modulated by the measured value, as illustrated in Figure 8.2.3-c. This imposed signal is the starting point for the elimination of unknown variables.

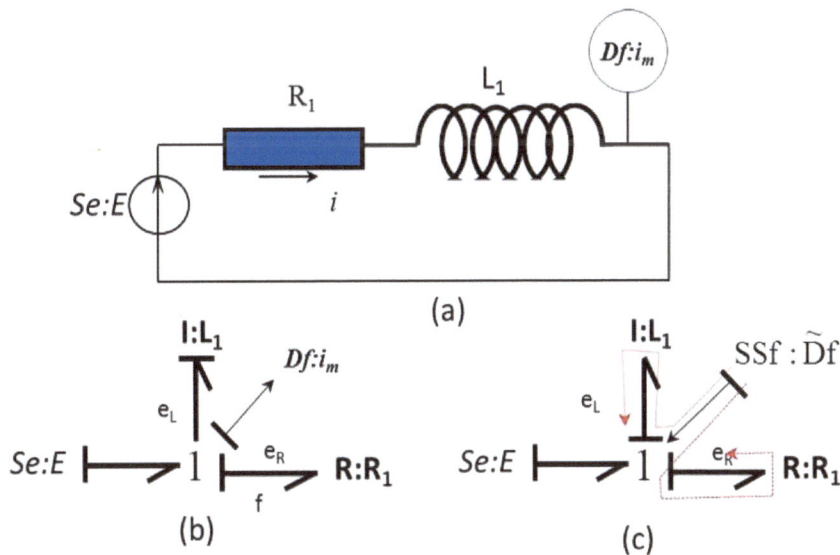

FIGURE 8.2.3. Electrical system (a) and its BG model in integral causality (b) with a flow sensor. and in derivative causality with a dualized flow sensor (c)

The determination of the ARRs on a BGM is based on elimination of the unknown variables using covering causal paths (from unknown variables to known ones) by using properties contained in the structural constraints of junctions 0 and 1. The candidate ARRs are generated from equations of the power balance which are represented in a BG by junction equations [19], [20]:

(8.2.4)
$$\sum b_i e_i = 0 \text{ for "1" junction}$$

(8.2.5)
$$\sum b_i f_i = 0 \text{ for "0" junction}$$

with i the number of the links connected to the junction and $b_i = \pm 1$ following the half-arrow orientation. From the BG model given by Figure 8.2.3b, the candidate ARR can be written as:

$$\sum b_i.e_i = Se - e_R - e_L = ARR_1$$

ARRs generation consists in eliminating unknown variables e_i by following the causal paths (as shown by the dashed lines in Figure 8.2.3c) from an unknown variable to a known one. e_L and e_R are then calculated (eliminated) using the following paths: $Df : i_m \rightarrow L_1 \frac{di_m}{dt} \rightarrow e_L$ and $Df : i_m \rightarrow R_1 i_m \rightarrow e_R$. We obtain the ARR given in Eq. 8.2.6

$$(8.2.6) \qquad\qquad Se - L_1 \frac{di_m}{dt} - R_1.i_m = ARR_1 = F(Df, R_1, L_1, Se)$$

The residual obtained by evaluating ARR_1 is sensitive to the faults which may affect the components (associated with their constraints) covered by the causal path during the elimination process, i.e. Df, R_1, L_1 and Se in our case. However, the elimination of the unknown variables on the considered causal constraint is not always possible. Consider the RLC electrical system given by Figure 8.2.4a with initially only one current sensor $Df : i_m$ and its BGM in integral causality given in Figure 8.2.4b.

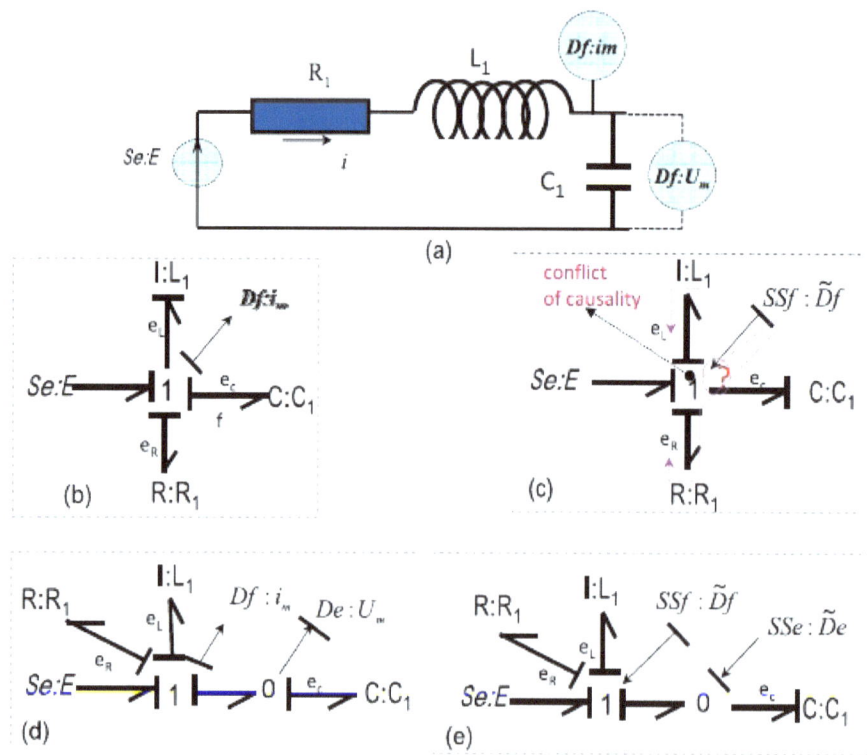

FIGURE 8.2.4. Electrical system (a) with a Bond Graph model in integral causality (b), with a conflict of causality (c) causally correct after dualizing the sensor (d) and causally correct after adding a new sensor (e).

In the presence of a C element (Figure 8.2.4c), a conflict of causality appears on the bond graph when trying to put both dynamic elements in derivative causality. It means that C-element has to stay in integral causality. ARR will depend on the initial effort $e_{C(0)}$: the subsystem is then non monitorable. To make it monitorable, a voltage sensor (for instance $Df : U_m$) must be added (Figure 8.2.4d) to break a loop in the oriented graph. The system is then observable but non redundant.

The ARRs generation algorithm can be summarized by the followings steps:

(1) Assign a preferred derivative causality on nominal BG; if it is possible (the model is over-constrained), then proceed with the following steps;

(2) Choose a junction from a bond graph model in derivative causality. Find the corresponding ARR by writing its characteristic equations and by identifying the specific observable sub-graphs. The latter can be found using different techniques proposed in [21] to generate ARRs from a bond graph model using covering causal paths. The goal is to study all of the causal paths relating the considered junction to the sources and the sensors.

(3) Move to the following junction. If the second ARR is independent of the first one, keep it, otherwise move to another junction,

(4) Repeat point (3) until all distinct signatures are obtained. This procedure is implemented in a software developed by one of the authors as a toolbox in Symbols2000 [22].

8.2.3.1. Detection and isolation procedure. Residuals lead to the formulation of a binary coherence vector $C = [c_1 \ c_2 \ \ldots \ c_n]$, whose elements, c_i $(i = 1, \ldots, n)$, are determined from a decision procedure, φ, which generates the alarm conditions. A simple decision procedure can be used for instance, $C = \varphi \ (r_1, r_2 \ldots, r_n)$, whereby each residual, r_i is tested against a threshold ε, fixed according to parameter uncertainties, sensor noises etc. In the simple decision procedure, ε can be taken equal to twice the standard deviation σ of the residual in an normal operating mode: $(\varepsilon = \pm 2\sigma)$.

Robust decision procedures minimize misdetection and false alarms by processing the residual noises. The detection procedure tests each residual r_i, against a fixed or adaptive threshold ε_i, to generate a coherence vector C. The elements of C, $c_i(i = 1, \ldots, n)$, are determined from :

$$c_i = \begin{cases} 1 \text{ if } r_i \text{ is bounded by } \varepsilon_i \\ 0 \text{ otherwise} \end{cases}$$

A fault is detected when $C \neq [\ 0 \ \ 0 \ \ \ldots \ \ 0\ ]$ which means that, at least one residual exceeds its threshold. Fuzzy logic reasoning can be used for robust alarm detection [23].

But usually a diagnostic result is a set of fault candidates F_c among the set of faults $F : F_c \subseteq F$.

The diagnostic system cannot distinguish between all fault candidates if the observed behavior occurs for more than one fault. The isolability can be, in this case, improved by adding more sensors, for example. It is the aim of the decision phase to try to isolate the fault. It rests, in this case, on logical analysis or pattern recognition approaches using a Fault Signature Matrix (FSM), which describes the participation of various components (physical devices, sensors, actuators and controllers, phenomena) in each residual r_i. The elements of FSM, say S, are determined as follows:

$$S_{ij} = \begin{cases} 1 \text{ if } r_i \text{ is sensible to the } f_j \\ 0 \text{ otherwise} \end{cases}$$

The fault f_j is isolated if its signature is different from the others. When a fault is detected and isolated, the type of fault can be determined using detailed models and signatures.

Because of the functional aspect of the Bond Graph, the generated formal ARRs can be associated systematically to the faults which may affect the considered system. Indeed, if the fault indicator is determined from zero junction representing the mass conservation, this residual is null if there is no leakage (conservative law is respected).

Therefore, compared with the other fault diagnosis methods, the merits of the Bond Graph based method are as follows:

(1) only one representation is used for modeling, formal ARRs generation and diagnosability analysis [22].

(2) formal ARRs and dynamic model (under state equation format) can be generated automatically using dedicated software [24], [25].

(3) the model is based on an energetic approach, which means that the topological, the physical and instrumentation architecture are clearly displayed in the graph.

(4) the structural diagnosability can be determined from the graphical model (no need of numerical calculation).

(5) because of modular and functional aspect of the Bond Graph, the ARRs are systematically associated to specific (sensor, actuator and other hardware component) faults which may affect the system.

   However some of drawbacks can be cited:

   • the derived properties are available only structurally (to be verified numerically).

- the model is developed under lumped parameter approximation.

Inspection of ARRs with respect to the known variables and component parameters they link leads to a so-called fault signature matrix (FSM). Looking at ARRs, an occurrence matrix can be set up with one row for each known variable or component parameter and one column for each residual. A known variable or component parameter present in an ARR is indicated by $'1'$, its absence by $'0'$. That is, the resulting matrix shows which components contribute to which residuals. The rows in a structural Fault Signature Matrix (FSM) display the fault signatures of components. We propose in this chapter to use Fuzzy logic for robust decision making, by using both the logic allowing to build the FSM and the robust thresholds determined from the BGM, as an expert system. In fact, in Fuzzy logic, the two valued logic 0 and 1 are expanded to any continuous valued logic among the closed interval $[0, 1]$ by taking into account a certain threshold chosen from experimentation, and considering the modeling uncertainties, unknown inputs and measurement noises. Fuzzy Logic Methodology is described in more details in the next section.

8.3. Fuzzy logic methodology

To illustrate the fuzzy logic approach, let consider a basic example dealing with the evaluation of the speed of a car by taking into account the temperature and the cloud cover. Humans say, for example, things like "if it is sunny and warm today, I will drive fast" rather than "if the temperature is equal to $70°F$ and if the cloud cover is equal to 20% then I will drive at $70mph$". Fuzzy logic is a mean to represent information that is approximate rather than fixed and exact. A fuzzy logic system (see Figure 8.3.1) rests on three components namely, the fuzzifier, the inference engine implemented on a fuzzy rule base and the defuzzifier [27].

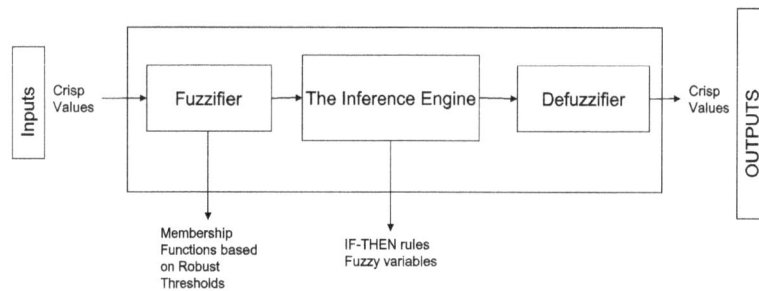

FIGURE 8.3.1. Fuzzy Logic System.

The fuzzifier receives in input crips variables which represent precise quantities and uses membership functions to assign to each variable a linguistic value. For the car example, the linguistic variables and their associated linguistic values are:

- Temp: freezing, cool, warm, hot
- Cloud Cover: overcast, partly cloudy, sunny
- Speed: slow, fast

Membership functions define fuzzy sets for the linguistic variables. If X is the universe of discourse and its elements are denoted by x, then a fuzzy set A in X is defined as a set of ordered pairs:

$$A = \{x, \mu_A(x) | x \in X\}$$

$\mu_A(x)$ is called the membership function (or MF) of x in A. The membership function maps each element of X to a membership value between 0 and 1. It is worth noting that the intersection of two fuzzy sets A and B is specified in general by a binary mapping T, which aggregates two membership functions as follows: $\mu_{A \cap B}(x) = T(\mu_A(x), \mu_B(x))$. Like fuzzy intersection, the fuzzy union operator is specified in general by a binary mapping S: $\mu_{A \cup B}(x) = S(\mu_A(x), \mu_B(x))$. The membership functions of variables used for the car example (temp, cloud and speed) are given on Figure 8.3.2 and Figure 8.3.3. They define four fuzzy sets for the variable Temp, three for the variable Cover and two for the variable Speed. If the crips value of the temperature is equal to $65°F$, the linguistic values (or fuzzy set values) returned by the fuzzifier are: $hot = 0$, $warm = 0.7$, $cool = 0.2$, $freezer = 0$.

Temp: {Freezing, Cool, Warm, Hot}

Cover: {Sunny, Partly, Overcast}

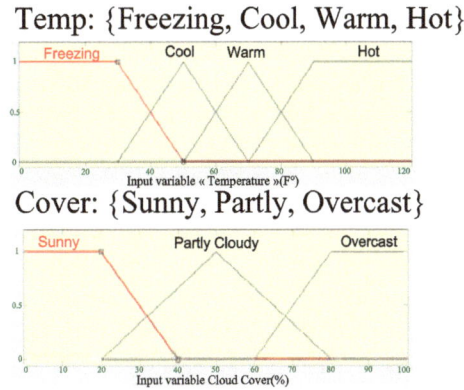

FIGURE 8.3.2. Membership functions: Input variables

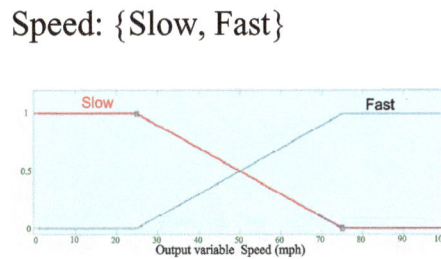

Speed: {Slow, Fast}

FIGURE 8.3.3. Membership functions: output variable

The **inference Engine** simulates the human reasoning by using IF-THEN rules provided by experts and stored in a knowledge base. In its simplest form, a fuzzy if-then rule follows the pattern: Ïf x is A then y is Bẅhere:

- 'if x is A' is the antecedent and 'y is B' is the consequent.
- A and B are linguistic values defined by fuzzy sets in the universes of discourse X and Y.
- x is the input variable and y is the output variable.

Note that in the if-then rule, the word 'is' gets used in two entirely different ways depending on whether it appears in the antecedent or in the consequent part of the rule. In MATLAB terms, it corresponds to the distinction between a relational test using "==" and a variable assignment using the "=" symbol. A less confusing way of writing the rule would be: "If x==A then y=B". The input to the rule is a linguistic value obtained from the crisp value of variable x through the fuzzification process. The result of the rule is a fuzzy set assigned to the output variable y. There are several ways to define the result of a rule. The rule is executed applying a fuzzy implication operator, whose arguments are the antecedent's value and the consequent's fuzzy set values. Given the consequent of each rule (a fuzzy set) and the antecedent value, the fuzzy implication operator is applied to obtain a new fuzzy set. Two of the most commonly used implication operators are the minimum, which truncates the consequent's membership function, and the product, which scales it. Let now apply this process on the car example, to determine the speed through the two following rules, in the case where the temperature is equal to $65°F$ and the cloud cover is equal to 20%:

- R1: If it is Sunny and Warm then drive Fast
- R2: If it is Cloudy and Cool then drive Slow

Thanks to the fuzzifier, the membership values are obtained from the crips values (see Figure 8.3.4). As a result, the antecedent of each rule becomes:

- R1: Sunny=0.8 and Warm=0.7
- R2: Cloudy=0.2 and Cool=0.4

As the two parts of the antecedent are joined by a conjunction ('and'), a AND operation: the minimum, is applied to both membership values to obtain the membership value 0.7, for the antecedent part of R1 and 0.2 for the antecedent part of R2. If the antecedent were joined by a disjunction 'or', a OR operation: the maximum, would be applied.

The antecedent value is used by the implication operator to define the fuzzy set of the output. We chose, for our example to truncate the consequent's membership function by applying the Mamdani Minimum implication operator defined as: $\phi[\mu_A(x), \mu_B(y)] \equiv \mu_A(x) \wedge \mu_B(y)$. In other words, the antecedent value determines the truncation threshold. Because decisions are based on the testing of all of the rules, the fuzzy sets that represent the results of the rules are then aggregate into a single fuzzy set. The input of the aggregation process is the list of truncated output functions returned by the implication process for each rule. The output of the aggregation process is one fuzzy set for each output variable (see Figures 8.3.4 and 8.3.5).

The **defuzzifier** combines the results of all the rules to give to the output variable (speed) a crips value. Because decisions are based on the testing of all of the rules, the fuzzy sets that represent the results of the rules are then aggregate into a single fuzzy set. The input of the aggregation process is the list of truncated output functions returned by the implication process for each rule. The output of the aggregation process is one fuzzy set for each output variable (see Figures 8.3.4 and 8.3.5). Several methods exist to calculate the crips value of the output. Among them, the "centroid" method is very popular, in which the "center of mass" of the fuzzy set resulting of the aggregation procedure is calculated (see Figures 8.3.5 and 8.3.6).

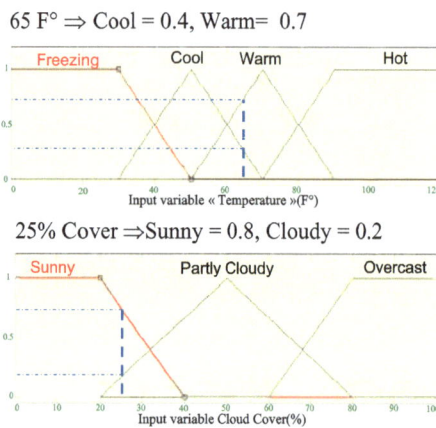

FIGURE 8.3.4. Fuzzification

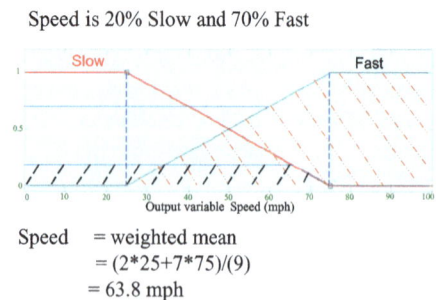

FIGURE 8.3.5. Defuzzification

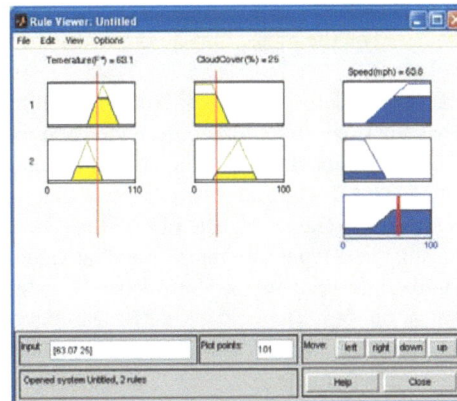

FIGURE 8.3.6. Defuzzification and rule viewer

8.3.1. Fuzzy logic characteristics. As shown by the previous example, fuzzy logic is concerned with the continuous transition from truth state to falsity state, as opposed to the discrete true/false transition in binary logic. It deals with many sorts of vagueness and uncertainty in data and information about a specific problem. Most engineering applications of fuzzy logic belong to "Linguistic Mathematics". In the meantime, it maintains a formal structure of logical operations as well as a rigorous axiomatic framework like classical binary logic. The possibility theory of fuzzy logic provides a measure of the potential ability of a subset in belonging to another subset [**34**]. It can be shown that probability theory is a special case of possibility theory. In this subsection, we propose an overview of the main characteristics of the Fuzzy Logic:

- The mathematical concepts behind fuzzy reasoning are very simple. This is why it is conceptually easy to understand.
- Fuzzy logic is based on natural language. The basis for fuzzy logic is the basis for human communication.
- Fuzzy logic is tolerant of imprecise data. Everything is imprecise if you look closely enough, but more than that, most things are imprecise even on careful inspection. Fuzzy reasoning builds this understanding into the process rather than tacking it onto the end.
- Fuzzy logic is a convenient way to map an input space to an output space. Mapping input to output is the starting point for everything. Consider the following examples:
  - With your specification of how hot you want the water, a fuzzy logic system can adjust the faucet valve to the right setting
  - With information about how fast the car is going and how hard the motor is working, a fuzzy logic system can shift gears for you.

8.3.2. Fuzzy logic in the context of diagnosis. The essence of fault diagnosis based on Fuzzy Logic is to accumulate the operating experience to organize the control structure to improve control performance automatically [**26**]. Especially, it appears very useful when the system parameters are uncertain or imprecise or when the system is too complex to be analyzed by mathematical methods. Nevertheless, because this methodology does not use structural information but the expert's experience, its several steps in design are difficult to generalize. In contrast, BGM contains exactly the structural information with which to generalize design steps for fuzzy logic robust decision making based detection. This is why, we proposed in this chapter to combine BGM and Fuzzy Logic in order to detect more accurately the fault when it appears.

8.3.2.1. *The Fuzzification.* By referring to the fuzzy logic system given on Figure 8.3.1, the fuzzifier receives in input the crips values of the residuals generated y the ARRs. A qualitative category is defined for each one of them based on the threshold values for the inputs (residuals) according to the residual sensitivity to certain faults and a degree of detectability for the outputs (faults having the same signature in the FSM). The shape of these functions can be diverse, but in our work, only triangles and trapezoids are used. Indeed, a residual will be coherent with the model of the system if it is null or inferior to a chosen threshold and it represents a fault indicator that reflects the faulty situation of the monitored system. The residual can be also positive or negative. This is why, we associate for example for each value taken by the residual a membership function corresponding to "Negative", "Zero" or "Positive" value and the point definitions of each MF is based on the robust thresholds and experimentations in both normal and faulty situations.

8.3.2.2. Base rule. In our context, a rule derives operating knowledge from given residuals deduced from the BGM and is generated from the expert knowledge. In our case, from the FSM logic and experimentation which have been done by introducing some different faults and analyzing the effect on each residual. The rule based structure is made using the Fuzzy Logic toolbox of MATLAB. The Fuzzy outputs are distinguished in three levels (Insensitive, Uncertain, and Sensitive) in order to be used by the diagnosis. The Fuzzy Rule-Based algorithm embedded in MATLAB Fuzzy Logic toolbox has the following steps: the inputs are transformed into memberships of fuzzy sets by fuzzifying functions. This information is given to the inference engine. Subsequently, membership values are transformed into required output variables by a defuzzification step. The Fuzzy Rule Base is characterized by the construction of a set of linguistic rules based on expert knowledge. The expert knowledge is usually in the form of IF-THEN rules, which can be easily implemented by fuzzy conditional statements.

8.3.2.3. Defuzzification. The MFs of the output have always the same shape and configuration in our risk model: the risk of any problem has the same ranks for the MFs of the output: "Insensitive", "Uncertain" and "Sensitive" and always without overlapping. In order to obtain a percentage of detectability of each set of faults, the output is defuzzyfied. The equations of the straight lines of each MF of the output are calculated. The calculations for each of the MFs are done by using the formalism described by Mamdani. In fact, Mamdani FIS is the most known or used in developing fuzzy models. The output of the system is generally defuzzified resulting in fuzzy sets which are combined using aggregation operators from the consequent of each rule of the input as used by [28]. Basically, the comparison of the outputs allows the identification of the higher value corresponding to the fault signature detected in the system.

8.4. Application

The robust decision making approach for fault detection developed in this chapter is applied to a traction system of a mobile robot named Robotino (see Figure 8.4.1). Robotino is an autonomous omni directional three wheel drive mobile robot. For actuation, three DC motors with each fitted with two sensors (measuring the current and the angular speed) and a reducer, are used. For our experimentation, we focus on the electromechanical traction system of Robotino composed of the DC motor, the two sensors and the wheel. Furthermore, we apply different kind of faults to Robotino in order to demonstrate the consistency of our approach.

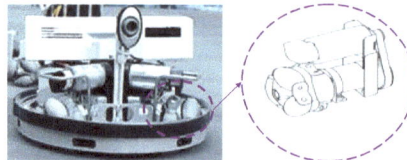

$$L_a = 8.9\text{mH} \quad ; m = 43.1\text{mV} \; ; J_s = 63 \times 10^{-3} \text{Kg.m}^2$$
$$R_a = 8.13 \; \Omega \; ; r = 16 \quad ; R_s = 0.02 \text{ Nm.sec.rad}^{-1}$$
$$R_e = 47 \; \mu\text{Nm.sec.rad}^{-1} \quad ; \quad J_e = 7.95 \times 10^{-6} \text{Kg.m}^2$$

FIGURE 8.4.1. Traction System in Omnidirectional Mobile Robot: Robotino

8.4.1. System Bond Graph Modelling. The Figure 8.4.2 (resp. Figure 8.4.3) gives the BG model of the Robotino traction system in integral (resp. derivative) causality. The electric power is provided by the electrical part of a DC motor which is equivalent to an input voltage source U in serial with a resistance R_a and an inductance L_a. The electrical current is measured by the sensor $Df : i$. The gyrator element GY describes the power transformation from the electrical part of the DC motor to its mechanical part which is characterized by its rotor inertia J_e, its viscous friction parameter R_e and the detector $Df_2 : \dot{\theta}_m$. The mechanical gear which links the mechanical and wheel parts is represented by a transformer element TF. The wheel is characterized by its inertia J_s, its viscous friction parameter R_s and a contact effort F_x which is calculated by using the so called Pacejka magic formula for the estimation of the longitudinal force [33].

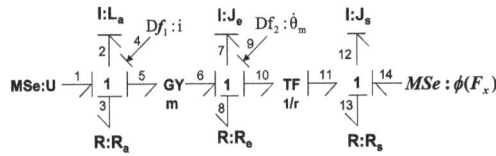

FIGURE 8.4.2. BG Model in preferred integral causality

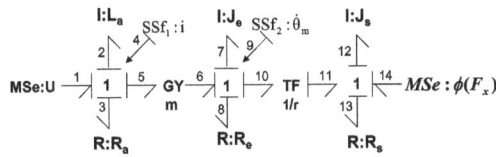

FIGURE 8.4.3. BG Model in derivative causality

8.4.2. ARRs Generation. From the model of Figure 8.4.3, the constitutive relations of the junctions, the gyrator and of the transformers are as follows:

$$\text{Junction } 1_1 : \begin{cases} e_4 = e_2 + e_3 + e_5 - e_1 \\ f_4 = f_1 = f_2 = f_3 = f_5 \end{cases}$$

$$\text{Junction } 1_2 : \begin{cases} e_9 = e_7 + e_8 + e_{10} - e_6 \\ f_9 = f_7 = f_8 = f_{10} = f_6 \end{cases}$$

$$\text{Junction } 1_3 : \begin{cases} e_{11} = e_{12} + e_{13} - e_{14} \\ f_{11} = f_{12} = f_{13} = f_{14} \end{cases}$$

$$\text{Gyrator } GY : m : \begin{cases} e_6 = \frac{1}{m}.f_5 \\ e_5 = \frac{1}{m}.f_6 \end{cases}$$

$$\text{Transformer } TF : r : \begin{cases} e_{11} = \frac{1}{r}.e_{10} \\ f_{10} = \frac{1}{r}.f_{11} \end{cases}$$

The elements R, C, I of the BG introduce the following behavioral equations:

$$\begin{aligned} \text{Element } R : R_a \quad & e_3 = R_a.f_3 \\ \text{Element } I : L_a \quad & f_2 = \frac{1}{L_e}.\int e_2 dt \Rightarrow e_2 = L_e.\dot{f}_2 \\ \text{Element } R : R_e \quad & e_8 = F_e.f_8 \\ \text{Element } I : J_e \quad & f_7 = \frac{1}{J_e}.\int e_7 dt \Rightarrow e_7 = J_e.\dot{f}_7 \\ \text{Element } I : J_s \quad & f_{12} = \frac{1}{J_s}.\int e_{12} dt \Rightarrow e_{12} = J_s.\dot{f}_{12} \\ \text{Element } R : R_s \quad & e_{13} = F_s.f_{13} \end{aligned}$$

The set $K = \left\{ U, \phi(F_x), i, \dot{\theta}_m \right\}$ is the set of the known variables corresponding to the sources and the detectors. From this set, the following equations are introduced:

$$\begin{aligned} \text{Source } MS_e : U \quad & e_1 = U, f_1 = 0 \\ \text{Detector } SSf_1 : i \quad & f_4 = i, e_4 = 0 \\ \text{Detector } SSf_2 : \dot{\theta}_m \quad & f_9 = \dot{\theta}_m, e_9 = 0 \\ \text{Source } MS_e : \phi(F_x) \quad & e_{14} = \phi(F_x), f_{14} = 0 \end{aligned}$$

Two ARRs are established linking only the known variables and the parameters. They are:

$$ARR_1 : U - R_a.i - L_a.\frac{d}{dt}(i) - m.\dot{\theta}_m = 0$$
$$ARR_2 : m.i - J_e\frac{d}{dt}(\dot{\theta}_m) - R_e\dot{\theta}_m - \frac{J_s}{r^2}.\frac{d}{dt}(\dot{\theta}_m) - \frac{R_s}{r^2}.\dot{\theta}_m + \frac{1}{r}.\phi(F_x) = 0$$

The Fault Signature Matrix is then constructed by stacking the coherence vectors as seen in Table 2. The matrix is extended by two additional columns. The first one with the heading $'M'$ indicates whether a fault can be detected. The second additional column with the heading $'I'$ indicates whether a fault can be isolated.

	R_1	R_2	M	I
L_a	1	0	1	0
R_a	1	0	1	0
MS_e	1	0	1	0
m	1	1	1	0
SSf_1	1	1	1	0
SSf_2	1	1	1	0
R_e	0	1	1	0
J_e	0	1	1	0
J_s	0	1	1	0
R_s	0	1	1	0
F_x	0	1	1	0

TABLE 2. Fault Signature Matrix

Using the procedure of input and output uncertainties modeling developed in [**29**], the following thresholds can be generated directly from the graphical model using the causal paths:

$$a_1 = \max\left(-R_a\zeta_i - L_a\frac{d\zeta_i}{dt} - m\zeta_{\dot{\theta}_m}\right);$$
$$= R_a\max(\zeta_i) + L_a\max\left(\frac{d\zeta_i}{dt}\right) + m\cdot\max\left(\zeta_{\dot{\theta}_m}\right);$$
$$a_2 = \max\left(-m\zeta_i - R_e\zeta_{\dot{\theta}_m} - J_e\frac{d\zeta_{\dot{\theta}_m}}{dt} - \frac{J_s}{r^2}\frac{d\zeta_{\dot{\theta}_m}}{dt} - \frac{R_s}{r^2}\zeta_{\dot{\theta}_m}\right);$$
$$= m\cdot\max(\zeta_i) + R_e\max\left(\zeta_{\dot{\theta}_m}\right) + J_e\max\left(\frac{d\zeta_{\dot{\theta}_m}}{dt}\right)$$
$$+ \frac{J_s}{r^2}\max\left(\frac{d\zeta_{\dot{\theta}_m}}{dt}\right) + \frac{R_s}{r^2}\max\left(\zeta_{\dot{\theta}_m}\right).$$

where ζ_i and $\zeta_{\dot{\theta}_m}$ are the measurement errors on the current and velocity detectors respectively. And where:

$$\max\left(\frac{d\zeta_{\dot{\theta}_m}}{dt}\right) = \frac{2\cdot\max\left(\zeta_{\dot{\theta}_m}\right)}{\Delta t}$$
$$\max\left(\frac{d\zeta_i}{dt}\right) = \frac{2\cdot\max(\zeta_i)}{\Delta t}$$

and Δt is the sampling time.

8.4.3. Fuzzy Logic approach. The rule based structure is made using the Fuzzy Logic toolbox for MATLAB. The Fuzzy outputs are associated with three qualitative values (Insensitive, Uncertain, and Sensitive) in order to be used for the diagnosis. The Fuzzy Rule Base algorithm embedded in MATLAB Fuzzy Logic toolbox consists of the following steps: the inputs are transformed into memberships of fuzzy sets by fuzzifying functions. This information is given to the inference engine. Then, membership values are transformed into required output variables by a defuzzification step. The main objective is to ensure more accurately the degree of detectability of the output variables of the fuzzy system, which correspond to the signature of a fault occurring in such parameters. The residuals are the input variables considered in the system by taking into account the thresholds of each residual and different ranges that can be reached.

Three types of membership functions having three membership functions in each namely combinations of trapezoidal and triangular members are considered. Linguistic variables such as "Negative", "Zero", "Positive" and "Insensitive", "Uncertain", "Sensitive" have been considered for evaluating respectively residuals and the degrees of detectability of different faults.

Some of the fuzzy rules (see Figure 8.4.4) are activated according to the information acquired by the fault signature matrix logic. The outputs of the activated rules are weighted by fuzzy reasoning and the degrees of detectability of different faults are calculated. The parameters defining the input and output functions are seen in Figure 8.4.5 where a_1 and a_2 represent the robust thresholds for each residual.

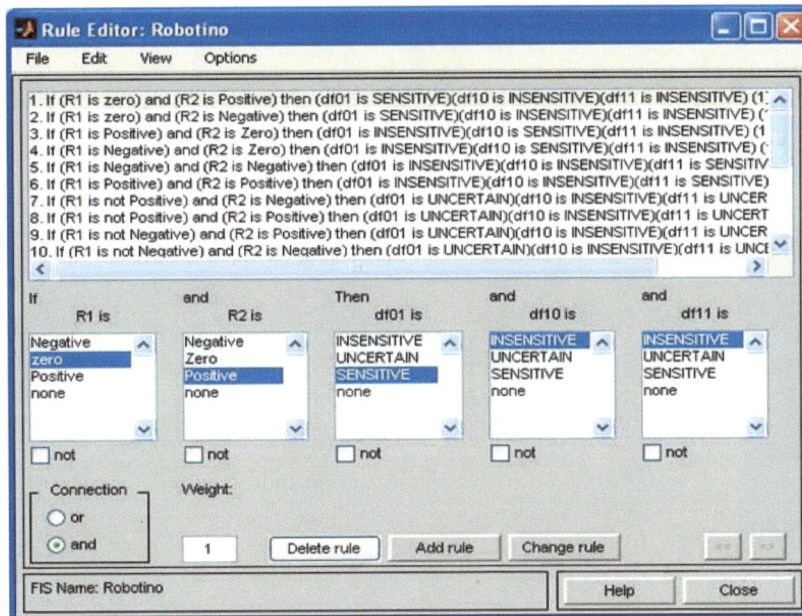

FIGURE 8.4.4. Overview of some rules used in the Fuzzy Mechanism

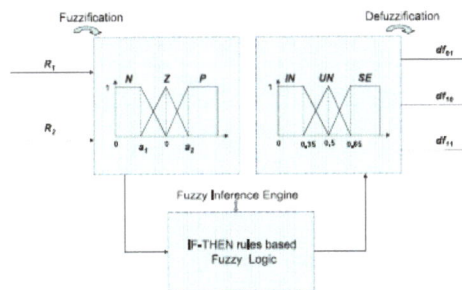

FIGURE 8.4.5. Fuzzy Logic Approach and Membership Functions

8.4.4. Experimental Results. The SIMULINK diagram (see Figure 8.4.11) consists of two parts; the first one is the Robotino model with its inputs and the second is the Fuzzy Block Controller which takes into account the residuals deduced from BG as explained before and provides the outputs namely the degrees of detectability of each residual. This shows that, the comparison between these different outputs allows a robust decision making in terms of which signature is observed and therefore a fault is detected. Figure 8.4.6 corresponds to the residuals' response in normal operation. This shows that, many fault scenarios have been tested. For instance, in Figure 8.4.7, the response of the residuals to a fault in the current sensor which is introduced during a 9 s time interval, from 5 s to 14 s is presented. The comparison of the three outputs of the Fuzzy system (see Figure 8.4.8) reveals that the signature observed is "f_{11}" which correspond to the introduced fault. Figure 8.4.9 shows the response of the residuals to a fault in the voltage which is introduced during the same time interval. The comparison of the three outputs of the Fuzzy system (see Figure 8.4.10) reveals that the signature observed is "f_{10}" which correspond to the voltage fault. Furthermore, whatever the values taken by the residuals, the fuzzy outputs allow the accurate detection of the signature corresponding to the fault when it occurs.

The Rule Viewer seen in Figure 8.4.12 depicts the fuzzy inference diagram for an FIS stored in a file. It is used to view the entire implication process from beginning to end. Indeed, we can move around the line indices that correspond to

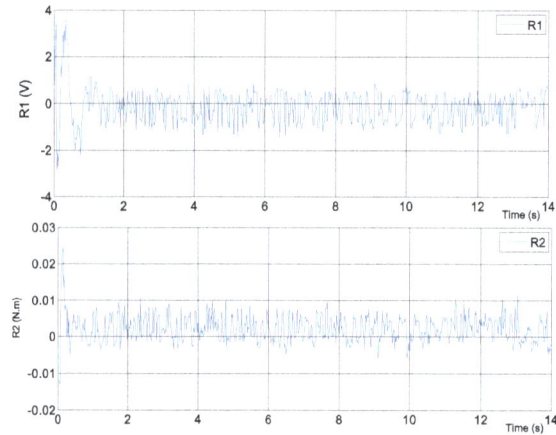

FIGURE 8.4.6. Residuals in normal functioning

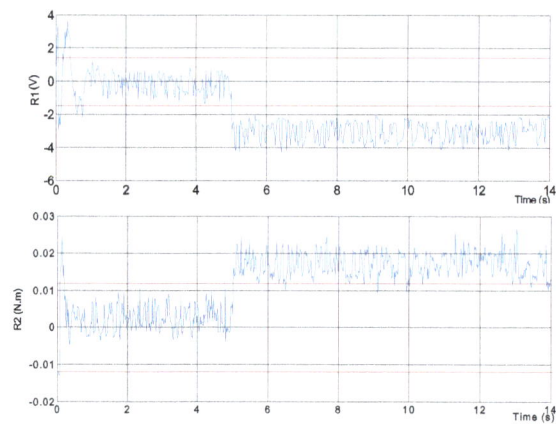

FIGURE 8.4.7. Residuals in faulty situation (current fault)

the inputs and then watch how the system readjusts and computes the new output. Finally, The Surface Viewer as seen in Figure 8.4.13 is a Graphical User Interface (GUI) tool that lets you examine the output surface of an FIS stored in a file for any one or two inputs. Since it does not alter the fuzzy system or its associated FIS structure in any way, it is a read-only editor.

8.5. Conclusion

A robust decision making approach for fault detection based on fuzzy evaluation of residuals has been elaborated and implemented for an electromechanical system. The main contribution of this study concerns the use of a behavioral model namely Bond Graph (BG) and the definition of membership functions by using the robust thresholds deduced from BG in order to detect more accurately a fault upon occurrence. The results of implementation in a real system reveal the quality of fuzzy logic in diagnosing the faults especially for the cases that crisp logic is not able to detect a fault accurately.

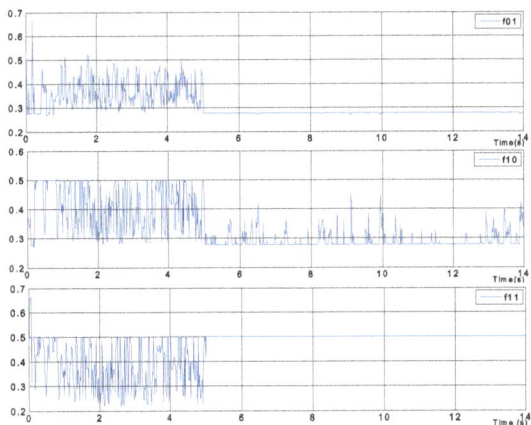

FIGURE 8.4.8. Fuzzy system outputs for current fault

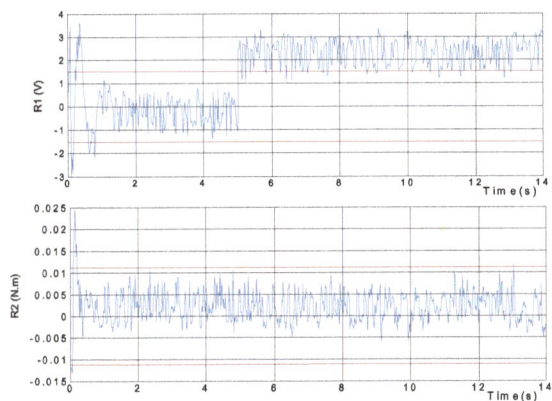

FIGURE 8.4.9. Residuals in faulty situation (Voltage fault)

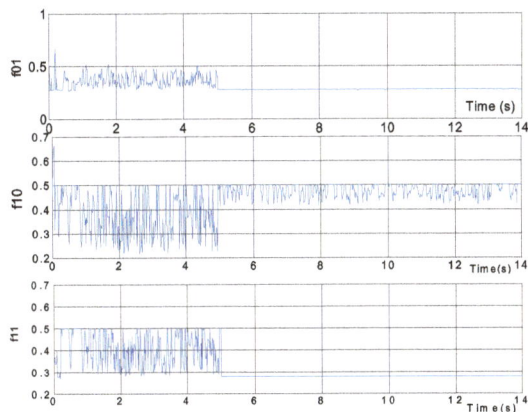

FIGURE 8.4.10. Fuzzy system outputs for voltage fault

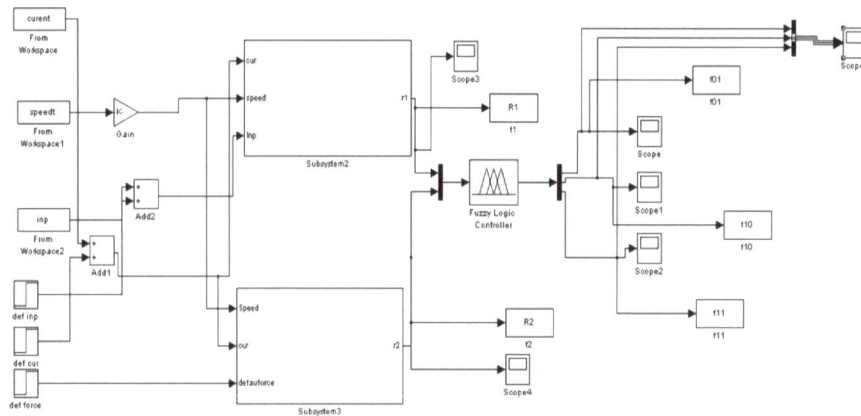

FIGURE 8.4.11. Block diagram showing the implementation of the developed approach in SIMULINK

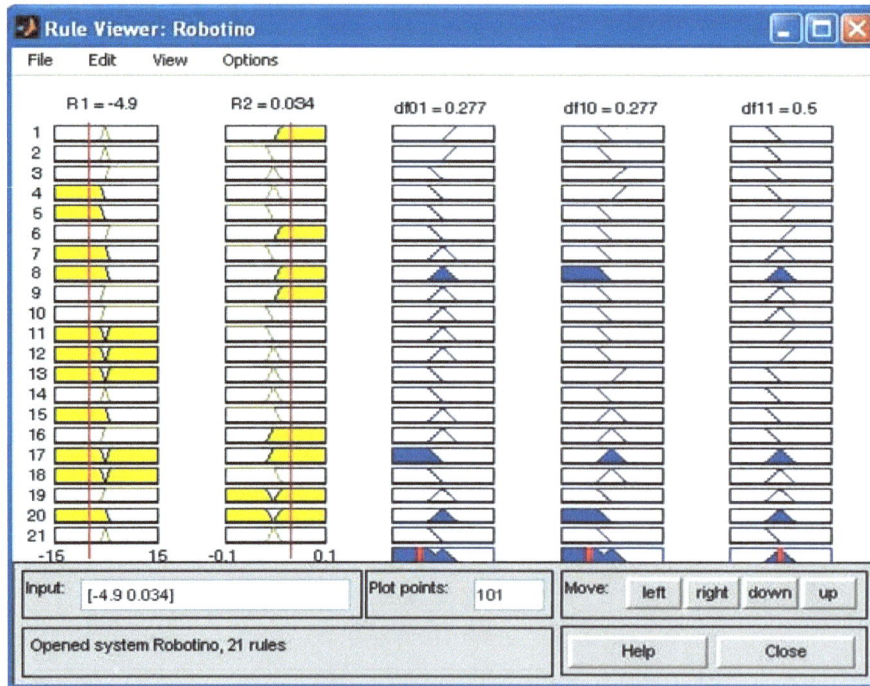

FIGURE 8.4.12. Fuzzy toolbox rule viewer

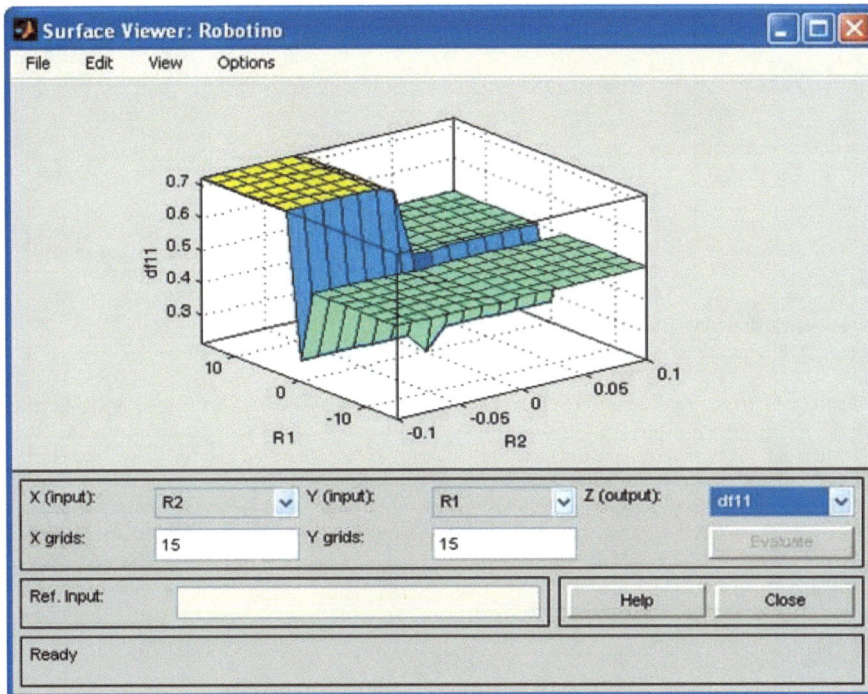

FIGURE 8.4.13. Fuzzy toolbox surface viewer

Bibliography

[1] N. Chatti, A-L. Gehin, R. Merzouki and B. OuldBouamama, *Online supervision of intelligent vehicle using functional and behavioral models*. IEEE Intelligent Vehicles Symposium (IV), 2011, pp.827-832.

[2] L.J. De Miguel and L.F. Blazquez, *Fuzzy logic-based decision-making for fault diagnosis in a DC motor*. Eng. Appl. Artif. Intell. 2005, vol.18, pp.423-450.

[3] O. Adort, D. Maquin, and J. Ragot, *Fault detection with model parameter structured uncertainties*. Proceeding Eur. Control Conf. Karlsruhe, Germany, 1999.

[4] M.A. Djeziri, R. Merzouki, B.Ould-Bouamama, and G. Dauphin-Tanguy, *Robust Fault Diagnosis by Using Bond Graph Approach*. Mechatronics, IEEE/ASME Transactions on Mechatronics. 2007, vol.12, pp.599-611.

[5] B.Ould-Bouamama, and A. Samantaray, *Model-based Process Supervision. A Bond Graph Approach*. Spring-Verlag, 2008.

[6] R. Merzouki, M.A. Djeziri, and B.Ould-Bouamama, *Intelligent monitoring of electric vehicle*. IEEE/ASME International Conference on Advanced Intelligent Mechatronics, 2009, pp.797-804.

[7] C.Boon-low, D.Wang, Arogeti, and M.Luo, *Quantitative Hybrid Bond Graph-Based Fault Detection and Isolation*. MIEEE Transactions on Automation Science and Engineering, 2010, vol.7, pp.558-569.

[8] B. Ould-Bouamama, K. Medjaher, M. Bayart, A.K. Samantaray, *Fault detection and isolation of smart actuators using bond graphs and external models*. Control Engineering and Practice, 2005, pp.159-175.

[9] P.J. Feenstra, P.J. Mosterman, G. Biswas, and P.C. Breedveld, *Bond graph modeling procedures for fault detection and isolation of complex flow processes*. Proc. ICBGM'01, Simulation Series, 2001, vol.33, pp.77-82.

[10] W. Bouallegue, S-B. Bouabdallah and M. Tagina, *Diagnosis of Bond Graph modeled uncertain parameters systems using residuals sensitivity*. IEEE International Conference on Systems Man and Cybernetics (SMC), 2010, pp.593-600.

[11] D. Henry and A. Zolghari, *Norm-based design of robust FDI schemes for uncertain systems under feedback control: Comparison of two approaches*. Control Eng. Pract, 2006, vol.14, pp.1081-1087.

[12] M.A. Djeziri, R. Merzouki, B.Ould-Bouamama, and G. Dauphin-Tanguy, *Bond Graph Model Based For Robust Fault Diagnosis*. Proceeding of the American Control Conference, New York City, USA, 2007, pp.3017-3022.

[13] L.J. De Miguel and L.F. Blazquez, *Fuzzy logic-based decision-making for fault diagnosis in a DC motor*. Eng. Appl. Artif. Intell., 2005, vol.18, pp.423-450.

[14] G. Dauphin-Tanguy, A. Rahmani, C. Sueur, *Formal determination of controllability/observability matrices for multi variables systems modeled by bond graph*. In International IMACS/SILE Symposium on Robotics, Mechatronics and Manufacturing Systems, 1992, pp.573-578.

[15] P. Gawthrop, *Bicausal bond graphs*. International conference on bond graph modeling and simulation. New York City, USA, 1995, pp.83-88.

[16] M. Blanke, M. Kinnaert, J. Lunze, M. Staroswiecki, J. Schreider, *Diagnosis and Fault-Tolerant Control*. Springer-Verlag New York, 2006, USA.

[17] M. Maurya, R. Rengaswamy, V. Venkatasubramanian, *A signed directed graph and qualitative trend analysis-based framework for incipient fault diagnosis*. Chemical Engineering Research and Design, 2007, vol.85, pp.1407-1422.

[18] M.-A. Djeziri, R. Merzouki, B. Ould-Bouamama, G. Dauphin-Taguy, *Robust fault diagnosis by using bond graph approach*. Mechatronics, IEEE/ASME Transactions on Mechatronics, 2007, vol.12, pp.599-611.

[19] A. Samantaray, K. Medjaher, B. Ould-Bouamama, M. Staroswiecki, G. Dauphin-Tanguy, *Diagnostic bond graphs for online fault detection and isolation*. Simulation Modelling Practice and theory, 2006, vol.14, pp.237-262.

[20] B. Ould-Bouamama, K. Medjaher, A. Samantaray, M. Staroswiecki, *Supervision of an industrial steam generator. part 1: Bond graph modelling*. Control Engineering Practice (CEP), 2005, vol.14, pp.71-86.

[21] M. Tagina, J. Cassar, G. Dauphin-Tangy, M. Staroswiecki, *Monitoring of systems modelled by bond graph*. International Conference on Bond Graph Modeling, 1995, pp.275-280.

[22] B. Ould-Bouamama, A. Samantaray, Staroswiecki, *Software for supervision system design in process engineering*. IFAC World Congress, 2006, pp.691-695.

[23] L.-J. De Miguel, L.-F. Blazquez, *Fuzzy logic-based decision-making for fault diagnosis in a dc motor*. Eng. Appl. Artif. Intell., 2005, vol.18, pp.423-450.

[24] A. Mukherjee, A. Samantaray, *System modelling through bond graph objects on symbols 2000*. In Int. Conf. Bond Graph Modeling and Simulation, 2001, vol.33, pp.164-170.

[25] J. F. Broenink, *Modelling, simulation and analysis with 20-sim*. Journal A Special Issue CACSD, 1997, vol.38, pp.22-25.

[26] G. Zhang, S.Ibuka, and K. Yasuoka, *Application of Fuzzy Data Processing for Fault Diagnosis of Power transformers*. Proceeding of IEEE Conference Publication, High Voltage Engineering Symposium, 1999, pp.22-27.

[27] L.A. Zadeh, *Fuzzy sets, Information and Control*. Information and control, 1965, vol.8, pp.338-353.

[28] T. Akter, and S.P. Simonovic, *Aggregation of Fuzzy Views of a Large Number of Stake Holders for Multi-objective Flood Management Decision Making*. Journal of Environmental Management, 2005, vol.77, pp.133-143.

[29] Y. Touati, R. Merzouki, and B. Ould-Bouamama, *Fault Detection and Isolation in Presence of Input and Output Uncertainties Using Bond Graph Approach.* The 5th International Conference on Integrated Modeling and Analysis in Applied Control and Automation, 2011, pp.221-227.

[30] B.P.Graham, and R.B. Newell, *Fuzzy identification and control of a liquid level rig.* Fuzzy Sets System, 1988, vol.26, pp.255-273.

[31] J. Thoma, *Bond graphs: introduction and applications.* Elsevier Science, 1975.

[32] W. Borutzky, A. Orsoni, and R. Zobel, *Bond graph modelling and simulation of mechatronics systems an introduction into the methodology.* In Proceedings of the 20th European Conference on Modelling and Similation, Bonn, Germany, 2006.

[33] H. B. Pacejka and R.S. Sharp, *Shear Force Developments by Pneumatic Tires in Steady State Conditions.* A Review of modelling Aspects. Vehicle Systems Dynamics, 1991, vol.20, pp.121-176.

[34] Y.M. Ali and L.C. Zhang, *Estimation of Residual Stresses Induced by Grinding Using a Fuzzy Logic Approach.* Journal of Materials Processing Technology, 1997, vol.63, pp.875-880.

[35] 2011, from http://www.intrade-nwe.eu.

[36] 30 June 2009, from http://www.netlib.org, .

CHAPTER 9

Contribution to the Dynamic Modeling and Control of an Hexapod Robot

Mahfoudi Chawki[1],
Faculty of Science and Technology, Ain Beida
University Larbi ben M'hidi, Algeria
mahfoudi.chawki@gmail.com

Djouani Karim
Laboratory LISSI
University Paris12, France
Djouani@univ-paris12.fr

Mohamed Bouaziz
Laboratory of mechanics
Polytechnic school, Algiers, Algeria
mbouaziz@enp.edu.dz

ABSTRACT. This chapter concerns real-times hexapod robot force control. Based on an operational trajectory planner, a computed torque control for each leg of hexapod robot is presented. This approach takes into account the force distribution on the robot legs in real time and the hexapod dynamic model. First, Kinematic and dynamic modeling are presented. Then, a methodology for the optimal force distribution is given. The issue of force distribution is expressed on the basis of nonlinear programming terms that take into consideration both the equality and the inequality of constraints. Subsequently, nonlinear inequalities of friction constraints can be replaced by a combination of linear equalities and inequalities [22]. The original constraining nonlinear programming problem is then transformed to a problem of a quadratic optimization. Therefore, the overall hexapod computed torque control is presented. Finally some simulations are also given in order to show the effectiveness of the proposed approach.

Keywords: Walking Robot, Quadratic Optimization, Dynamic Modeling and Control

9.1. Introduction

As a part of legged vehicles, hexapod robots can be used to work in space fields with rough ground, e.g. map building of an uneven terrain, to perform hazardous tasks as searching and removing landmines and also to collect volcano data.

As shown in [**29, 3, 11**], the interests of walking machines from the research and application point of views, are twofold. Firstly, the complexity nature of legged locomotion has been very attractive and challenging to many pioneering researches. However, due to the complexity of legged robots applications in real world are not significant. Major problems are concerned with real time dynamic control of the legged robot under several constraints. To overcome such problems, dynamic models should be integrated in every control strategy. The second interest reflects that legged robots are very effective as an efficient transportation device on irregular and/or soft terrain.

When studding legged vehicles, one may be concerned with the stability analysis of the motion. Thus, the corresponding mechanics in the case of three or more supporting legs, is considered as statically stable. Contrarily, in the case of two or fewer legs (monopodes, bipedes,..), the system is considered dynamically stable.

---

[1]Corresponding author

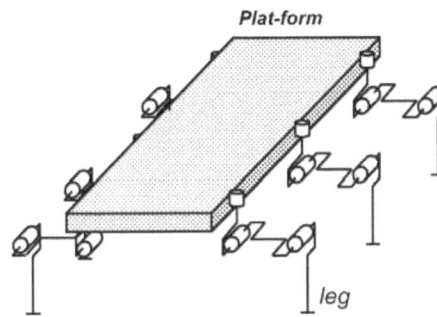

FIGURE 9.1.1. **View of the hexapod**

Before we address the hexapod robot's dynamic modeling it is helpful to have an overall view of how the robot is controlled. In the task planning stage, a trajectory planner is used to determine a path that guides the hexapod from its initial position to a given final position. Then, a gait, which gives the position and events for placing and lifting the robot's legs is selected [**25, 32**]. The inverse kinematic model is then used in order to compute the desired trajectory (positions and velocities) in joint space. A joint computed torque control strategy is used for the hexapod real-time control . The proposed approach is based on the computation of the legs force distribution . The existence of three actuated joints in each leg of the hexapod robot generates a redundant actuation that leads to more active joints (18) than the number of degrees of freedom of the robot platform (6 dof) as shown in Figure (9.1). During the setting equation of the force distribution, we found that the number of unknown variables is not in agreement with the number of equation of the system; which means that the system solution is not unique. Moreover, some physical constraints such as the friction and the nature of contact, etc, have not been taken into account. In addition, joints torque saturation has also not been considered. Thus, the Force Distribution Problem (FDP) can be formulated as a nonlinear constraining programming problem by considering constraints of both nonlinear equality and inequality. Several resolution approaches have been proposed to deal with such a problem [1]-[21], whose we have emerged the following four main methods:

(1) Linear Programming (LP) method [**2**],[**5**]
(2) Compact Dual Linear Programming (CDLP)method [**9**],
(3) Quadratic Programming (QP) Method [**12**], [**13**]
(4) Analytical Method [**24**] [**27**]

A comparative study for the four referred methods can be found in [**22**]. On the other hand, some researchers have proposed an optimal force distribution scheme with the collaboration of multiple robots by applying a combination of CDLP and QP methods[**15**]. In FDP solving system and according to some criteria, the following physical aspects of the robot crawling must be considered.

(1) preventing legs from slipping
(2) avoiding discontinuities of the foot forces.
(3) avoiding hard impact when the contact between the foot and the ground start to establish, by making the foot forces of the swing leg increase smoothly starting from zero value.

In this paper, we propose to solve the problem of determining the real time force distribution for an hexapod robot, by applying the approach proposed in [**22**] for a four- legged robot It consists of using a combination of QP method with a technical reduction of problem size. The robot crawling is divided into 3 phases. In the first phase, only 3 legs are supporting the robot, for instance legs 1-2-3, leading to a force distribution problem with 9 unknown variables. Furthermore, in the second phase, all the six legs are supporting the robot, leading to a force distribution problem with a number of 18 unknown variables. So, in order to reduce the number of unknown variables, we consider that the contact forces on the legs 1-2-3 can be deduced from the first phase by introducing a continuous, decreasing function varying from 1 to 0. Thus, the dimension of the problem, in the second phase, can be reduced to 9 unknown variables. The treatment of third phase which is related to the robot functioning legs 4, 5, 6 is similar to the first one. In the whole

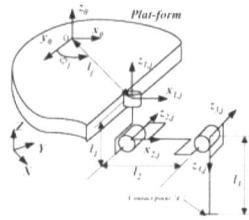

FIGURE 9.2.1. **geometrical model**

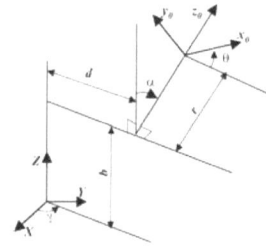

FIGURE 9.2.2. **geometrical parameters**

three phases, the force distribution problem still the same and is solved with the same algorithm. The main idea is to reduce the problem size by transforming the nonlinear constraints into linear ones, while respecting some physical considerations referred above. This chapter has been organized as follows: Direct and inverse geometrical models of the hexapod are presented in section 2 . In section 3 the dynamic model of an hexapod robot is derived. Section 4 is concerned with the treatment of the force distribution problem. Section 5 is dedicated to the simulation of a real time control of Problem reduction and optimal solution. In section 6, the simulation results are presented.

9.2. Geometrical Modeling

Before starting to present the direct and inverse geometrical model , let us first describe succinctly the hexapod architecture. Only one leg modeling is considered here because robot hexapod legs are all identical, the j th architecture leg (j=1 to 6) is presented in Figure (9.2.1). Each leg " j" is fixed to robot platform by a revolute joint located at a certain distance l_j from the center of gravity of the platform that represents the Body. The angle ϕ_j represents the coordinate frame orientation $(x_{1,j}, y_{1,j}, z_{1,j})$ situated at the first articulation of the leg and the fixed coordinate frame of the body (x_0, y_0, z_0). In case where the robot is walking, it is considered as an arborescent robot with some closed loops. To study this kind of robots we use the method defined by Wissama and Klifinger [1].

The transformation matrix between two coordinate frame attached to ith joint's and the (i-1)th joint's is given by figure (9.2.2):

(9.2.1) $$^{i-1}\boldsymbol{T}_i = R(Z, \gamma)T(Z, b)R(X, \alpha)T(X, d)R(Z, \theta)T(Z, r)$$

Thus: $^{i-1}\boldsymbol{T}_i =$

$$\begin{pmatrix} C\gamma_i C\theta_i - S\gamma_i C\alpha_i S\theta_i & -C\gamma_i S\theta_i - S\gamma_i C\alpha_i C\theta_i & S\gamma_i C\alpha_j & d_i C\gamma_i + r_i S\gamma_i S\alpha i \\ S\gamma_i C\theta_i + C\gamma_i C\alpha_i S\theta_i & -S\gamma_i S\theta_i + C\gamma_i C\alpha_i C\theta_i & -C\gamma_i S\alpha_i & d_i S\gamma_i - r_i C\gamma_i S\alpha i \\ S\gamma_i S\theta_i & S\gamma_i C\theta_i & C\alpha_i & r_i C\alpha_i + bi \\ 0 & 0 & 0 & 1 \end{pmatrix}$$

Table 1 describes the transformation process from the ground coordinate frame (X, Y, Z) to the coordinate frame located at the contact point "'4'" of each robot leg .

frame	α	d	θ	r	b	γ
plat-form	α	d	θ	r	h	β
liaison"1"	0	l_j	$\theta_{1,j}$	0	0	ϕ_j
liaison"2"	$-\pi/2$	0	$\theta_{2,j}$	0	$-l1$	0
liaison"3"	0	l_2	$\theta_{3,j}$	0	0	0
contact Pt"4"	0	l_3	$\theta_{4,j}$	0	0	0

TABLE 1. **geometrical parameters**

The transformation allowing the position location of the point contact 4 of any leg in the ground absolute coordinate frame can be given as:

(9.2.2) $$^R\boldsymbol{T}_4 = {}^R \boldsymbol{T}_0^0 \boldsymbol{T}_1^1 \boldsymbol{T}_2^2 \boldsymbol{T}_3^3 \boldsymbol{T}_4$$

When both position and orientation angle of the last coordinate frame that is fixed to the end of the leg "j"' are known, we use Paul's method [21]. It provides joint's coordinates values $\theta_{i,j}$ $(i = 1, 2, 3)$ $(j = 1, .., 6)$ as follows:

(9.2.3)
$$\begin{cases} \theta_{1,j} = \arctan(S1, C1) \\ \theta_{2,j} = \arctan(S2, C2) \\ \theta_{3,j} = \arctan(S3, C3) \end{cases}$$

with:
$$\begin{cases} S1 = S\phi X_0 - C\phi Y_0 \\ C1 = S\phi Y_0 - C\phi X_0 - l_j \end{cases}$$

$$\begin{cases} X_0 = (e - f)P_{x,j} + (e + f)P_{y,j} + gP_{z,j} \\ \quad -gh - dS\theta \\ Y_0 = (-e - f)P_{x,j} + (-e + f)P_{y,j} + gP_{z,j} \\ \quad -gh + dS\theta \\ e = C\beta C\theta \\ f = S\beta C\alpha S\theta \\ g = S\alpha S\theta \end{cases}$$

$$\begin{cases} S2 = \frac{XZ + Y\sqrt{X^2 + Y^2 - Z^2}}{X^2 + Y^2} \\ C2 = \frac{XZ - Y\sqrt{X^2 + Y^2 - Z^2}}{X^2 + Y^2} \end{cases} \quad \begin{cases} X = 2Z_2 X_1 \\ Y = -2Z_2 Y_1 \\ Z = W^2 - X_1^2 \\ \quad -Y_1^2 - Z_1^2 \end{cases}$$

$$\begin{cases} Z_2 = l_2 \\ W = -l_3 \\ X_1 = S\beta S\alpha P_{x,j} - C\beta S\alpha P_{y,j} + C\alpha P_{z,j} \\ \quad -hC\alpha - r + l_1 \\ Y_1 = C(\phi + \theta_{j,1})(X_0) + S(\phi + \theta_{j,1})(Y_0 - lC\theta_{j,1}) \end{cases}$$

$$\begin{cases} S3 = \frac{X_1 C2 + Y_1 S2}{W} \\ C3 = \frac{X_1 S2 - Y_1 S2 + Z_2}{W} \end{cases}$$

Remark: S*=sin(*); C*=cos(*); $(P_{x,j}, P_{y,j}, P_{z,j})$, are the coordinates of the end point 4 formulated in (x_0, y_0, z_0).

9.3. Hexapod Dynamics Model

9.3.1. Introduction. The robot dynamics is given by [28] [1] [31]:

(9.3.1)
$$\Gamma = f(\theta, \dot{\theta}, \ddot{\theta})$$

Where, θ, $\dot{\theta}$, $and\ \ddot{\theta}$ are the generalizes coordinates, speeds and acceleration, respectively.
The explicit form of Eq.9.3.1 can be expressed as follows for any leg "j":

(9.3.2)
$$\Gamma = M(\theta)\ddot{\theta} + C(\theta, \dot{\theta})\dot{\theta} + Q(\theta) + J^T f$$

where $M \in \Re^{3 \times 3}$, $C \in \Re^{3 \times 1}$, $Q \in \Re^{3 \times 1}$ and $J^T f \in \Re^{3 \times 1}$.

- $M(\theta)$, matrix $(n \times n)$ representing the inertia of the robot, which is deduced from the kinetic energy:

$$E = \frac{1}{2}\dot{\theta}^T M \dot{\theta}$$

- $C(\theta, \dot{\theta})$, a vector $(n \times 1)$ representing Coriolis torques and centrifugal forces, so that:

$$C\dot{\theta} = \dot{M}\dot{\theta} - \frac{\partial E}{\partial \theta}$$

- $Q(\theta) = [Q_1, ..., Q_n]$, a vector of gravity torque and forces, so that:

$$Q_i = \frac{\partial U}{\partial \theta_i}$$

- f the reaction of the ground
- E and U are respectively the kinetic and the potential energy of the system

The Eq.9.3.1 can be rewritten as:

(9.3.3)
$$\Gamma = M(\theta)\ddot{\theta} + H(\theta,\dot{\theta}) + J^T f$$

with:

$$H(\theta,\dot{\theta}) = C(\theta,\dot{\theta})\dot{\theta} + Q(\theta)$$

Then, we can write:

$$H(\theta,\dot{\theta}) = \Gamma \quad ; \quad if\ \ddot{\theta} = 0\ and\ f = 0$$

So if we use Newton-Euler formalism with $\ddot{\theta} = 0\ and\ f = 0$ we obtain the value of $H(\theta,\dot{\theta})$. This transformation is quite important and enables to:

- extract the acceleration vector $\ddot{\theta}$ by:

$$\ddot{\theta} = M^{-1}(\theta)(\Gamma - J^T f - H(\theta,\dot{\theta}))$$

- avoid the computation of the vector $C(\theta,\dot{\theta})$ which has redundant algorithms.

9.3.2. Newton-Euler Formalism. *Remark*: The contact forces on the ground, f_x, f_y, f_z (Figure 9.3.2.1) are 0 if the leg is lifted and $\neq 0$, otherwise.

The Newton-Euler algorithm can be established as follows [1]:

9.3.2.1. *Velocities and accelerations calculations.* Let C_i be any link of a leg "j", Figure

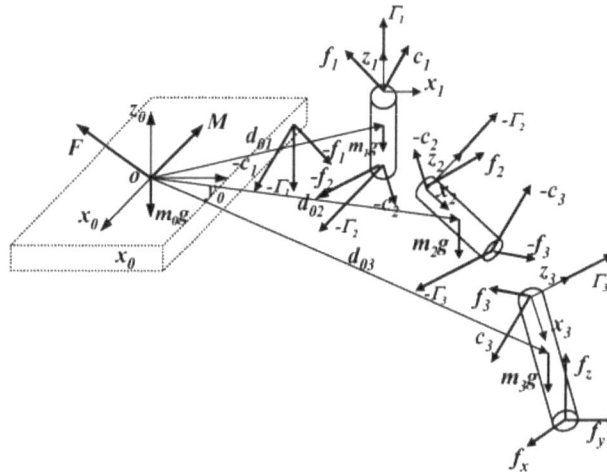

FIGURE 9.3.1. **Forces acting on the hexapod system**

(9.3.4)
$$P_{oi} = P_{o,i-1} + P_{i-1,i}$$

and

(9.3.5)
$$d_{oi} = P_{oi} + d_{i,i}$$

after derivation:

(9.3.6)
$$\dot{P}_{oi} = \dot{P}_{o,i-1} + \omega_{i-1} \wedge P_{i-1,i} + \sigma_i \dot{\theta}_i Z_i$$

In our case, we have only rotational articulations so $\sigma_i = 0$, then:

$$v_i = \dot{d}_{oi} = \dot{P}_{oi} + \omega_{i-1} \wedge d_{i,i}$$

The second derivation gives:

(9.3.7)

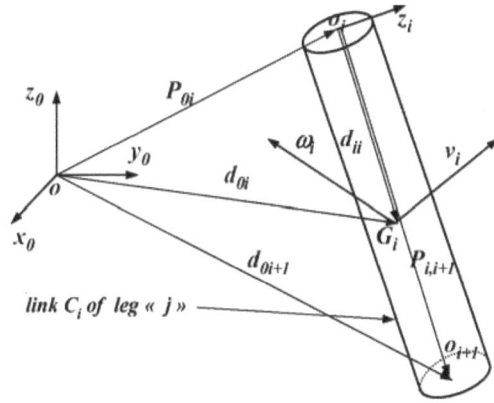

FIGURE 9.3.2. **Kinematics parameters representation of a link "i" in a leg "j"**

$$\ddot{\boldsymbol{P}}_{oi} = \ddot{\boldsymbol{P}}_{o,i-1} + \boldsymbol{a}_{i-1} \wedge \boldsymbol{P}_{i-1,i} + \omega_{i-1} \wedge (\omega_{i-1} \wedge \boldsymbol{P}_{i-1})$$
$$+ \sigma_i (2\dot{\theta}_i \omega_{i-1} \wedge \boldsymbol{Z}_i + \ddot{\theta}_i \boldsymbol{Z}$$

And

(9.3.8) $$\gamma_i = \dot{\boldsymbol{v}}_i = \ddot{\boldsymbol{d}}_{oi} = \ddot{\boldsymbol{P}}_{oi} + \boldsymbol{a}_{i-1} \wedge \boldsymbol{d}_{i,i} + \omega_i \wedge (\omega_{i-1} \wedge \boldsymbol{d}_{ii}),$$

where, γ_i represents the gravity center acceleration.

Therefore we can define the global acceleration with the following relation:

(9.3.9) $$\alpha_i = \ddot{\boldsymbol{P}}_{oi} + g\boldsymbol{Z}_0$$

where, g is the acceleration of gravity with:

(9.3.10) $$\boldsymbol{a}_i \wedge \star + \omega_i \wedge (\omega_i \wedge \star) = \underline{\boldsymbol{b}}_i \star$$

In condensed form:

(9.3.11) $$\underline{\boldsymbol{b}}_i = \widehat{\boldsymbol{a}} + \widehat{\omega}.\widehat{\omega}$$

Where,

$$\widehat{\boldsymbol{a}} = \begin{pmatrix} 0 & -a_z & a_y \\ a_z & 0 & -a_x \\ -a_y & a_x & 0 \end{pmatrix}$$

$$\widehat{\omega} = \begin{pmatrix} 0 & -\omega_z & \omega_y \\ \omega_z & 0 & -\omega_x \\ -\omega_y & \omega_x & 0 \end{pmatrix}$$

$\widehat{\boldsymbol{a}} \, and \, \widehat{\omega}$: represent the skew symmetric matrices.

Eq.9.3.9 can be represented in condensed form:

(9.3.12) $$\alpha_i = \sigma_i \boldsymbol{t}_i + r_{i-1}$$

with:

$$\begin{cases} \boldsymbol{t}_i = 2\dot{\theta}_i \omega_{i-1} \wedge \boldsymbol{Z}_i + \ddot{\theta}_i \boldsymbol{Z}_i \\ r_{i-1} = \alpha_{i-1} + \underline{\boldsymbol{b}}_{i-1} \boldsymbol{P}_{i-1,i} \end{cases}$$

In our case, $\sigma_i \boldsymbol{t}_i = 0$ because $\sigma_i = 0$, then we obtain:

(9.3.13) $$\gamma_i + g\boldsymbol{Z}_0 = \alpha_i + \underline{\boldsymbol{b}}_i \boldsymbol{d}_{ii}$$

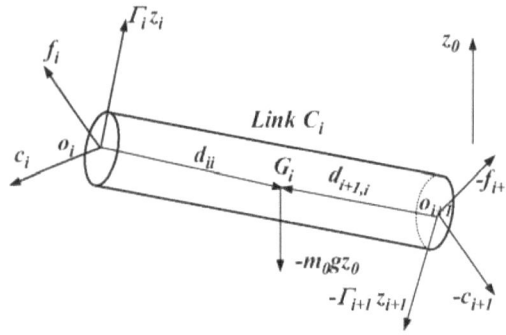

FIGURE 9.3.3. **Equilibrium of a link "i" in a leg "j"**

9.3.2.2. *Forces and torques computation.*

- Let $\boldsymbol{c}_i$, the interaction torque exerted by the link C_{i-1} on the link C_i Figure 9.3.2.2.
- $\boldsymbol{f}_i$, the interaction force exerted on o_i by the link C_{i-1} on the link C_i.
- $\boldsymbol{F}^i = \boldsymbol{f}^i + \sigma_i \Gamma^i \boldsymbol{Z}_i$, the force exerted, on o_i by the ith actionnary on the link C_i, where Γ^i is a scalars, representing the moment of the motor.
- $\boldsymbol{C}^i = \boldsymbol{c}^i + \overline{\sigma}_i \Gamma^i \boldsymbol{Z}_i$: the moment exerted, by the link C_{i-1} and the ith actionnary on the link C_i.

The equilibrium of the link C_i can be represented by:

(9.3.14)
$$\boldsymbol{F}^i_{res} = \boldsymbol{F}^i - \boldsymbol{F}^{i+1} - m_i g \boldsymbol{Z}_0 = m_i \gamma_i$$

(9.3.15)
$$\boldsymbol{C}^i_{res} = \boldsymbol{C}^i - \boldsymbol{C}^{i+1} - \boldsymbol{d}_{i,i} \wedge \boldsymbol{F}^i + \boldsymbol{d}_{i+1,i} \wedge \boldsymbol{F}^{i+1}$$

and

(9.3.16)
$$\boldsymbol{C}^i_{res} = \varphi^i \boldsymbol{a}_i - \omega_i \wedge (\varphi^i \omega_i)$$

Where , φ^i is the inertia matrices of the link C_i. from Eq.9.3.14:

(9.3.17)
$$\boldsymbol{F}^i = \boldsymbol{F}^{i+1} + m_i(\alpha_i + \underline{\boldsymbol{b}}_i \boldsymbol{d}_{ii})$$

Then, from Eq.9.3.15 and (9.3.16) we obtained:

(9.3.18)
$$\boldsymbol{C}^i = \boldsymbol{C}^{i+1} + \varphi^i \boldsymbol{a}_i + \omega_i \wedge (\varphi^i \omega_i) + \boldsymbol{d}_{ii} \wedge (\boldsymbol{F}^i - \boldsymbol{F}^{i+1}) + \boldsymbol{P}_{i,i+1} \wedge \boldsymbol{F}^{i+1}$$

9.3.2.3. *The kinetic energy computation.* The total kinetic energy of the system is given as follows:

(9.3.19)
$$E = \sum_{j=1}^{n} E_j$$

where E_j, kinetic energy of the link C_j, expressed by the following equation:

(9.3.20)
$$\boldsymbol{E}_j = \frac{1}{2}(\omega_j^T \varphi^j \omega_j + m_j \boldsymbol{V}_{Gj}^T \boldsymbol{V}_{Gj}),$$

where ω_j is the instantaneous rotation velocity of the link "C_j" expressed in the coordinate frame "j". $\boldsymbol{V}_{Gj}$ is the linear velocity of the gravity center of the link "C_j" expressed in the coordinate frame "j".

As depicted in Figure 9.3.2.3,

(9.3.21)
$$\boldsymbol{V}_{Gj} = {}^j \boldsymbol{V}_j + {}^j \omega_j \boldsymbol{d}_{jj}$$

as we know:

(9.3.22)
$$\boldsymbol{k}_j = \varphi^j - m_j \widehat{\boldsymbol{d}}_{jj} \widehat{\boldsymbol{d}}_{jj}$$

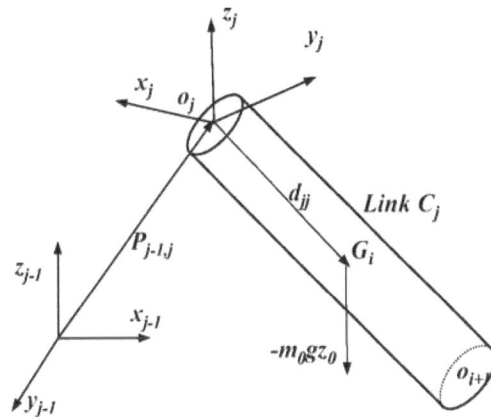

FIGURE 9.3.4. **Modeling a link of a leg**

Eq.9.3.20 can be transformed as:

$$(9.3.23) \qquad \boldsymbol{E}_j = \frac{1}{2}(\omega_j^T \, {}^j\boldsymbol{k}_j \, {}^j\omega_j + m_j \, \boldsymbol{V}_j^T \, {}^j\boldsymbol{V}_j + 2m_j \, {}^j\boldsymbol{d}_{jj}^T({}^j\boldsymbol{V}_j \wedge^j \omega_j))$$

with:

$$(9.3.24) \qquad {}^j\omega_j = {}^j\boldsymbol{A}_{j-1} \, {}^{j-1}\omega_{j-1} + \overline{\sigma}_j \dot{\theta}_j^j \boldsymbol{Z}_j = {}^j\omega_{j-1} + \overline{\sigma}_j \dot{\theta}_j \, {}^j\boldsymbol{Z}_j$$

and,

$$(9.3.25) \qquad {}^j\boldsymbol{V}_j = {}^j\boldsymbol{A}_{j-1} ({}^{j-1}\boldsymbol{V}_{j-1} + {}^{j-1}\omega_{j-1} \wedge \boldsymbol{P}_{j-1,j}) + \sigma_j \dot{\theta}_j \, {}^j\boldsymbol{Z}_j$$

${}^j\boldsymbol{A}_{j-1} \in \Re^{3\times 3}$: represented the orientation matrix .

9.3.2.4. *Elements M_{ij} of the matrix $\boldsymbol{M}$.* :
In the end we computed the matrix $\boldsymbol{M}$ from Eq. 9.3.19 and 9.3.23:
The elements M_{ii} of the matrix $\boldsymbol{M}$ is equal to the coefficient of $\dot{\theta}_i^2/2$ in the expression of kinetic energy, and the elements M_{ij} , if $i \neq j$ is equal to the coefficient of $\dot{\theta}_i \dot{\theta}_j$. $M_{11} = Iz_1 + S2^2 Ix_2 + C2^2 Iy_2 + S23^2 Ix_3 + C23^2 Iy_3 + m_3 C2^2 l_2^2 + m_3 C23 \, C2 l_2 l_3 + Ia_1$
$M_{12} = M_{13} = 0$
$M_{22} = Iz_2 + Iz_3 + m_3 S3^2 l_2^2 + m_3 C3^2 l_2^2 + m_3 C3 \, l_2 l_3 + Ia_2$
$M_{23} = Iz_3 + 1/2 \, m_3 C3 l_2 l_3$
$M_{33} = Iz_3 + Ia_3$
With, $(Ia_1, Ia_2 \text{ and } Ia_3)$ represent the motors inertias. These results are obtained with the diagonal inertia matrix of the legs links.
At the end of computation for all legs, we can describe the equilibrium of the plat-form as following, Figure9.3.2.4.

- let $\gamma_{0,0}$ and $\omega_{0,0}$ respectively the plat-form desired linear acceleration and angular acceleration in the coordinates frame (x_0, y_0, z_0).
- $\boldsymbol{F}_{1,j}^0$, the force applied by the leg "j" at the articulation "1" on the plat-form "0".
- $\boldsymbol{C}_{1,j}^0$, the moment applied by the jth leg in the articulation "1" on the plat-form .
- $\boldsymbol{P}_{01,j}^0$, the distance between the articulation "1" of the jth leg and the origin of the coordinate frame (x_0, y_0, z_0) expressed in the same coordinate frame.

The application of the dynamic fundamental principle, at the mass center of the platform provides the following matrix equation:

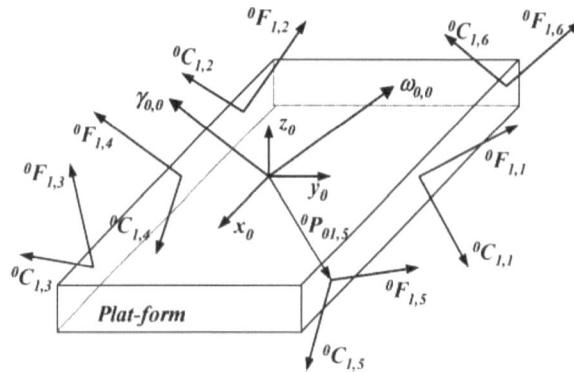

FIGURE 9.3.5. **Equilibrium of the plat-form**

(9.3.26)
$$\begin{pmatrix} m_0 \mathbf{I}_3 & \mathbf{0} \\ \mathbf{0} & \underline{\varphi}^0 \end{pmatrix} \begin{pmatrix} \gamma_{0,0} \\ \omega_{0,0} \end{pmatrix} + \begin{pmatrix} -m_0 \mathbf{g}_0 \\ \omega_{0,0} \wedge (\underline{\varphi}^0 \omega_{0,0}) \end{pmatrix} = \begin{pmatrix} \mathbf{F} \\ \mathbf{M} \end{pmatrix}$$

Where,

- m_0 and $\underline{\varphi}^0$ are respectively the mass and the inertia matrix of the plat-form.
- $\mathbf{I}_3$, the identity matrix (3×3).
- $\mathbf{g}_0$: gravity vector

$\mathbf{F}$ and $\mathbf{M}$ are defined as follows:

(9.3.27)
$$\begin{cases} \mathbf{F} = \sum_{j=1}^{6} \mathbf{F}_{1,j}^0 \\ \mathbf{M} = \sum_{j=1}^{6} (\mathbf{C}_{1,j}^0 + \mathbf{P}_{01,j}^0 \wedge \mathbf{F}_{1,j}^0) \end{cases}$$

9.4. Force Distribution Problem

9.4.1. Problem Formulation. The Figure9.4.1 shows the system of forces acting on an hexapod robot. To simplify this graphical scheme, we have represented only components of the force. In the major part of cases, the existing rotational torques at the feet are negligible. Let (x_0, y_0, z_0) the robot fixed coordinate frame in which the body of the robot is located in the (x_0, y_0) plane and $(x_{1,j}, y_{1,j}, z_{1,j})$ represent the coordinate frame fixed at the end of the leg "j", with the z axis normal to the support surface of the foot which is assumed to be parallel to the (x_0, y_0) plane. $\mathbf{F} = [F_X F_Y F_Z]^T$ and $\mathbf{M} = [M_X M_Y M_Z]^T$ denote respectively the robot body force vector and moment vector, which results from the gravity and the external force acting on the robot body. $f_{x,j}, f_{y,j}$, and $f_{z,j}$ are defined as the components of the force acting on the supporting foot "j" in the directions of x_0, y_0 and z_0, respectively. The number of supporting feet n, can vary between 3 and 6 for an hexapod robot. The robot's quasi-static force/moment equation can be written as:

(9.4.1)
$$\begin{cases} \sum_{j=1}^{n} \mathbf{f}_j = \mathbf{F} \\ \sum_{j=1}^{n} \mathbf{OP}_j \wedge \mathbf{f}_j = \mathbf{M} \end{cases}$$

where $\mathbf{OP}_j$ represents the position vector of the contact point of the junction of the leg "j" with the body gravity center. The general form of the matrix of this equation can be written as follow:

(9.4.2)
$$\mathbf{A}\,\mathbf{G} = \mathbf{W}$$

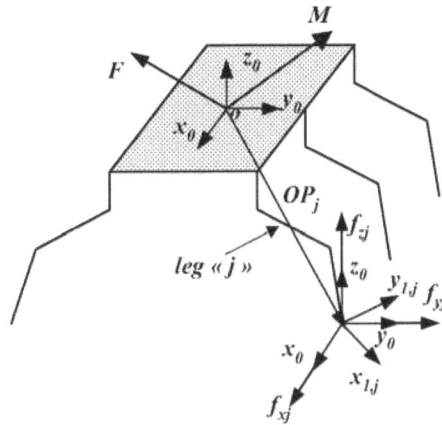

FIGURE 9.4.1. **Orientation of coordinate frame**

with :

$$\begin{cases} \boldsymbol{G} = [\boldsymbol{f}_1^T \boldsymbol{f}_2^T \cdots \boldsymbol{f}_n^T]^T & \in \Re^{3n} \\ \boldsymbol{f}_j^T = [f_{x,j}\ f_{y,j}\ f_{z,j}]^T & \in \Re^3 \\ \boldsymbol{W} = [\boldsymbol{F}^T \boldsymbol{M}^T]^T & \in \Re^6 \end{cases}$$

$$\boldsymbol{A} = \begin{pmatrix} \boldsymbol{I}_3 & \cdots & \cdots & \boldsymbol{I}_3 \\ \boldsymbol{B}_1 & \cdots & \cdots & \boldsymbol{B}_n \end{pmatrix} \quad \in \Re^{6 \times 3n}$$

$$\boldsymbol{B}_j \equiv \widehat{\boldsymbol{OP}}_j \equiv \begin{pmatrix} 0 & -P_{z,j} & P_{y,j} \\ P_{z,j} & 0 & -P_{x,j} \\ -P_{y,j} & P_{x,j} & 0 \end{pmatrix} \quad \in \Re^{3 \times 3}$$

where $\boldsymbol{I}_3$ is the matrix of identity and $\mathbf{G}$ is the foot force vector, corresponding to three ($\boldsymbol{G} \in \Re^9$) or six ($\boldsymbol{G} \in \Re^{18}$) supporting legs. $\boldsymbol{A}$ is matrix coefficient that is related to supporting feet positions , and $\boldsymbol{B}_j$ is a skew symmetric matrix consisting of $(P_{x,j}, P_{y,j}, P_{z,j})$, that represents the coordinate position of the supporting foot "j" in (x_0, y_0, z_0). $\boldsymbol{W}$ is the vector of a total body force / moment. It is obvious that Eq.9.4.2 is an under-determined system of equations and it does not accept a unique solution, which means that forces of the feet can have many solutions in accordance with the equation of equilibrium. Nevertheless, the feet forces must satisfy the requirements of the following physical constraints, otherwise they will be useless:

(1) no slipping supporting feet when the robot is walking and it shows up in the following constraint:

(9.4.3)
$$\sqrt{f_{x,j}^2 + f_{y,j}^2} \le \mu\, f_{z,j}$$

where μ is the friction static parameter of the ground.

(2) knowing that the feet forces are generated from the corresponding joints actuators, we must consider the physical limits of the joint torques which leads to:

(9.4.4)
$$-\tau_{jmax} \le \boldsymbol{J}_j^T\ {}^j\boldsymbol{A}_{0j} \begin{pmatrix} f_{x,j} \\ f_{y,j} \\ f_{z,j} \end{pmatrix} \le \tau_{jmax}$$

for $(j = 1, ..., n)$, where $\boldsymbol{J}_j \in \Re^{3 \times 3}$ is the Jacobian of the leg "j", $\tau_{jmax} \in \Re^{3 \times 1}$ represents the vector of maximum joint torque of the leg "j" , and $\boldsymbol{A}_{0j} \in \Re^{3 \times 3}$ is the orientation matrix of $(x_{1,j}, y_{1,j}, z_{1,j})$ with respect to (x_0, y_0, z_0).

(3) In order to get a whole and complete contact between the leg foot and the ground, there must exist a $f_{z,j}$ such way that we can write:

(9.4.5)
$$f_{z,j} \geq 0$$

Now, we proceed with problem size reduction. Firstly we apply the linearization procedure and then it is solved for the hexapod case. It is evident that it is not that easy to solve this kind of nonlinear programming problem to determine the real time feet force distribution with its constraints complexity.

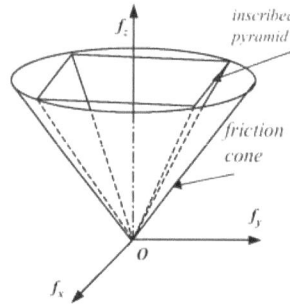

FIGURE 9.4.2. **Conservative inscribed pyramid**

9.4.2. Problem Size Reduction. Eq.9.4.3 represents a friction cone formulation, Figure 9.4.2. To overcome the nonlinearities induced by the equation written below, most of the researchers replaced this friction cone by the inscribed pyramid [**5, 10, 12**]. Consequently the expression of nonlinear friction constraints can be approximated by the linear inequalities that follow:

(9.4.6)
$$f_{x,j} \geq \acute{\mu} f_{z,j}, \quad f_{y,j} \geq \acute{\mu} f_{z,j}, \quad j = 1, ..., n$$

where $\acute{\mu} = \frac{\sqrt{2}\mu}{2}$ is for the inscribed pyramid representation . Thus, by substituting the nonlinear constraint of Eq.9.4.3 by the linear one of Eq.9.4.6 the initial nonlinear constraining programming problem becomes a linear programming problem [**5**],[**9**], and [**12**]. The slipping phenomena can be minimized by optimizing the ratio of tangential to normal forces at the feet. In [**14**] the authors proved that for the case of multi-legged robots, all ratios (at the feet) are equal to the global ratio that is be defined as the ratio of the tangential to normal forces at the robot body. This result guided Liu and Wen [**24**] to find the relationship among the feet forces and to have the possibility to transform the initial friction constraints from nonlinear inequalities to a set of linear equalities. The advantage of these methods relies in the fact that parts of feet forces component satisfy the global ratio relationship and gives the possibility to the other components to fulfill the linear inequality constraints requirements as Eq.9.4.6. For example, defining $f_{x,j}$ $(j = 1; ; n)$ and $f_{y,j}$ $(j = 1; ; n)$, for the foot "j" chen et al. [**22**] show that:

(9.4.7)
$$f_{x,j} = k_{xz} f_{z,j} \quad , (i = 1, ..., n)$$

(9.4.8)
$$f_{y,j} \leq \mu^\star f_{z,j} \quad , (i = 1, ..., n)$$

where $k_{xz} = \frac{F_X}{F_Z}$ is the forces global ratio at the body of the robot in the x_0 and z_0 direction . $\mu^\star$ is the given friction constraints coefficient. Finally, and on the basis of a nonlinear programming problem, the force distribution expression is transformed to a linear one by making the substitution of Eq.9.4.3 with both Eqs (9.4.7) and (9.4.8).

9.4.3. Problem Transformation and Continuous Solution. In modeling the behavior of the hexapod robot in walking motion, we suppose that the robot is supported by three legs at the time [**30, 4, 8**]. This means that discontinuity avoidance has to be taken into consideration and this later should be of two types: the first is to ensure that the foot force of the swing leg transits continuously while the leg moves freely until it touches the ground; the second type is to make forces of the feet raises smoothly from zero value, to prevent from impact arising from the swing leg placing . The hexapod robot crawling steps are as follow:

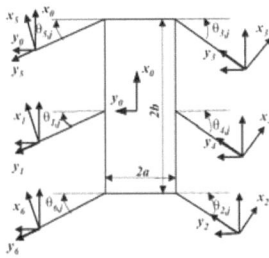

FIGURE 9.4.3. **bottom View of the hexapod**

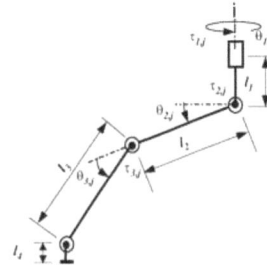

FIGURE 9.4.4. **basic mechanism of the leg**

- $[t_1; t_2]$ indicates the period of time from which the swinging legs (4,5 or 6) are raised, to the time t_2 up till they are positioned.
- $[t_2; t_3]$ indicates the time period from the time t_2 at which the legs (4,5,6) are positioned , to the time t_3 till legs numbered legs (1,2,3) (next swing legs) are lifted .
- $[t_3; t_4]$ indicates the time period from the time t_3 at which the legs (1,2,3) (the swing leg) is lifted, to the time t_4 till legs numbered (1,2,3) are positioned.

In the time period $(t_1; t_2)$ $n = 3$, so A and G become respectively a matrix of 6×9 and a vector of 9×1. Eq.9.4.2 is made up of six equations with nine unknown variables. By bringing together Eq.9.4.7 and Eq.9.4.2 we obtain a system of nine equations.

(9.4.9) $$\overline{A}\,G = \overline{W}$$

with:

$$\overline{A} = \begin{pmatrix} I_3 & & & I_3 & & & I_3 & & \\ B_1 & & & B_2 & & & B_3 & & \\ 1 & 0 & -k_{xz} & 0 & 0 & 0 & 0 & 0 & 0 \\ 0 & 0 & 0 & 1 & 0 & -k_{xz} & 0 & 0 & 0 \\ 0 & 0 & 0 & 0 & 0 & 0 & 1 & 0 & -k_{xz} \end{pmatrix}$$

$$G = \begin{pmatrix} f_1 \\ f_2 \\ f_3 \end{pmatrix} \qquad \overline{W} = \begin{pmatrix} F \\ M \\ 0 \\ 0 \\ 0 \end{pmatrix}$$

Using some rows combination of the matrix $\overline{A}$, Eq.9.4.9 can be written as:

(9.4.10) $$\widehat{A}\,G = \widehat{W}$$

With:

$$\widehat{A} = \begin{pmatrix} 1 & 0 & 0 & 1 & 0 & 0 & 1 & 0 & 0 \\ 0 & 1 & 0 & 0 & 1 & 0 & 0 & 1 & 0 \\ 1 & 0 & -k_{xz} & 0 & 0 & 0 & 0 & 0 & 0 \\ 0 & 0 & 0 & 1 & 0 & -k_{xz} & 0 & 0 & 0 \\ -P_{y,1} & P_{x,1} & 0 & -P_{y,2} & P_{x,2} & 0 & 1 & 0 & -k_{xz} \\ P_{z,1} & 0 & -P_{x,1} & P_{z,2} & 0 & -P_{x,2} & -P_{y,3} & P_{x,3} & 0 \\ 0 & 0 & 0 & 0 & 0 & 0 & P_{z,3} & 0 & -P_{x,3} \\ 0 & -P_{z,1} & P_{y,1} & 0 & -P_{z,2} & P_{y,2} & 0 & -P_{z,3} & P_{y,3} \end{pmatrix}$$

$$\widehat{W} = \begin{pmatrix} F_X \\ F_Y \\ 0 \\ 0 \\ M_Z \\ M_Y \\ 0 \\ M_X \end{pmatrix}$$

where $\widehat{A} \in \Re^{8 \times 9}$ is the resulting matrix of $\overline{A}$ after combination. $G \in \Re^8$ is the foot force vector. $\widehat{W} \in \Re^8$ is the resulting vector of W after combination. Thus, the force distribution problem is subjected to the inequality constraints expressed by Eqs (9.4.4),(9.4.5)and (9.4.7). In the time period $(t_2; t_3)$, G is a 18 dimensional vector, and A is a matrix of 6×18. For the sake of continuity of solution, the foot forces of the legs (1,2,3) denoted by $f_j = [f_{x,j} \ f_{y,j} \ f_{z,j}]^T$ should be changed smoothly from $f_j(t_2)$ to 0, (j=1,2,3). Therefore, the foot force of the legs $f_j \ (j = 1, 2, 3)$ in the time period $(t_2; t_3)$ can be expressed as:

(9.4.11) $\qquad\qquad\qquad f_j = \delta(t) f_j(t_2) \quad (j = 1, 2, 3)$

where $\delta(t)$ can be any chosen continuous scalar function that can vary from $\delta(t_2) = 0 \quad to \quad \delta(t_3) = 1$. Therefore, the original problem with 18 unknown variables $[f_1, f_2, ..., f_6] \in \Re^{1 \times 18}$ is limited to 9. Thus, f_j (j=1,2,3) are known in this period of time and Eq.9.4.3 can be similarly written as:

(9.4.12)

$$\begin{pmatrix} I_3 & I_3 & I_3 \\ B_4 & B_5 & B_6 \end{pmatrix} \begin{pmatrix} f_4 \\ f_5 \\ f_6 \end{pmatrix} = \begin{pmatrix} F_X \\ F_Y \\ F_Z \\ M_X \\ M_Y \\ M_Z \end{pmatrix} - \begin{pmatrix} I_3 & I_3 & I_3 \\ B_1 & B_2 & B_3 \end{pmatrix} \begin{pmatrix} f_1 \\ f_2 \\ f_3 \end{pmatrix}$$

The general form of Eq.9.4.12:

(9.4.13) $\qquad\qquad\qquad\qquad \widetilde{A}G = \widetilde{W}$

with :

$$\begin{cases} \widetilde{W} = W - A_i V \\ V = [f_1 f_2 f_3]^T \end{cases}$$

Using the same process as in the first period (t_1, t_2) the combination of both Eq.9.4.13 and Eq.9.4.7 leads to a system of eight independent equations with the same nine unknown variables of Eq.9.4.10, with $\widehat{\widetilde{A}}$ the resulting matrix of $\widetilde{A}$ after combination, and $\widehat{\widetilde{W}}$ the resulting vector of $\widetilde{W}$ after combination.

9.5. Quadratic Problem Formulation and Solution

The resulting solution of the inverse dynamic equations of a hexapod robot is not unique, yet it can be selected in an optimal way And that can be done by a minimization of some objective function . The applied approach intends to minimize the robot weighted torque summation, which is reflected in the objective function that follow [13, 15, 7]:

(9.5.1) $\qquad\qquad\qquad\qquad f_G = p^T G + \dfrac{G^T Q G}{2}$

with:

$$p^T = [\widehat{\tau}_1^T] J_1^T \ ...., \widehat{\tau}_n^T J_n^T] \quad \in \Re^{3n}$$

$$Q = \begin{pmatrix} J_1 q_1 J_1^T & \cdots & 0 \\ \vdots & \ddots & \vdots \\ 0 & \cdots & J_n q_n J_n^T \end{pmatrix} \quad \in \Re^{3n \times 3n}$$

where $\hat{\tau}_j$ represents the torque of the joint that is due to the weight and inertia of the leg "j", $\mathbf{J}_j$ is the leg "j" Jacobian, and $\mathbf{q}_j$ represents a definite diagonal weighting matrix of positive nature of the leg "j". We notice that this objective function is absolutely convex. Because the time allowed to get a solution is independent of an initial guess, a quadratic programming is greater in speed and in a got solution quality than a linear programming [13]. The general linear quadratic programming problem of the force distribution for an hexapod robot legs can be now formulated mathematically as follow:

$$(9.5.2) \qquad \mathbf{p}^T \mathbf{G} + \frac{\mathbf{G}^T \mathbf{Q} \mathbf{G}}{2}$$

$$(9.5.3) \qquad \widehat{\mathbf{A}} \widehat{\mathbf{G}} = \widehat{\mathbf{W}}$$

$$(9.5.4) \qquad \mathbf{B} \widehat{\mathbf{G}} \le \mathbf{C}$$

where $\mathbf{G} \in \Re^9$ is a vector of feet forces design variables . it should be emphasized that , Eq.9.5.3 represents Eq.9.4.10, and Eq.9.5.4 is the resulting inequality constraints of a series combination of Eq.9.4.4, Eq.9.4.5 and 9.4.8 where

$$\mathbf{B} = [\mathbf{B}_1^T \ \mathbf{B}_2^T \ \mathbf{B}_3^T \ \mathbf{B}_4^T]^T \quad \in \Re^{9 \times 24}$$
$$\mathbf{C} = [\tau_{1max} \ \tau_{2max} \ \tau_{3max} \ -\tau_{1max} \ -\tau_{2max} \ -\tau_{3max} \ 0\,0\,0\,0\,0\,0]^T \quad \in \Re^{24}$$

with

$$\mathbf{B}_1 = \begin{bmatrix} \mathbf{J}_1^T \mathbf{R}_1 & 0 & 0 \\ 0 & \mathbf{J}_2^T \mathbf{R}_2 & 0 \\ 0 & 0 & \mathbf{J}_3^T \mathbf{R}_3 \end{bmatrix} \in \Re^{9 \times 9}$$

$$\mathbf{B}_2 = \begin{bmatrix} -\mathbf{J}_1^T \mathbf{R}_1 & 0 & 0 \\ 0 & -\mathbf{J}_2^T \mathbf{R}_2 & 0 \\ 0 & 0 & -\mathbf{J}_3^T \mathbf{R}_3 \end{bmatrix}$$

$$\mathbf{B}_3 = \begin{bmatrix} 0 & 0 & -1 & 0 & 0 & 0 & 0 & 0 & 0 \\ 0 & 0 & 0 & 0 & 0 & -1 & 0 & 0 & 0 \\ 0 & 0 & 0 & 0 & 0 & 0 & 0 & 0 & -1 \end{bmatrix}$$

$$\mathbf{B}_4 = \begin{bmatrix} 0 & 1 & -\mu^\star & 0 & 0 & 0 & 0 & 0 & 0 \\ 0 & 0 & 0 & 0 & 1 & -\mu^\star & 0 & 0 & 0 \\ 0 & 0 & 0 & 0 & 0 & 0 & 0 & 1 & -\mu^\star \end{bmatrix}$$

The solution of this system is developed in [25].

9.6. Computed-Torque Control

Suppose that a desired trajectory $\theta_d(t)$ has been selected for the arm motion. To ensure trajectory tracking by the joint variable, errors [28, 1, 30, 31].

$$(9.6.1) \qquad \mathbf{e}(t) = \theta_d - \theta(t)$$

differentiate $\mathbf{e}(t)$ to obtain:

$$(9.6.2) \qquad \dot{\mathbf{e}}(t) = \dot{\theta}_d - \dot{\theta}(t)$$

Then, the overall robot arm input becomes:

$$(9.6.3) \qquad \Gamma = \mathbf{M}(\ddot{\theta} + \mathbf{k}_v \dot{\mathbf{e}} + \mathbf{k}_p \mathbf{e}) + \mathbf{H}(\theta, \dot{\theta}) + \mathbf{J}^T \mathbf{f}$$

This controller is shown in Figures 9.6.1, and 9.6.2

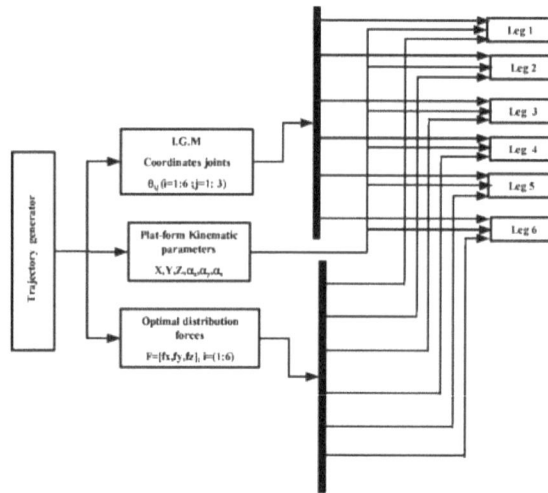

FIGURE 9.6.1. **Control model of the hexapod**

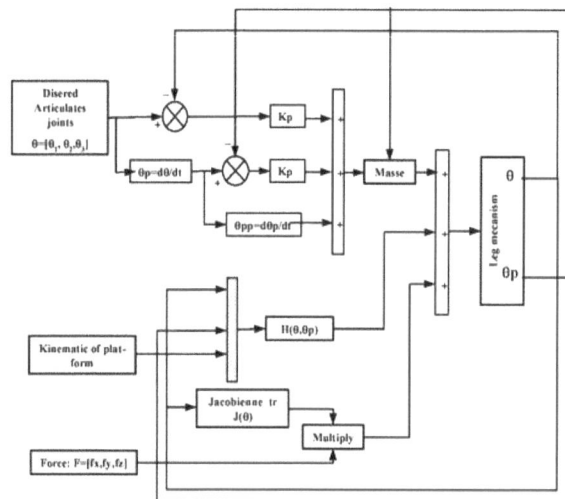

FIGURE 9.6.2. **Control model of a leg "j"**

9.6.1. Choice of PD Gains. It is usual to take the $n \times n$ diagonal matrices so that:

$$\boldsymbol{k}_v = diag[k_{vi}], \quad \boldsymbol{k}_p = diag[k_{pi}]$$

and $k_{pi} = \omega_n^2$, $k_{pi} = 2\xi\omega_n$ with ξ the damping ratio and ω_n the natural frequency. The PD gains are usually selected for critical damping $\xi=1$. Then, to avoid exciting the resonant mode, we should select natural frequency to half the resonant frequency $\omega_n < \omega_r/2$.

9.6.2. Simulation results. In order to show the effectiveness of the proposed approach, some simulations were conducted using Matlab. We consider that the hexapod robot is crawling in a linear trajectory (Y=3X), on an ground situated, in the X-Y plane. Furthermore, the force tensor acting at the body center are (Fx = -3 [N], Fy = 5 [N], Fz = -50 [N], Mx = 0 [Nm], My = 2 [Nm], Mz = 1 [Nm]). The sizes and parameters of the Hexapod robot are shown in

Figures 9.4.3 and 9.4.4, where a = 0.25 [m], b = 0.6 [m], $l_1 = 0.05$ [m], $l_2 = 0.20$ [m], $l_3 = 0.30$ [m] and $l_4 \simeq 0$ [m]. There are three actuated joints $\theta_{1,j}$, $\theta_{2,j}$, $and\,\theta_{3,j}$ in the leg "j", for (j=1,...,6). The masses and the inertia of links are respectively (m1 = 0.1, m2 = 0.07, m3 = 0.03 [kg]) and (Ix1 = 1.36, Iy1 = 0.297, Iz1 = 1.6, Ix2 = 2.1, Iy2 = 2.29, Iz2 = 0.33, Ix3 = 0.001, Iy3 = 0.05, Iz3 = 0.05 [kg cm^2]); The simulation is presented for two cycles of walking corresponding to 15 seconds with 0.01 S as the step. In this time the hexapod cross 0.14 [m]. The lifted legs do a cycloid trajectory.

9.7. Conclusion

The chapter presents a complex mathematical formalism useful in dynamic studies of mobile robots. This formalism is based on the Newton Euler method, completed by the Lagrange formalism and adapted to the complex study of the mobile spatial systems. We presented the problem of the force distribution that can be initiated and applied on the hexapod robot. At first, the hexapod inverse and direct geometric models were developed and presented. After working on reducing the size of the problem and proceeding to the necessary transformation, the initial problem is then solved in terms of the quadratic programming problem. In the end an algorithm of the computed-torque control was established to ensure trajectory tracking by the joint variable, errors. Finally, some simulations were given in order to show the effectiveness of the proposed approach.

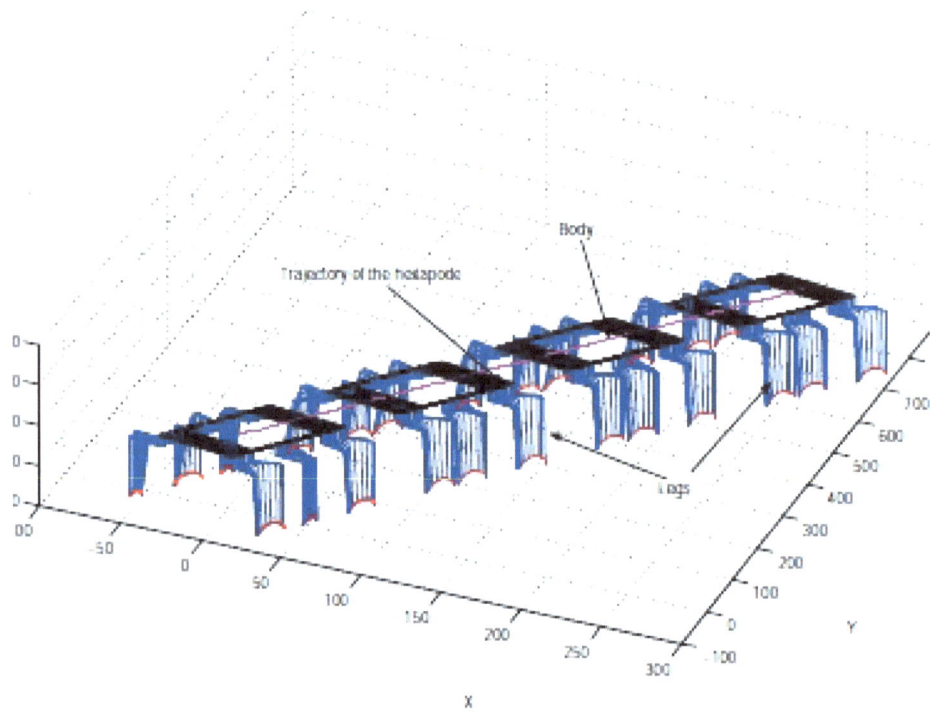

FIGURE 9.7.1. **View of crawling hexapod**

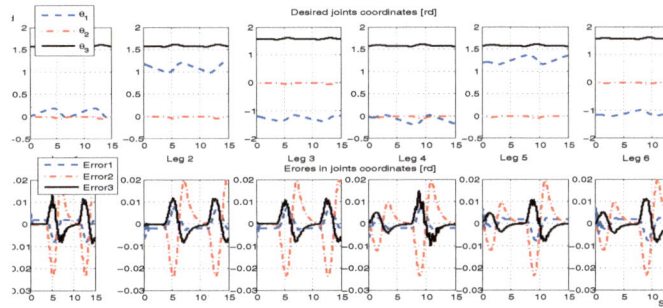

FIGURE 9.7.2. **Errors in coordinates joints**

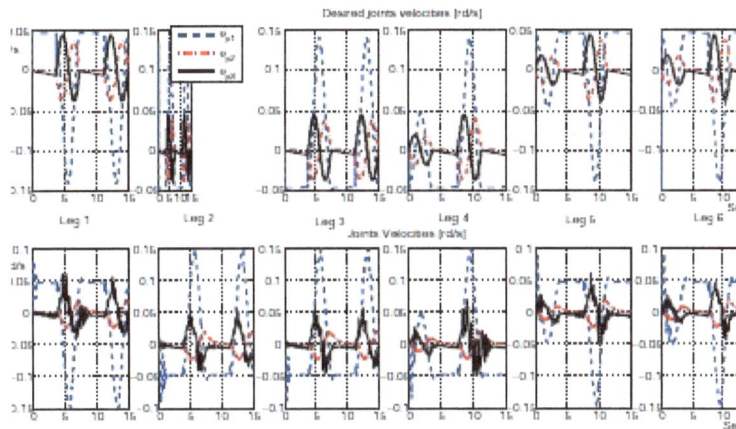

FIGURE 9.7.3. **Velocities coordinates joints**

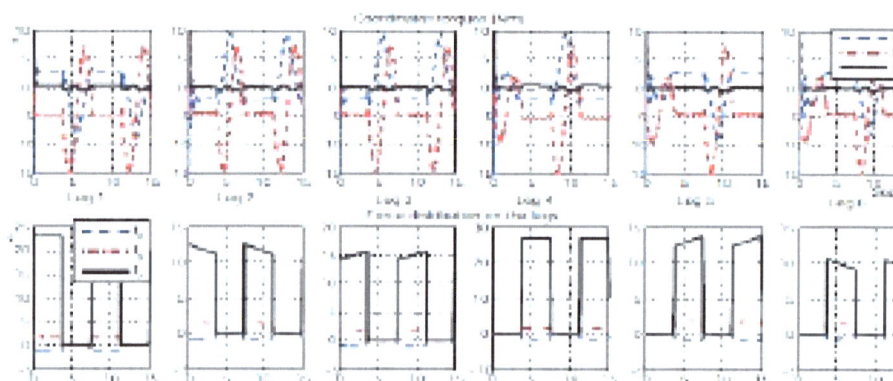

FIGURE 9.7.4. **Articulations torques and reactions of the ground**

Bibliography

[1] W. Khalil and E.Dombre, *modelisation identification et commande des robots*, Hermes sciences, paris, 1999.

[2] D. E. Orin and S. Y. Oh, *Control of force distribution in robotic mechanisms containing closed kinematic chains*, Journal: Trans. of the ASME, J. of Dynamic Systems, Measurement, and Control, volume:102,134-141,1981.

[3] Graca, P.; Zimon, J, *Six- legged technical walking considering biological principles*, Electrodynamic and Mechatronics, 2009. SCE 11 '09. 2nd International Students Conference , 23-24,2009.

[4] Zoltan Pap et al, *Optimization of the Hexapod Robot Walking by Genetic Algorithm* , SISY 2010 o 2010 IEEE 8th International Symposium on Intelligent Systems and Informatics,September 10-11, 2010, Subotica, Serbia

[5] C. A. Klein and S. Kittivatcharapong, *Optimal force distribution for the legs of a walking machine with friction cone constraints*, IEEE Trans. on Robotics and Automation, volume:6, 73-85,1990.

[6] F. T.Cheng and D. E. Orin, *Optimal force distribution in multiplechain robotic systems*, Journal:IEEE Trans. on Systems, Man, and Cybernetics, volume:21, 13-24,1991.

[7] Xingjun Tan, Yujun Wang and Xinqiang He, *The Gait of a Hexapod Robot and Its Obstacle-surmounting Capability* , Proceedings of the 8th World Congress on Intelligent Control and Automation June 21-25 2011, Taipei, Taiwan.

[8] William A. Lewinger and Roger D. Quinn, *A Hexapod Walks Over Irregular Terrain Using a Controller Adapted from an Insect's Nervous System*, The 2010 IEEE/RSJ International Conference on Intelligent Robots and Systems October 18-22, 2010, Taipei, Taiwan.

[9] F. T.Cheng and D. E. Orin, *Efficient formulationof the force distribution equations for simple closed-chain*, Journal:IEEE Trans. on Systems, Man, and Cybernetics, volume:21, 25-32,1991.

[10] F. T. Cheng and D. E. Orin, *Efficient algorithm for optimal force distribution-the compact-dual LP method*, Journal:IEEE Trans. on Robotics and Automation, volume:6, 178-187,1990.

[11] Pei Chun Lin,Haldun Komsuoglu, and Daniel E. Koditschek, *A Leg Configuration Measurement System for Full-Body Pose Estimates in a Hexapod Robot*, Journal: IEEE Trans. on robotics, volume: 21, NO. 3, June 2005.

[12] M. A. Nahon and J. Angeles, *Optimization of dynamic forces in mechanical hands*, Journal:Trans. of the ASME, J. of Mechanical Design, volume:113,167-173, 1999.

[13] M. A. Nahon and J. Angeles, *Real-time force optimization in parallel kinematic chains under inequality constraints* Journal:IEEE Trans. on Robotics and Automac 1999 Cyber Scientific Machine Intelligence and Robotic Control, volume:12,87-94, 1999.

[14] J.F. Gardner, *Force distribution in walking machines over rough terrain*, Journal:Trans. of the ASME, J. of Dynamic Systems, Measurement and Control, volume:113, 754-758,1991.

[15] W. Kwon and B.C. Wen, *A new optimal force distribution scheme of multiple cooperating robots using dual method*, Journal of Intelligent and Robotic Systems, volume:21, 301-326,1998.

[16] Y. D. Shin and M. J. Chung, *Optimal force distribution by weak point force minimization in cooperating multiple robots*, in Proc. of the IEEE/RSJ Int. Workshop on Intelligent Robots and Systems, Osaka, Japan, 767-772,1991.

[17] Y. R. Hu and A. A.Goldenberg, *Dynamic control of multiple coordinated redundant robots*, Journal:IEEE Trans. on System, Man, and Cybernetics,volume:22,568-574,1992.

[18] S. Mukherjee and K. J.Waldron, *An exact optimization of interaction forces in three-fingered manipulation*, Journal:Trans. of the ASME, J. of Mechanical Design, volume:114, 48-54,1992.

[19] Gao Jianhua, *Design and Kinematic Simulation for Six-DOF Leg Mechanism of Hexapod Robot* , Proceedings of the 2006 IEEE International Conference on Robotics and Biomimetics , December 17 - 20, 2006, Kunming, China.

[20] L. T.Wang and M. J. Kuo, *Dynamic load-carrying capacity and inverse dynamics of multiple cooperating robotic manipulators*, Journal:IEEE Trans. on Robotics and Automation, volume:10,71-77, 1994.

[21] R.C Paul, *Robots manipulators, mathematics, programing and control* , MIT press, 1981.

[22] X. Chen and K.Watanabe, *Optimal force distribution for the legs of quadruped robot*, Journal:Machine inteligence and robotique control, volume:1,87-94,1999.

[23] J.F. Gardner, K. Srinivasane, *A solution for the Force distribution probleme in reduandanly actuated closed Kinematic chain*, Journal:Trans. of the ASME, J. of Dynamic Systems, Measurement,and Control, volume:112, 523-526,1990.

[24] H. Liu and B. Wen, *Force distribution for the legs of quadruped robot*, Journal: of Robotique Systemes, volume:14, 1-8,1997.

[25] C. Mahfoudi K Djouani S. Rechak and M. Bouaziz, *Optimal Force Distribution for the legs of an Hexapod robot* , Journal:2003 IEEE Conference on Controle application CCA 2003, volume:June 23-25, Istambul, Turkey, 1997.

[26] A. Fijany K. Djouani G. Fried and J. Pontnau, *New Factorization Technics and Fast Serial and Parallel Algorithms for Operational Space Control of Robot Manipulators", Proceedings of IFAC, 5th Symposium on Robot Control,* ,Nantes, France,813-820,1997.

[27] V. Kumar and K. J.Waldron, *forces distribution in walking vehicules*, Journal: Trans. of the ASME, J. of Mechanical Design, volume:112,90-99,1990.

[28] Q. Xiding and G. Yimin, *Analysis of the Dynamics of a six-legged Vehicle* , journal:The International Journal of Robotics Research, volume:14,1-8,1995.

[29] B. S. Lin and S. Song, *Dynamic Modeling, Stability, and Energy Efficiency of a Quadrupedal Walking Machine* , Journal of Robotics Systeme, volume:18,657-670,2001.

[30] P. Alexandre and A. Preumont, *On the gait control of a six-legged walking machine*, Journal: International Journal of Systemes Sience, volume:27(8), 713-721,1996.

[31] F. Pfeiffer J. Eltze and H. Weidman, *Six- legged technical walking considering biological principles*, Journal: Robotics and Autonomous Systems, volume:14, 223-232,1995.

[32] Richard J. Lock, Ravi Vaidyanathan, Stuart C. Burgess, and Roger D. Quinn, *Impact of Passive Stiffness Variation on Stability and Mobility of a Hexapod Robot* , IEEE/ASME International Conference on Advanced Intelligent Mechatronics, Singapore, July 14-17, 2009.

CHAPTER 10

Robotized Brachytherapy of Prostate

Vincent Coelen[1],
Polytech-Lille, LAGIS, UMR CNRS 8219
Avenue Paul Langevin, 59655 Villeneuve D'Ascq, France
vincent.coelen@polytech-lille.net

Rochdi Merzouki
Polytech-Lille, LAGIS, UMR CNRS 8219
Avenue Paul Langevin, 59655 Villeneuve D'Ascq, France
rochdi.merzouki@polytech-lille.fr

Eric Lartigau
Centre Oscar Lambret de Lille
Université Lille 2, France
59000 Lille Cedex, FRANCE
E-Lartigau@o-lambret.fr

ABSTRACT. In this chapter, a robotic concept for the prostatic brachytherapy is described. Based on adaptive tracking of mobile targets under ultrasound control using an industrial robot manipulator, the concept integrates two main tasks: mobile target tracking and on-line supervision. The mobile target corresponds to the track at the millimeter scale to the cancerous tissues. This task is done after detection of the prostate contour applied on acquired ultrasound images. The targets are selected safety with manual or automatic approaches, through a virtual environment, where their coordinates are transferred to the robot. The latter is controlled adaptively according to the target position, thus realizing the both brachytherapy motions, related to the needle insertion and the seed injection. The on-line supervision of the robotic concept is performed, based on quantitative approach, it improves the safety during the brachytherapy motions. A fault diagnosis strategy is applied to detect a possible change of the operation modes of the joint actuator. Experimental results are applied on a prostate phantom using one access point, they show the advantage of the described concept in reducing trauma and recovery times for real patients

Keywords: Robot manipulator, Prostatic brachytherapy, Target tracking and On-line supervision.

10.1. Introduction

During the last decade, new cases of the prostate cancer was increased significantly all over the world, where the incidence in France for example, was reached the 62 245 new cases in 2005, and there were 8937 deaths in 2006 [1]. Due to the increase of the incidence since 1980, the prostate cancer is considered as a public health problem. The World Cancer Report [13] published by the World Health Organization (WHO) [27] described 700 000 new cases of this cancer in the world, and also a net increase of incidence in the developed and emergent countries. Brachytherapy is a recent technique used to treat located cancerous tumors, consisting of injecting radioactive elements directly in a microscopic targeted area. This treatment is used partly in the treatment of prostate cancer, where the manual procedure involves inserting a set of needles through the abdomen to the prostate and loaded with the adequate dosimetry of radioactive material, using a static grid under ultrasound guidance. The treatment of the prostate cancer with this technique was known from 1910, but real development of prostate brachytherapy was started from 1970-1980 [2]. The aim of this technique is to implant permanent radioactive seeds into the prostate, in order to eliminate located cancer

---

[1]Corresponding author

cells. The applied technique on the patients is done under general anaesthesia. Seeds are hold in needles, which are inserted into the prostate across the perineum by the radiotherapist. To help the operator for the manual insertion of the needle, an external grid exists. The needles have to go through the grid before be inserted into the patient. The prostate and the needles are followed by an endo-rectal ultrasound probe. Compared to surgery, the classical brachytherapy gives approximately the same results in term of the accuracy, but it is better for sexual toxicity and less suffering for the patient [2]. In Europe, thousands of prostate implantation had been realized this last decade. In parallel, research has been conducted to introduce robotic assistance of brachytherapy of the prostate in order to improve the accuracy in term of target reachability and seed placement. Such a solution has an additional advantage of maintaining accurate repetitive movements.

Major developments in the robotized surgery had come by using human-machine interface like the *DaVinci* robot. It allows minimally invasive surgery, with accuracy, 3D and haptic feedback detailed in [4]. Others systems of robotized surgery are in the border with brachytherapy, in Trejos *et al.* [5], a robotic system is used to implant seeds in a phantom representing a chest, under both ultrasound and electromagnetic guidances.

Wei et al. in [22], presented a solution of inserting multiple needles simultaneously in the prostate is proposed. The benefits presented by the authors is a reduction in the time of the intervention, a reduction of edema and a reduction in the number of registration in the calculation of image-coupling robot. Other contributions to the automation of brachytherapy propose to use systems to overcome the static grid, using the robots' accuracy far greater than that of a human. For example, in Fichtinger et al. [6], the proposed system can vary both the position of the needle guide to reproduce virtually grid, and can change the orientation of the guide. In [16], automated architecture of brachytherapy is given in closed loop using industrial manipulator. The aim is to exploit the accuracy and mobility of a 6 degree of freedom (dof) robot for the both mission of insertion and injection. Phee et al. in [7] have proposed a technique to access the prostate through a single input point on the perineum, but it may cause a toxicity problem and suboptimal dose coverage. Patriciu et al. in [8] suggested an automatic system seed placement under Magnetic Resonance Imaging (*MRI*) guidance. A robot gripper has been managed to place seeds into a phantom. The *MRI* guidance is better accurate than ultrasound, but involve non-metallic components. Another independent system is the *Cyberknife* robot, which realize an external radiotherapy without physical contact with the patient and under indication in oncology [9]. This last technique uses an industrial robot of type of *KUKA*$^@$ [14], including a mini linear accelerator, with positioning and mobile tumor tracking system. There is no direct interaction between patient and unit. This system had managed to open new indications of incurable case in oncology field [12].

This chapter presents a concept of robotized brachytherapy of the prostate, based on adaptive control for target tracking using a manipulator robot under ultrasound guidance. The proposed concept combines between the accurate control of the brachytherapy needle for insertion and injection movements with on-line supervision. This latter is considered to monitor the whole robot functioning during the brachytherapy act. The model-based monitoring technique allows not only to detect abrupt faults but also to anticipate any deviations on the robot trajectories in real time, through a virtual and interactive environment.

10.2. Mobile Target Tracking

This section presents the methodology for controlling a manipulator robot in tracking on-line a mobile target at the millimeter scale. This methodology regroups three main steps: contour detection of the prostate, target definition and adaptive tracking of the target. For this, the following assumptions are considered:

- Ultrasound image processing is done in 2D using different transversal sequences, related to the depth of the ultrasound probe in the interacted environment;
- The dosimetry is not considered in this method, only the reachability of mobile targets with accuracy and safety is taken into account.

10.2.1. Contour detection. The contour detection algorithm is based on the active contour model algorithm [10], [11] and applied to an ultrasound image as it is the most used for the brachytherapy acts nowadays. The algorithm is given as follows:

- *Preprocessing*: When the ultrasound image is acquired, frequently, it contains a low level of contrast with noises, then preprocessing filter is used to prepare the image before the segmentation step. A median filter is used for ultrasound image preprocessing, according to the high frequency of the noises.

- *Segmentation*: Detection of the prostate contour is performed by using active contour called *snake* algorithm [**10**]. This segmentation is a semiautomatic technique, where its goal is to approximate the perimeter of a static object. The active contour is defined as a continuous deformation curve, represented parametrically as:

$$v(s) = \{x(s), y(s)\}\, ; s \in [0,1]$$

where $x(s)$ and $y(s)$ are the coordinates along the contour $v(s)$.

Position and deformation of the contour are defined by an energy function of 10.2.1:

(10.2.1)
$$E(s) = \int_0^1 [E_{int}(v(s)) + E_{img}(v(s))]\, ds$$

where E_{int} is the internal energy of the contour, it represents the flexibility and extensibility of the contour. E_{img} is the energy associated with the image, it forces the contour to move to the edge of object. Position and shape are determined by the energy minimization process, where the lower values of E_{img} should be at edge positions.

Internal energy is calculated as:

(10.2.2)
$$E_{int} = \alpha E_{cont} + \beta E_{curv}$$

Where:

E_{cont} is the continuity energy, where it forces the contour to keep continuous. In discrete case, it maintains the points at equal distance.

E_{curv} is the smoothness terms of the energy.

Then, image energy is defined as:

(10.2.3)
$$E_{img} = \gamma E_{grad} + \delta E_{flow}$$

E_{grad} is the gradient energy. This energy is mostly used in active contour because it is minimal at edge location. E_{flow} is the gradient flow energy, it guides the snake structure to the minimum value of E_{grad}.

α, β, γ and δ are parameters used to control the importance of each element of energy in the total energy.

- *Minimization*: Energy minimization could be done by calculating the energy of all the image points, but it takes long, then an iterative method is used to make the convergence of the prostate contour. This method requires an initial referred contour, which is close to the real prostate contour and done by an expert radiologist before the brachytherapy, during the pre-implant exam. Thus, the energy image is calculated directly after the preprocessing for all the image.

Results of the presented algorithm are given in Figure 10.2.1, which describes the steps of active contour detection algorithm. Figure 10.2.1-b shows the preprocessing image result applied on the initial image of Figure 10.2.1-a using a median filter, in order to exclude noises with high frequencies. Then, Figure 10.2.1-c shows a progressive convergence of the detection contour algorithm where the energy attraction is still actuate. In Figure 10.2.1-d, a final convergence of the algorithm is reached and noticed by stop motion of all considered contour samples. For this presented example of image, the segmentation time approached 0.225 sec.

Figure 10.2.1 shows the accuracy of the contour detection algorithm (red line) comparing to manual drawing contour done by an expert (green line). Finally, Figure 10.2.1 gives a real scale for comparison between the expert contour and the automatic detected contour.

10.2.2. Target definition. When the contour is well detected after the active algorithm convergence (Figure 10.2.2), a virtual grid (green points) is then automatically generated from referred origin position, and superposed on the ultrasound image. The distance between two successive points of this grid is about 5 mm. The accessible area is then defined according to the dynamic contour position and the prohibited area of the urethra (red zone) of Figure 10.2.2. Thus, the reached targets are selected manually or automatically. The coordinates of the targets are deduced in two dimensions at the image frame, while the third dimension coordinate corresponds to the insertion depth of the ultrasound probe. They are finally represented in robot world frame.

10.2.3. Adaptive target tracking . Hardware and software tools used for adaptive target tracking of the robotic concept of prostatic brachytherapy are developed in this section.

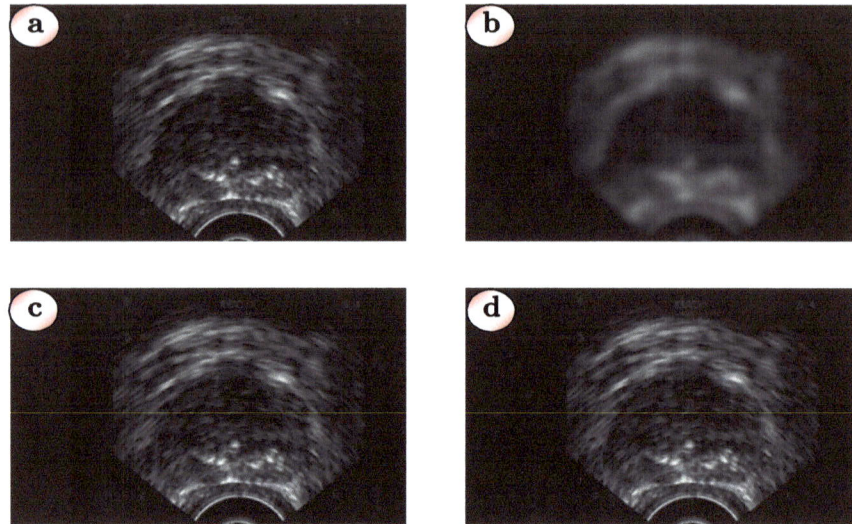

FIGURE 10.2.1. 2D contour detection of the prostate: (a): Initial image; (b): Preprocessing image; (c) During algorithm convergence; (d) Final convergence

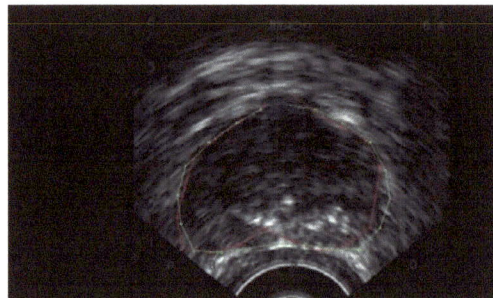

FIGURE 10.2.2. Superposition of the prostate contour of expert (green line) and detected contour (red line)

10.2.3.1. *Hardware components description.* The industrial robot manipulator used for this medical application is a KUKA KR6-ARC (Figure 10.2.5). It is developed to perform precision tasks such as welding or painting for industrial application. This robot consists of 6 degrees of freedom, with a redundant 7th linear axis (Figure 10.2.6-3).

It is controlled in joint or linear motions, through a Human Machine Interface (HMI) developed under WindowsXP Embedded (Figure 10.2.5). The robot operating system is VxWorks, with real-time kernel.

10.2.3.2. *On-line control.* The control phase is defined by the following steps: first, a trans-rectal ultrasound probe is used to acquire prostate images. Then, an active contour detection algorithm is implemented (Figure 10.2.6-1), in order to detect the dynamic prostate contours in 2D. This step is managed into virtual environment , composed by the industrial robot dynamic and its associated prostatic phantom (Figure 10.2.6-2). The identified targets are determined also in the virtual environment, using an adaptive virtual grid. These targets can be selected manually by a human operator or automatically after defining the safety areas (Figure 10.2.6-1). These targets are defined in image coordinates, then converted in the world robot frame as a centre of the tool coordinates (Figure 10.2.6-4). Robot controller defines the robot movements by using the inverse geometrical model (Figure 10.2.6-3). Communication between the virtual environment and the real robot is made by a local network.

The whole control is done through a virtual environment (Figure 10.2.6-2), by considering the dynamics of the robot and the targets. This is realized by the on-line data fusion of the joint positions of the robot and the targets

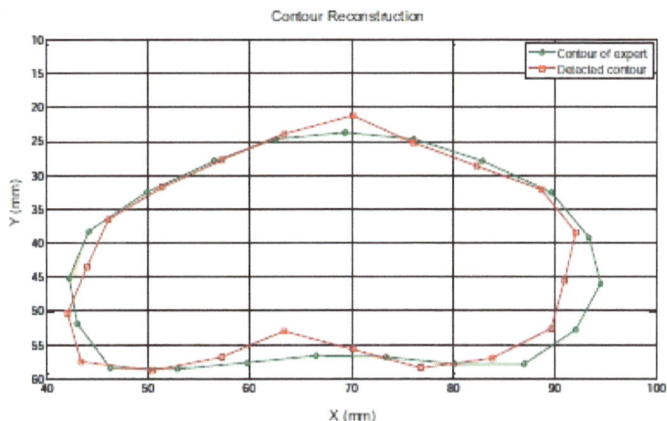

FIGURE 10.2.3. Comparison of expert contour and detected contour of prostate in real scale

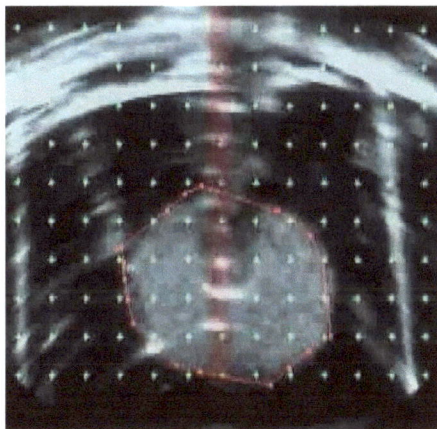

FIGURE 10.2.4. Target position on the grid

location. The proposed algorithm compares between experimental and theoretical positions of the targets and can correct adoptively the robot trajectory.

10.2.4. Experimental results. The experimental test bench of Figure 10.2.7 is composed by an industrial robot, equipped with an adaptive pneumatic gripper to guide the brachytherapy needle . An ultrasound phantom system of the prostate is used (Figure 10.2.7).

After contour detection of the prostate and the virtual grid reconstruction, two trajectories are planned in order to access twice time the prostate phantom after validation under the virtual environment. The trajectories are transmitted off-line to the robot kernel system via Ethernet connection where it repeats the desired kinematics. In Figure 10.2.8, we describe the 3D trajectories of the center of tool, related to the needle extremity, planned to reach the two targets on the prostate phantom. The distance on *X* is calculated experimentally and it is not automatically deduced from the 2D image processing. Superposition of simulated and real trajectories shows the accuracy rate of the target tracking, where in Figure 10.2.9, the errors on the three axis is presented.

In Figure 10.2.10, the needle insertion is presented for the first selected target on the prostatic phantom. In Figure 10.2.11, an extraction of the needle is shown. It is clearly visible the flexibility effect of the needle after insertion task, and what could be significant the influence of the environment in accessing the desired targets. Figure 10.2.12 and Figure 10.2.13 show respectively insertion and extraction of the brachytherapy needle according to the second trajectory.

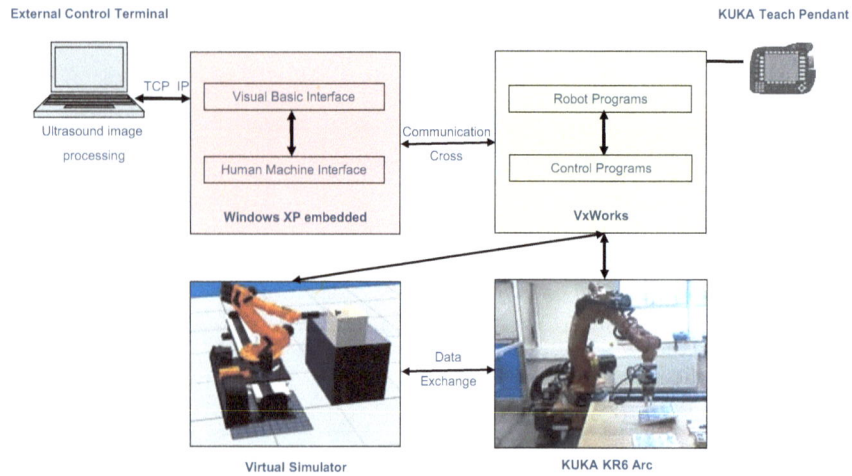

FIGURE 10.2.5. Image-robot architecture

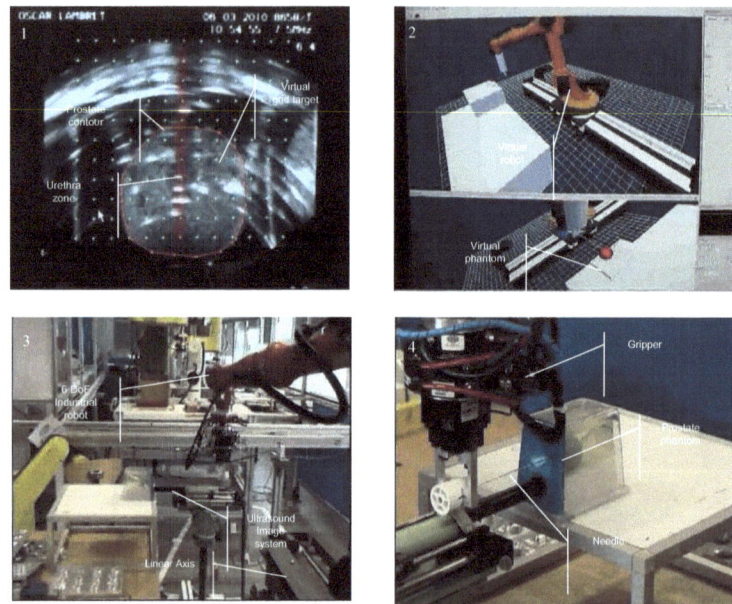

FIGURE 10.2.6. 1): Active contour detection; (2): Virtual environment of the prostatic brachyther-apy; (3): Industrial robot under ultrasound guidance; (4): Needle Insertion into the prostate target

10.3. On-line Supervision

In order to monitor the robotized brachytherapy system, it is important to track the healthy status of robot actuators and those of the medical gripper in real time , in order to detect and isolate some undesired faulty situations and to prevent reconfigurable scenarios.

In the literature, two types of methods are used for robust diagnosis of mechatronic systems, the qualitative and quantitative methods. The qualitative approach consists on dividing the parameter space into several classes that correspond to known operation modes and then to determine by learning the mathematical relationships between the effects (observations of experts, sensor measurements and statistics), and causes (faults). But often, it is generally not possible to identify all possible operation modes of the monitored system, due to imperfect knowledge of some

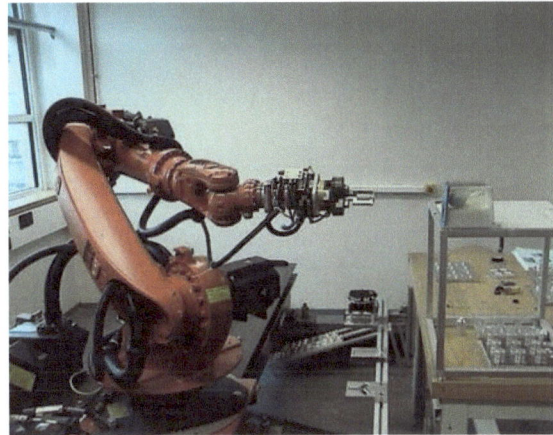

FIGURE 10.2.7. Experimental Test Bench for the prostatic Brachytherapy with robotic concept

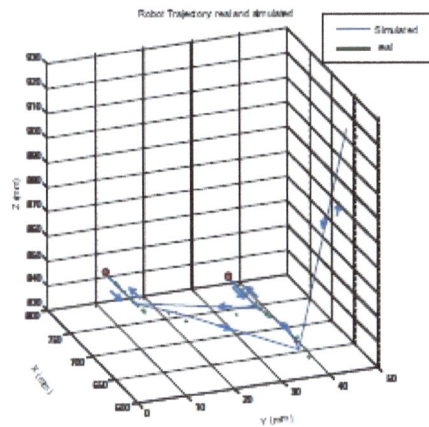

FIGURE 10.2.8. Simulated and real trajectories of the robot

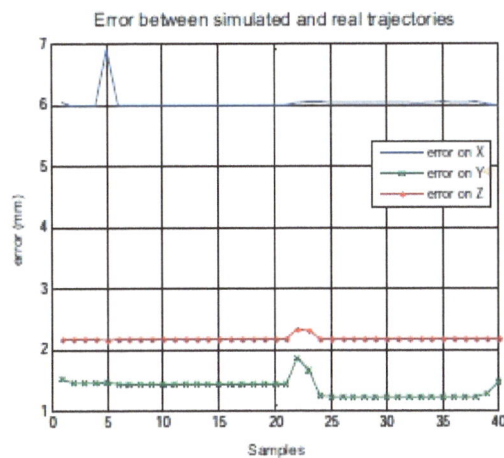

parameters and uncertainties on its model. Thus, the principle of qualitative methods is mainly based on solving a problem of pattern recognition. Several studies have been made in recent years using this qualitative approach, such as in Caccavale et al. [15], the authors used the neural networks method in order to estimate the input fault of a robot

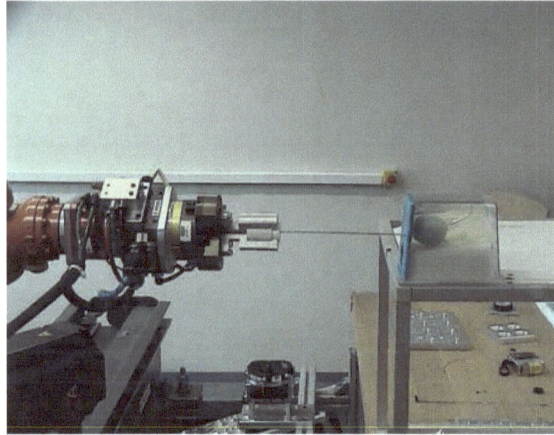

FIGURE 10.2.10. Needle insertion to the first target

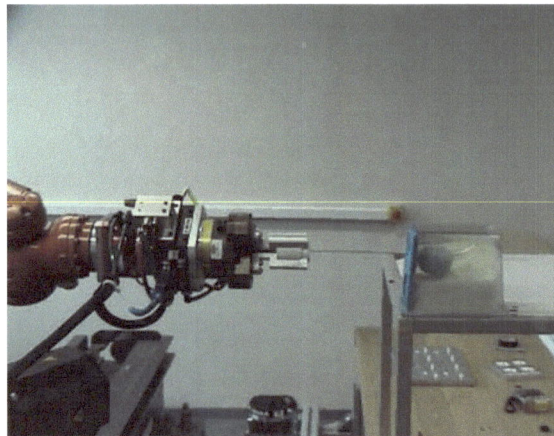

FIGURE 10.2.11. Needle extraction from the first target

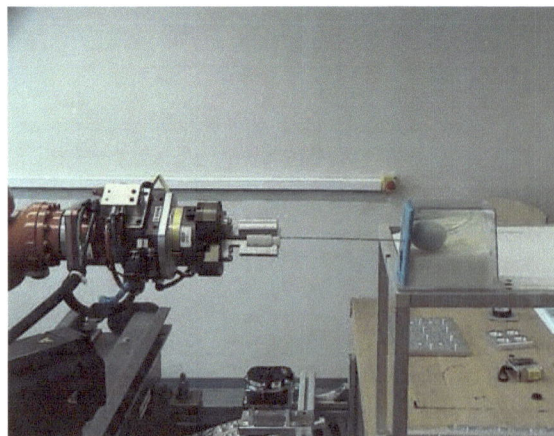

FIGURE 10.2.12. Needle insertion to the second target

manipulator. In Pettersson et al. [19], a pattern recognition techniques is applied for model-free execution monitoring in normal or faulty execution. Two model-based Fault Detection and Isolation (FDI) schemes for robot manipulators

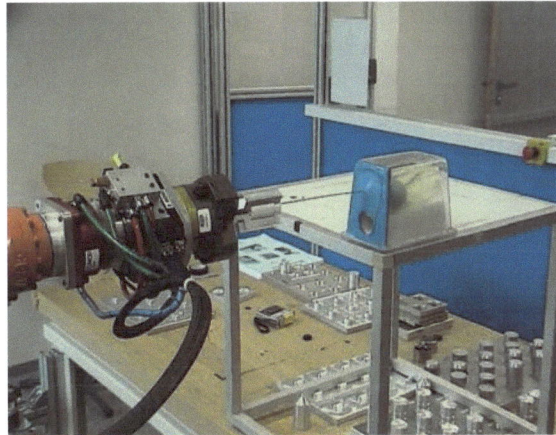

FIGURE 10.2.13. Needle extraction from the second target

using soft computing techniques, as an integrator of Neural Network (NN) and Fuzzy Logic (FL) is proposed in Yksel et al. [18]. Eski et al. in [17] shown experimental results for the detection of faults on actuators of robot manipulators using artificial intelligence based on neural networks. For quantitative approaches, also called model-based methods, the principle is to generate analytical redundancy relations, expressing the difference between the information issued from the real system and those generated by the system model given in normal operation. These relations also called residuals, characterize the mode of the system and are equal to zero in normal situation. Among the recent research literature, in [20], the authors used the observer principle in the generation of residuals, for detection of actuator faults for robot manipulator and in [21], a parity relation approach is used to fault diagnosis of manipulation robots using two distinguished methods.

In the following section, a dynamic model of electromechanical actuators located at the brachytherapy robot joints is used to develop a fault diagnosis strategy. The latter is considered to monitor the whole robot operations during the insertion and injection movements. The model-based monitoring technique allows not only to detect abrupt faults but also to anticipate any deviations on the robot trajectories in real time, through a virtual and interactive environment, used to make an on-line supervision of the robotized brachytherapy.

10.3.1. Supervision architecture. During the robotized brachytherapy operation, the robot trajectory needs to be planned with the seed dosimetry. In the presented methodology, a KUKA (KR6 Arc) manipulator robot is used to reach the desired millimetric targets. The programming language KUKA Robot Language (KRL) is used to create trajectories easily in industrial environment. The programs are done through the KUKA Control Software (KCS) which is running under operating system VxWorks. The Input/Output (I/O) connections are used to communicate with the program, such as the serial connection of type RS232. Although VxWorks is a real time operating system, the user can not influence the execution of a motion during the interpolation process.

For this case, another software package named Remote Sensor Interface (RSI) is installed. This package allows users to control the robot in real-time, by influencing the robot motion via a correction during the path execution. The add-on RSI-EthernetXML is considered to exchange data between the robot and a remote server using an Ethernet connection as shown in Figure 10.3.1. When the connection is established, the robot controller send messages every 12ms (interpolation cycle) and the remote computer must answer in a given time. All the communication packets are transmitted as Extended Markup Language (XML) strings, where it is possible to configure the content of these packets. In default configuration, the robot sends its actual joint and Cartesian positions, while the remote computer sends joint angle or position correction. Predefined objects in RSI software are used to send joints angles, position, current and torque of each joint actuator to the remote application for the supervision.

The remote computer is running under a Debian Squeeze GNU/Linux operating system with a real-time Xenomai Kernel. It is important for such accurate robot operation to use a real time kernel to ensure a robust execution of the task with respect to the timing.

In the supervision application (Figure 10.3.1), after a thread starts handling the communication with the robot, another part executes the monitoring algorithm. The thread communication opens a Transmission Control Protocol

(TCP) connection with the robot and start parsing the received messages using the libXML2 package (XML C parser and toolkit of Gnome) recommended by KUKA. The joint corrections are sent according to the planned path for the brachytherapy. Another task is executed in parallel for data collection, used to monitor on-line the joint actuators of the robot.

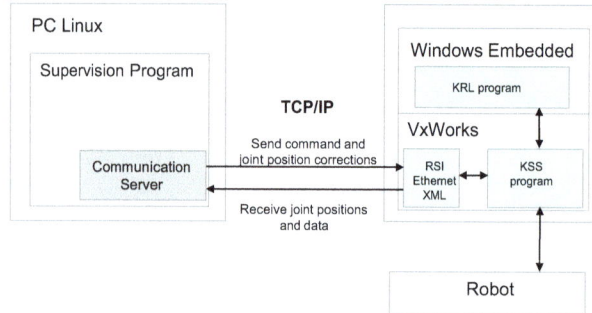

FIGURE 10.3.1. Supervision architecture for Robotized Brachyterapy of Prostate

10.3.2. Fault diagnosis strategy. The applied fault diagnosis methodology [**24**] is given by the following steps:

(1) Dynamic modeling of the non-linear and uncertain electromechanical joint actuator of the manipulator arm;
(2) Identification of the joint actuator parameters. Uncertainties are calculated using the statistical or interval approaches [**25**];
(3) Generation of Analytical Redundancy Relations (ARRs) for the uncertain system by decoupling the nominal and the uncertain parts. Residual corresponds to the nominal part of the ARR, while the threshold is deduced from the uncertain part of the ARR;
(4) Sensitivity analysis of the residual is calculated using the uncertain part of the ARR, allowing the calculation of the fault delectability indexes.

The considered electrical part of the electromechanical actuator corresponds to RL circuit of the jth DC motor, composed by: input voltage source U_{0_j}, electrical resistance R_{e_j}, inductance L_j and electromotive force feedback EMF, which is linear to the angular velocity of the rotor and equal to with k_{e_j} the EMF constant.

The mechanical part is characterized by its inertia, viscous friction f_{e_j}, a gear constant N_j and transmitted torque C_j used to drive the arm. The considered parameter uncertainties p_j are multiplicative noted δp_j. Thus, the jth dynamic of electromechanical joint actuator is expressed by the following equations [**26**]:

(10.3.1)
$$
\begin{cases}
U_{0_j} = R_j.I + L_j.\dot{I} + k_{e_j}.\dot{\theta}_e \\
\quad + \delta_{R_j}.R_j.I + \delta_{L_j}.L_j.\dot{I} + \delta_{k_{L_j}}.k_{e_j}.\dot{\theta}_e \\
k_{e_j}.I = J_{e_j}.\ddot{\theta}_e + f_{e_j}.\dot{\theta}_e + N_j.C_j - \delta_{k_{e_j}}.k_{e_j}.I \\
\quad + \delta_{J_{e_j}}.J_{e_j}.\ddot{\theta}_e + \delta_{f_{e_j}}.f_{e_j}.\dot{\theta}_e + \delta_{N_j}.N_j.C_j
\end{cases}
$$

The residual for each dynamic is calculated after ARRs generation as follows

(10.3.2)
$$
\begin{cases}
r_{1_j} = U_{0_j} - R_j.I - L_j.\dot{I} - k_{e_j}.\dot{\theta}_e \\
r_{2_j} = k_{e_j}.I - J_{e_j}.\ddot{\theta}_e - f_{e_j}.\dot{\theta}_e - N_j.C_j
\end{cases}
$$

while the thresholds are calculated from all parameter uncertainties as follows

(10.3.3)
$$
\begin{cases}
S_{1_j} = \left|\delta_{R_j}.R_j.I\right| + \left|\delta_{L_j}.L_j.\dot{I}\right| + \left|\delta_{k_{e_j}}.k_{e_j}.\dot{\theta}_e\right| \\
S_{2_j} = \left|\delta_{J_{e_j}}.J_{e_j}.\ddot{\theta}_e\right| + \left|\delta_{f_{e_j}}.f_{e_j}.\dot{\theta}_e\right| + \left|\delta_{N_j}.N_j.C_j\right| + \left|\delta_{k_{e_j}}.k_{e_j}.I\right|
\end{cases}
$$

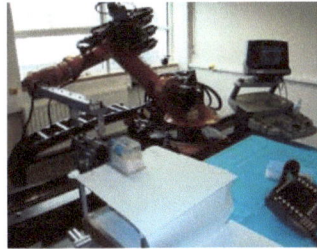

FIGURE 10.3.2. Robotic System for Brachytherapy of the Prostate

10.3.3. Experimental results. Experimental test bench of Figure 10.3.2 is composed by an industrial robot, equipped with an adaptive gripper to guide the brachytherapy needle.

These experiments focus on fault diagnosis of actuator $N°5$ (i.e. $j = 5$) for the KUKA KR6 Arc of Figure 10.3.2 used for prostatic brachytherapy motions of Figure 10.3.3.

On the Linux computer dedicated for supervision of Figure 10.3.1, the data exchange with the robot is used for on-line residual calculation and visualization via a simulator interface of Figure 10.3.6. The robustness of the residual for the mechanical part r_{2_j} of equation 10.3.2 is supported by the identified threshold, based on the uncertain part of the ARR S_{2_j} (10.3.3).

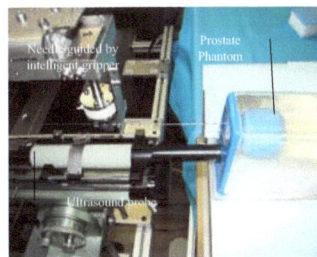

FIGURE 10.3.3. Insertion and Injection movements for the Brachytherapy of the prostate under ultrasound control

The parameters and uncertainties for actuator 5 are identified in this following Table.1:

Parameters	Nominal Values	Uncertainties	Values
J_{e5}	$0.000797 \ kg.m^2$	δJ_{e5}	0.00007
f_{e5}	$0.000015 \ N.m.s/rad$	δf_{e5}	0.0000015
R_{e5}	$1.44 \ \Omega$	δR_{e5}	0.144
L_5	$0.0147 H$	δL_5	0.00147
N_5	0.23	δN_5	0.023

TABLE 1. Experimental model parameters

The desired and planned trajectory for the robot according the centre of tool located at the extremity of the needle is given in Figure 10.3.4.

For normal situation (Healthy), the applied current on actuator $N°5$ in order to generate the trajectory of Figure 10.3.4 is represented in Figure 10.3.5.

The electric power generates a mechanical power represented by a torque effort of Figure 10.3.7 transmitted to the arm and a velocity of Figure 10.3.8.

In the normal operating mode, where any fault or disturbance are affecting the actuator $N°5$, the following residual for the mechanical part is obtained for the concerned trajectory. It is shown that the variation of the residual is encapsulated inside the adaptive thresholds (Figure 10.3.9).

For the faulty situation, the transmitted torque is stopped from 13.7 sec to 17.3 sec (Figure 10.3.10)

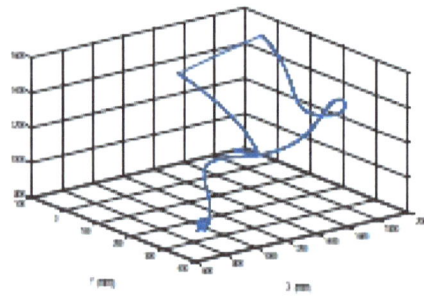

FIGURE 10.3.4. Planned robot trajectory

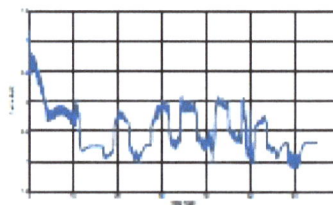

FIGURE 10.3.5. Current control signal for actuator 5

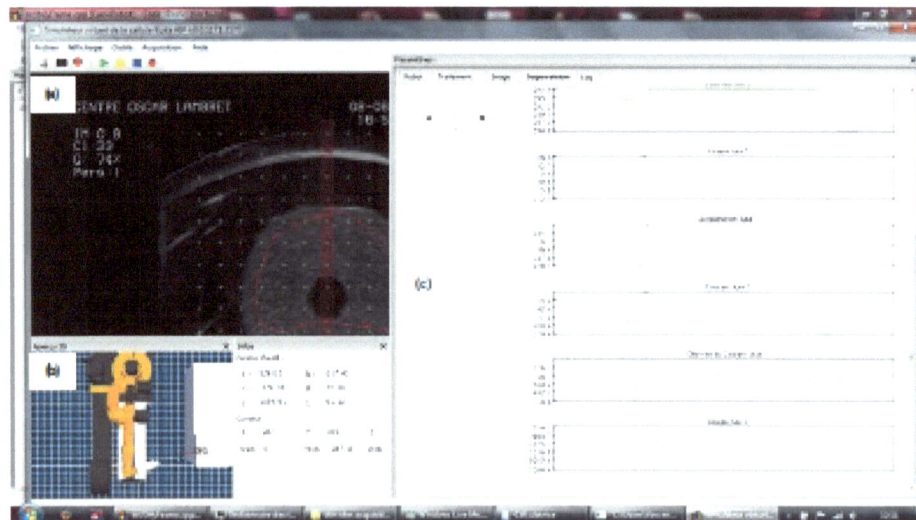

FIGURE 10.3.6. Real data acquisition with residuals deduction during needle insertion and injection operations

It is clearly noticed that for the time interval the residual of the mechanical part is sensitive to this abrupt fault, where it evaluates out the residuals (Figure 10.3.11).

10.4. Conclusion

This chapter presents and illustrates a technique of robotized brachytherapy of the prostate using an industrial robot manipulator. Industrial robots have proven themselves in the field of production. It is interesting to explore their technical speed, repeatability and accuracy in medical applications. Of course, such a robotic concept requires a perfect

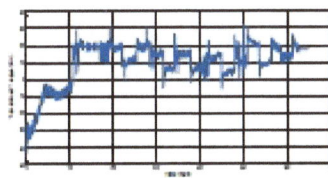

FIGURE 10.3.7. Transmitted torque in normal situation

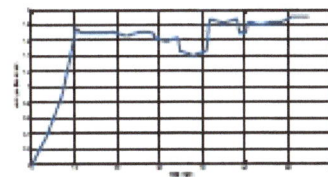

FIGURE 10.3.8. Angular position of joint 5

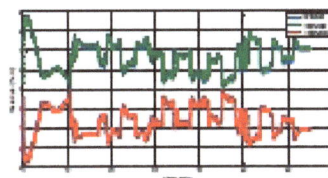

FIGURE 10.3.9. Residual of mechanical part of actuator 5 in normal situation

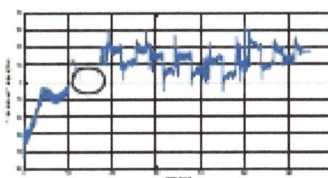

FIGURE 10.3.10. Transmitted torque for actuator 5 in faulty situation

surveillance of the almost constitutive components, during the interaction with the alive environment. Thus, a strategy of supervision is made to detect in real time any changes in the trajectories of the robot during the brachytherapy act.

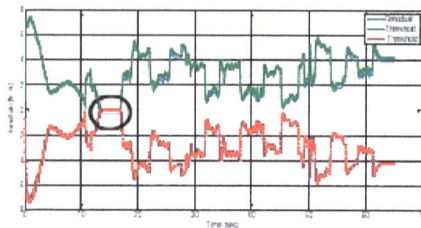

FIGURE 10.3.11. Residual of mechanical part of actuator 5 in faulty situation

Bibliography

[1] S. Guérin, F. Doyon, 'The frequency of cancer in France in 2006, mortality trends since 1950, incidence trends since 1980 and analysis of the discrepancies between these trends'. Bull Cancer vol. 96 N°1, January 2009.

[2] J-M. Cosset, T. Flam, N. Thiounn, J-C. Rosenwald, D. Pontvert, M. Timbert, S. Solignac, L. Chauveinc, 'brachytherapy for prostate cancer: old concept, new techniques.' Bull Cancer, 93 (8) : 761-6, 2006.

[3] B.S. Hilaris, M.E. Batata, L.L. Anderson, B. Pierquin, J.F. Wilson, D. Chassagne, 'Modern brachytherapy.' New York: Masson, 1987.

[4] C. Iselin , F. Fateri, A. Caviezel, J. Schwartz, J. Hauser. 'Usefulness of the DaVinci robot in urologic surgery'. La Revue médicale suisse, 3136, 2010.

[5] A-L. Trejos, A.W. Lin, M.P. Pytel, R.V. Patel, R.A. Malthaner. 'Robot-assisted minimally invasive lung brachytherapy.', Int Journal of Medical Robotics, Computer Assistance Surgery, 3: 41-51,2007.

[6] G. Fichtinger, J-P. Fiene, C. W. Kennedy, G. Kronreif, I. Iordachita, D. Y. Song, E. C. Burdette, P. Kazanzides. 'Robotic assistance for ultrasound-guided prostate brachytherapy'. Medical Image Analysis, 12 p535–545, 2008.

[7] L. Phee, J. Yuen, D. Xiao, C.F. Chan, H. Ho, C.H. Thang, P.H. Tan, C. Cheng, W.S. Ng., 'Ultrasound guided robotic biopsy of the prostate. Int J Human Robot, 3: 463-483, 2006.

[8] A. Patriciu, D. Petrisor, M. Muntener, D. Mazilu, M. Schär, D.Stoianovici. 'Automatic brachytherapy seed placement under MRI guidance.', IEEE Transactions on Biomedical Engineering, vol 54, n°8, August 2007.

[9] P-Y. Bondiau, K. Bénézery, V. Beckendorf, D. Peiffert, J.-P. Gérard, X. Mirabel, A. Noël, V. Marchesi, T. Lacornerie, F. Dubus, T. Sarrazin, J. Herault, S. Marcié, G. Angellier, E. Lartigau, 'CyberKnife® robotic stereotactic radiotherapy: technical aspects and medical indications', Cancer/Radiothérapie, Volume 11, Issues 6-7, pp. 338-344, 2007.

[10] M. Kass, A. Witkin and D. Terzopoulos , 'Snakes: Active contour models', International Journal of Computer Vision, pp. 321-331, 1988.

[11] H.M. Ladak, F. Mao, Y. Wang, D.B. Downey, D.A. Steinman, A. Fenster, 'Prostate Boundary Segmentation from 2D Ultrasound Images', Med. Phys.:27, 1777-1788, 2000.

[12] T. Lacornerie, X. Mirabel, E. Lartigau. 'The Cyberknife R : Experience of Centre Oscar- Lambret' Cancer/Radiothérapie 13, 391–398, 2009.

[13] World Cancer Report 2008. Boyle P and Levin B. International Agency for Cancer Research. Lyon 2008.

[14] www.KUKA-robotics.com

[15] F. Caccavale, P. Cilibrizzi, F. Pierri, L. Villani, 'Actuators fault diagnosis for robot manipulators with uncertain model', Control Engineering Practice, Vol. 17, pp. 146– 157, 2009

[16] V. Coelen, R. Merzouki, X. Liem, E. Lartigau, B. Ould-Bouamama, 'Contribution to a new robotic concept of prostate brachytherapy', IEEE International Conference on Robotics and Biomimetics (ROBIO), pp. 763 - 768, 2010

[17] I. Eski, S. Erkaya, S. Savas, S.Yildirim,'Fault detection on robot manipulators using artificial neural networks', Robotics and Computer-Integrated Manufacturing, Vol.27, pp. 115-123, 2011

[18] T. Yksel, A. Sezgin, 'Two fault detection and isolation schemes for robot manipulators using soft computing techniques', Applied Soft Computing, Vol. 10, pp. 125–134, 2010

[19] O. Pettersson, L. Karlsson, and A. Saffiotti, 'Model-Free Execution Monitoring in Behavior-Based Robotics', IEEE TRANSACTIONS ON SYSTEMS, MAN, AND CYBERNETICS-PART B: CYBERNETICS, Vol. 37, N°. 4, 2007

[20] V.F. Filaretova, M.K. Vukobratovic, A.N. Zhirabok, 'Observer-based fault diagnosis in manipulation robots', Mechatronics Vol. 9, pp. 929-939, 1999

[21] V.F. Filaretova, M.K. Vukobratovic, A.N. Zhirabok, 'Parity relation approach to fault diagnosis in manipulation robots', Mechatronics, Vol. 13, pp. 141–152, 2003

[22] Z. Wei, M. Ding, D. Downey, A. Fenster. 'Dynamic intraoperative prostate brachytherapy using 3D TRUS guidance with robot assistance'. Proceedings of the 2005 IEEE Engineering in Medicine and Biology 27th Annual Conference, Shanghai, China, 2005.

[23] Y. Yu, T.K. Podder, Y.D. Zhang, W.S. Ng, V. Misic, J. Sherman, L. Fu, D. Fuller, D.J. Rubens, J.G. Strang, R. A. Brasacchio, E.M. Messing. 'Robot-Assisted Prostate brachytherapy'. Book Chapter, Lecture Notes in Computer Science, Springer Berlin / Heidelberg, ISBN: 978-3-540-44707-8, 2007.

[24] M. A. Djeziri, R. Merzouki, B. Ould-Bouamama, 'Robust Monitoring of Electric Vehicle with Structured and Unstructured Uncertainties', IEEE Transactions on Vehicular Technology, Vol. 58, pp. 4710-4719, 2009.

[25] M. A. Djeziri, R. Merzouki, B. Ould-Bouamama, G. Dauphin-Tanguy, 'Robust Fault Diagnosis by Using Bond Graph Approach', IEEE/ASME Transactions on Mechatronics, Volume 12, Issue 6, pp. 599 – 611, 2007.

[26] R. Merzouki, K. Medjaher, M. A. Djeziri, B. Ould-Bouamama, 'Backlash Fault Detection in Mechatronics System', MECHATRONICS Journal, Vol. 17, pp 299-310, 2007.

[27] World Health Organization official website: www.who.int.

CHAPTER 11

Design of Hybrid Hyper-Redundant Robot Manipulator

Abdelhakim Chibani[1],
Polytech-Lille, LAGIS, CNRS UMR 8219
Avenue Paul Langevin, 59655 Villeneuve D'Ascq, France
abdelhakim.chibani@polytech-lille.fr

Chawki Mahfoudi
Faculty of Technology Ain Beida
University Larbi ben M'Hidi, Algeria
mahfoudi.chawki@gmail.com

Rochdi Merzouki
Polytech-Lille, LAGIS, CNRS UMR 8219
Avenue Paul Langevin, 59655 Villeneuve D'Ascq, France
rochdi.merzouki@polytech-lille.fr

Abdelouaheb Zaatri
Departement of mechanics
University Mentouri,Constantine, Algeria
zaatri@hotmail.com

ABSTRACT. The hybrid hyper-redundant robots have a very large number of controlled degrees of freedom. This work presents geometric and kinematic modeling using to design an hybrid hyper-redundant robot. The modular structure studied here is considered as a very redundant mechanism that leads to complicated kinematic model. Because the mechanical architecture of the manipulator is simplified, geometric resolution of the problem is made analytically and a specific solution to the case of hybrid hyper-redundant robot formed by the stacking of parallel mechanisms is obtained. It is based on inverse kinematics of each module which the Jacobian matrix (6×6) represents the relationship between the speed of the end effectors (linear and angular velocities) and the active joint velocities. Finally, a case study consisting of solving geometric and kinematic analysis of an 18-degrees of freedom (DOF) hyper-redundant manipulator is presented.

Keywords: Gough-Stewart platform, parallel manipulator, hybrid robot, hyper-redundant robot, kinematic modeling.

11.1. Introduction

Although parallel robots are more rigid and more accurate than series, it still suffer from a small working space and a reduced manipulability [2]. A hybrid robot (series-parallel) is a mechanism that combines the advantages of both types of manipulators [25]. If the number of independent parameters exceeds the minimum number of degrees of freedom necessary to accomplish a particular task, the robot manipulator is called redundant. Thus, hyper-redundant manipulators have a very large number of controlled degrees of freedom [7]. This kinematic redundancy is even more important in the case of spatial manipulator. In bio-inspired structures, such hyper-redundant robot can be similar in morphology and operation to snake, tentacles and elephant trunk [9][26][27]. The industrial field, such structures are present in many applications such as the hybrid robot LX4 of Logabex society [1], paint robots and those used in the

[1]Corresponding author

field of a Variable Geometry Truss (VGT) robots. Considering the performance introduced by this kind of manipulators in different applications, several authors have addressed the problem of modeling in its three aspects: geometric, kinematic and dynamic. We cite for example the work of O. Ibrahim and W. Khalil [4][5] who presented a recursive solution for the inverse modeling dynamic of hybrid robots built by a series of parallel modules. L. Romdhane [25] studied design of a hybrid serial-parallel Gough Stewart like mechanism. J. Gallardo and al. [13] presented a design of a novel hyper-redundant manipulator. The same authors studied the kinematics of modular spatial hyper-redundant manipulators formed from RPS-type limbs [14]. J. Mintenbeck and al. [9] described a design modeling and control of hyper-redundant parallel mechanism. It is a biomimetic inspired snake robot and the performance in redundancy is dedicated for human surgery. In this same context, N. Iwatsuki and al. [29] studied the kinematic analysis and motion control of hyper-redundant robot, designed using a serial connection of parallel mechanisms. Also, in the field of biomimetic robotics, few authors have exploited the kinematic redundancy of hyper-redundant mechanisms to develop new technologies such as the works of D. Chablat [30] et G. Guallot [10] dealing with design and control of an Eel robot. A. A. Ramadan and al. [18] presented the development process of a serial-parallel structure as a micromanipulator. The concept of hyper-redundant mechanism was also used in the design of humanoid robots as in [28] and [19], who have adopted a flexible spine imitating human morphology. Just recently, C. Liang and al. [11] have proposed a new waist-trunk system built by stacking of parallel modules. In this paper, a modeling development for hybrid hyper-redundant design is presented. After the development of geometric and kinematic model, a CAD design of parallel mechanism corresponding to a Stewart Gough platform is established. A virtual simulator is developed in OpenGL environment enabled to view the robot and to illustrate the kinematics. The hybrid structure is a concatenation of n parallel modules. The platform of each module is connected to its base by m kinematic chains. We define a frame $\mathbb{R}_k$ fixed to the platform of each module k and $\mathbb{R}_{bk}$ fixed at its base. The base of module k being fixed to the platform of module $k - 1$, therefore there is a constantly changing between $\mathbb{R}_{k-1}$ and $\mathbb{R}_{bk}$. Figure (11.1.1) shows an overall description of a module of hybrid (serial and parallel) hyper-redundant robot.

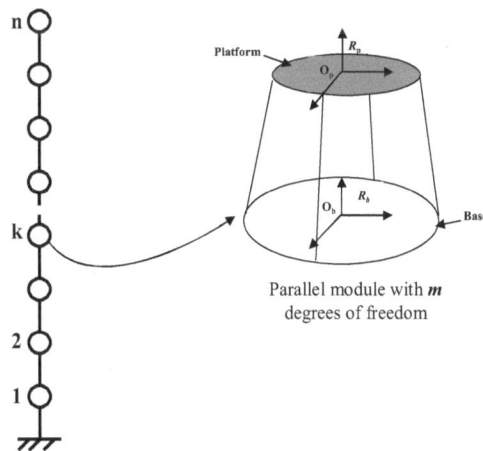

FIGURE 11.1.1. Description of hybrid hyper-redundant robot

Each module of the hybrid structure is composed of six identical kinematic chains linked to a mobile platform by spherical joints and by universal joints at the base (Figure 11.1.2). The attachment points on both sides of the module k and the corresponding axes are shown in Figure (11.1.3) Prismatic actuator is used to vary the length of leg segment. The minimum equivalent tree structure is obtained by isolating the platform [3]. We obtain for each module, a structure composed of six actuated prismatic joints and 12 passive rotary joint. The kinematic chain used to link the base and platform are serial structures composed of three joints and three body. In Figure (11.1.4) is represented the axes corresponding to the different joints.

Table (1) describes the Denavit Hartenberg [17] geometric parameters of corresponding to transformations along the chain i.

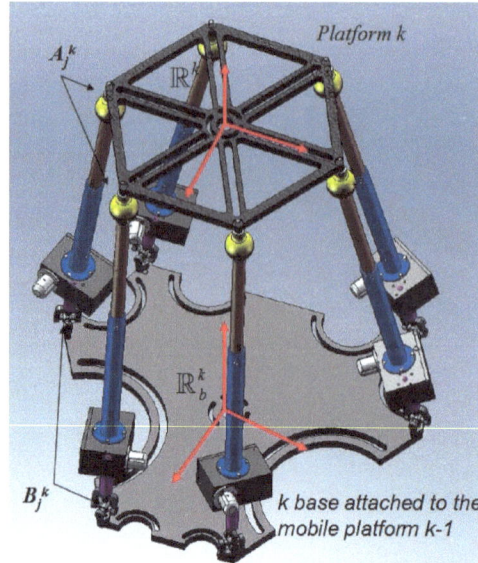

FIGURE 11.1.2. Description of parallel robot with six degrees of freedom

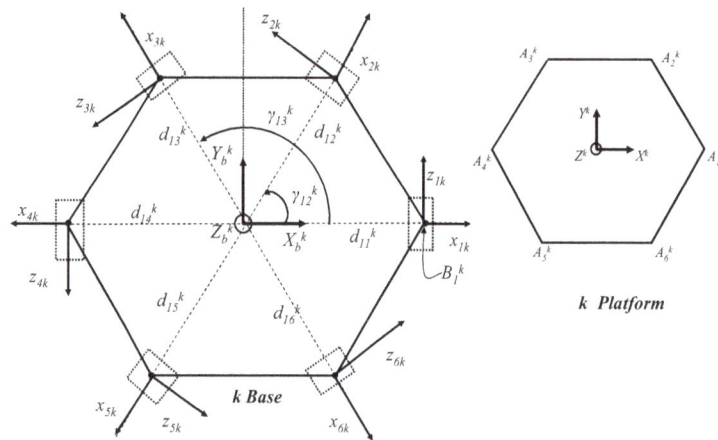

FIGURE 11.1.3. Geometric configuration of the base and the platform of each module k

frame	a_{ji}	μ_{ji}	σ_{ji}	α_{ji}	d_{ji}	θ_{ji}	r_{ji}	b_{ji}	γ_{ji}
$1_{i,k}$	0	0	0	$-\pi/2$	$d_{1i,k}$	$q_{1i,k}$	0	$b_{1i,k}$	$\gamma_{1i,k}$
$2_{i,k}$	$1_{i,k}$	0	0	$\pi/2$	0	$q_{2i,k}$	0	0	0
$3_{i,k}$	$2_{i,k}$	1	1	$\pi/2$	0	0	$q_{3i,k}$	0	0

TABLE 1. **Geometric parameters of the kinematic chain** i $(i = 1..6)$

The parameters $(\gamma_j , b_j , \alpha_j , \theta_j , r_j)$ (Tab.1) are used to define the frame $\mathbb{R}_j$ in the antecedent frame $\mathbb{R}_{j-1}$. The homogeneous transformation matrix is given by:

$$(11.1.1) \qquad\qquad {}^{i}\boldsymbol{T}_{j} = \begin{bmatrix} {}^{i}\boldsymbol{A}_{j} & {}^{i}\boldsymbol{P}_{j} \\ \boldsymbol{0}_{1\times 3} & 1 \end{bmatrix}$$

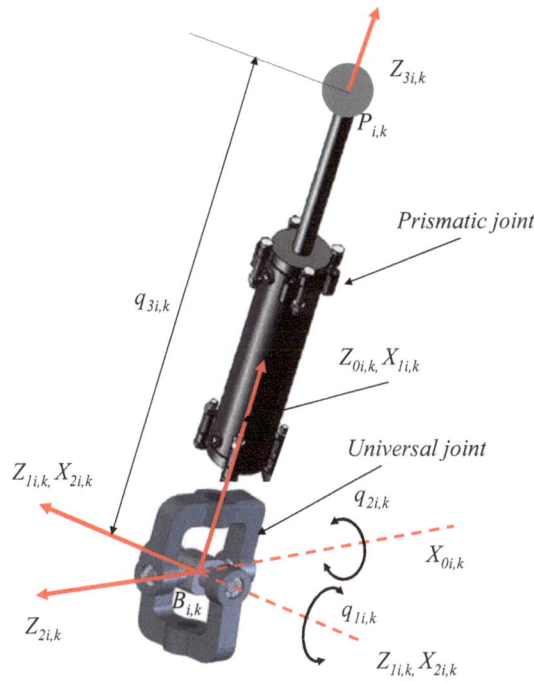

FIGURE 11.1.4. Location of joint frames of the kinematic chain i of the module k

(11.1.2) $$ {}^{i}\boldsymbol{A}_j = [\,{}^{i}\boldsymbol{s}_j \quad {}^{i}\boldsymbol{n}_j \quad {}^{i}\boldsymbol{a}_j\,] $$

${}^{i}\boldsymbol{A}_j$: is the matrix (3×3) which defines the orientation of $\mathbb{R}_j$ in $\mathbb{R}_i$
${}^{i}\boldsymbol{P}_j$: is the position vector (3×1) that defines the origin of $\mathbb{R}_j$ in $\mathbb{R}_i$

11.2. Geometric Model

11.2.1. Geometric model of a module k**.** The direct geometric model of a module k of the hybrid structure, expresses the coordinates describing the operational situation (position and orientation) of the mobile platform relative to the base according to its actuated joint variables. This model is given by the following relation:

(11.2.1) $$ \boldsymbol{X}_k = f(\mathbf{q}_{a,k}) $$

$\mathbf{q}_{a,k}$: is the column matrix $(N \times 1)$ of actuated joint variables and X_k the column matrix of operational coordinates of the module k, The inverse geometric model of parallel robots is generally easier to obtain than the direct model, it allows to express the actuated joint variables depending on the situation of the platform in the operational space and it is expressed by the following relation:

(11.2.2) $$ \mathbf{q}_{a,k} = f^{-1}(\boldsymbol{X}_k) $$

11.2.2. Direct geometric model of the kinematic chain i **of module** k**.** This model gives the coordinates of ${}^{i}\boldsymbol{P}_k$ expressed in $\mathbb{R}_{bk}$ system coordinates in function of joint variables $(q_{1i,k}, q_{2i,k}, q_{3i,k})$ $i = (1..6)$. This is done using the relationship (11.1.1) to set the frame $\mathbb{R}_{3i,k}$, whose origin is ${}^{i}\boldsymbol{P}_k$, which corresponds to the end frame of the kinematic chain i, in the basis frame $\mathbb{R}_{b,k}$:

(11.2.3) $$ {}^{b,k}\boldsymbol{T}_{3i,k} = {}^{b,k}\boldsymbol{T}_{1i,k} \quad {}^{1i,k}\boldsymbol{T}_{2i,k} \quad {}^{2i,k}\boldsymbol{T}_{3i,k} $$

with:

$$
{}^{b,k}\boldsymbol{A}_{3i,k} = \begin{bmatrix} C\gamma_{1i}C_{1i}C_{2i} - S\gamma_{1i}S_{1i} & C\gamma_{1i}S\theta_i - S_{1i} & C\gamma_{1i}C_{1i}S_{2i} + S\gamma_{1i}C_{1i} \\ S\gamma_{1i}C_{1i}C_{2i} - C\gamma_{1i}S_{1i} & S\gamma_{1i}S\theta_i - S_{1i} & S\gamma_{1i}C_{1i}S_{2i} - C\gamma_{1i}C_{1i} \\ S_{1i}C_{2i} & C_{1i} & -S_{1i}S_{2i} \end{bmatrix}
$$

(11.2.4)

$$
{}^{b,k}\boldsymbol{P}_{3i,k} = \begin{bmatrix} C\gamma_{1i}C_{1i}S_{2i}q_{3i} + S\gamma_{1i}C_{2i}q_{3i} \\ C\gamma_{1i}C_{1i}S_{2i}q_{3i} + S\gamma_{1i}C_{2i}q_{3i} \\ b_{1i} - S_{1i}S_{2i}S_{2i}q_{3i} \end{bmatrix}
$$

The solution of the direct geometric model of a module k of the hybrid structure is not unique since for a given joint configuration variables the platform can take many different situations.

11.2.3. Inverse geometric model of the kinematic chain i of module k. This model gives the joint variables $(q_{1i,k}, q_{2i,k}, q_{3i,k})$ (for $i = 1..6$) according to the Cartesian coordinates of the point $\boldsymbol{P}_{ik}$ expressed in the frame R_{bk}. Attachment points $\boldsymbol{P}_{ik}$ with the platform in relation to the reference base for this module R_{bk} is denoted ${}^i\boldsymbol{P}_k$. The intermediate vector ${}^{Bi,k}\boldsymbol{P}_{i,k}$, defines the point $\boldsymbol{P}_{ik}$ in the frame R_{Bik} whose origin is B_i (Figure 11.2.1).

(11.2.5) $${}^{Bi,k}\boldsymbol{P}_{i,k} = {}^{Bi,k}\boldsymbol{A}_{b,k} \ {}^{b,k}\boldsymbol{P}_{i,k} + {}^{Bi,k}\boldsymbol{P}_{b,k}$$

Where the coordinates of point ${}^{b,k}\boldsymbol{P}_{i,k}$ are determined by the relationship:

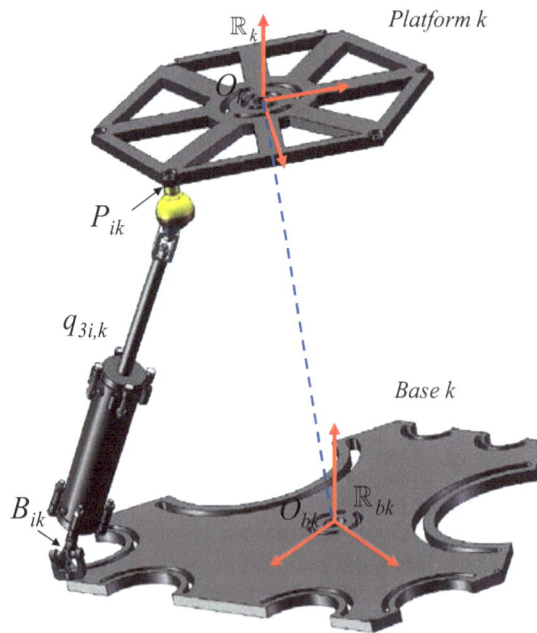

FIGURE 11.2.1. Geometrical representation of the active joint variable $q_{3i,k}$

(11.2.6) $${}^{b,k}\boldsymbol{P}_{i,k} = {}^{b,k}\boldsymbol{A}_{b,k} \ {}^k\boldsymbol{P}_{i,k} + {}^{b,k}\boldsymbol{P}_{b,k}$$

${}^{b,k}\boldsymbol{A}_{b,k}$ and ${}^{b,k}\boldsymbol{P}_{b,k}$ are matrices which define the orientation and position of the origin of the frame $\mathbb{R}_{bk}$ in $\mathbb{R}_k$ and ${}^k\boldsymbol{P}_{i,k}$ attachment points to the platform relative to reference $\mathbb{R}_k$.

(11.2.7) $${}^{b,k}\mathbf{A}_k = Rot(z, \phi) \times Rot(y, \theta) \times Rot(x, \psi)$$

(11.2.8)
$$^{b,k}\boldsymbol{P}_k = [\boldsymbol{P}_{x,k} \quad \boldsymbol{P}_{y,k} \quad \boldsymbol{P}_{z,k}]^T$$

(11.2.9)
$$^{b,k}\boldsymbol{P}_k = [^k\boldsymbol{P}_{xi,k} \quad {}^k\boldsymbol{P}_{yi,k} \quad {}^k\boldsymbol{P}_{zi,k}]^T$$

By expression of components of the vector $(BP)_{i,k}$, we can easily determine the prismatic variables $q_{3i,k}$:

(11.2.10)
$$q_{3i,k} = \sqrt{^{Bi,k}\boldsymbol{P}^2_{xi,k} + {}^{Bi,k}\boldsymbol{P}^2_{yi,k} + {}^{Bi,k}\boldsymbol{P}^2_{zi,k}}$$

where: $^{Bi,k}\boldsymbol{P}_{xi,k}$, $^{Bi,k}\boldsymbol{P}_{yi,k}$, $^{Bi,k}\boldsymbol{P}_{zi,k}$ are the coordinates of point $P_{i,k}$ in the frame $\mathbb{R}_{Bi,k}$. The joint variables of the universal joint $q_{1i,k}$ et $q_{2i,k}$ are determined by applying the method of Paul [**15**]:

(11.2.11)
$$\begin{cases} ^{Bi,k}\boldsymbol{P}_{xi,k}\boldsymbol{S}_{1i,k} + {}^{Bi,k}\boldsymbol{P}_{yi,k}\boldsymbol{C}_{1i,k} = 0 \\ ^{Bi,k}\boldsymbol{P}_{xi,k}\boldsymbol{C}_{1i,k}\boldsymbol{C}_{2i,k} + {}^{Bi,k}\boldsymbol{P}_{yi,k}\boldsymbol{S}_{1i,k}\boldsymbol{C}_{2i,k} + {}^{Bi,k}\boldsymbol{P}_{zi,k}\boldsymbol{S}_{2i,k} = 0 \\ -^{Bi,k}\boldsymbol{P}_{xi,k}\boldsymbol{C}_{1i,k}\boldsymbol{S}_{2i,k} - {}^{Bi,k}\boldsymbol{P}_{yi,k}\boldsymbol{S}_{1i,k}\boldsymbol{S}_{2i,k} + {}^{Bi,k}\boldsymbol{P}_{zi,k}\boldsymbol{C}_{2i,k} = -q_{3i,k} \end{cases}$$

The solutions of equation system (11.2.11) are given by:

(11.2.12)
$$q_{1i,k} = atan2(^{Bi,k}\boldsymbol{P}_{yi,k}, {}^{Bi,k}\boldsymbol{P}_{xi,k}) \quad \text{ou} \quad q_{1i,k}{}' = q_{1i,k} + \pi$$

and

(11.2.13)
$$q_{2i,k} = atan2(\frac{-q_{3i,k}(^{Bi,k}\boldsymbol{P}_{xi,k}\boldsymbol{C}_{1i,k} + {}^{Bi,k}\boldsymbol{P}_{yi,k}\boldsymbol{S}_{1i,k})}{Det}, \frac{q_{3i,k}\,{}^{Bi,k}\boldsymbol{P}_{zi,k}}{Det})$$

such as

$$Det = {}^{Bi,k}\boldsymbol{P}^2_{zi,k} + (^{Bi,k}\boldsymbol{P}_{xi,k}\boldsymbol{C}_{1i,k} + {}^{Bi,k}\boldsymbol{P}_{yi,k}\boldsymbol{S}_{1i,k})^2$$

The condition required for the application of equation (11.2.13) is $Det \neq 0$. So that for Det is zero, it is necessary that $^{Bi,k}\boldsymbol{P}^2_{zi,k}$ and $(^{Bi,k}\boldsymbol{P}_{xi,k}\boldsymbol{C}_{1i,k} + {}^{Bi,k}\boldsymbol{P}_{yi,k}\boldsymbol{S}_{1i,k})$ are zero. This condition is easily verified because $^{Bi,k}\boldsymbol{P}^2_{zi,k}$ is positive. Indeed if it is zero, this means that the leg is collinear with the plane of the base which is not realistic. Articular variable $q_{1i,k}$ has two theoretical solutions, but only one solution will be considered because the value of $q_{1i,k}$ is between zero and π.

11.2.4. Geometric model of the hyper-redundant structure. The operational coordinates of the origins of intermediate platforms $[X_k, Y_k, Z_k, \phi_k, \theta_k, \psi_k]$ are determined by an interpolation of the operational coordinates of the n platform. X_k, Y_k, Z_k represents the Cartesian coordinates of the origin of the frame $\mathbb{R}_k$ in $\mathbb{R}_{k-1}$ and ϕ_k, θ_k, ψ_k the orientation of $\mathbb{R}_k$ in the frame $\mathbb{R}_{k-1}$.

We calculate the operational coordinates through the following steps (Figure 11.2.2):

- Find a polynomial interpolation points corresponding to (X_0, Y_0, Z_0) and (X_n, Y_n, Z_n). This polynomial equation is the desired curve of the axis of the robot.
- Subdivide of the curve in n stages.
- Determine the coordinates of the origin of intermediate platforms.

11.3. Kinematic Model

11.3.1. Kinematic Model of a leg *i* of a module *k*. Kinematic model of a kinematic chain i, gives the linear velocity of point $P_{i,k}$ depending on the speed of the joints of the kinematic chain i $(q_{1i}, q_{2i}, q_{3i})_k$:

(11.3.1)
$$^{b,k}\boldsymbol{v}_{Pi,k} = {}^{b,k}\boldsymbol{J}_{3i,k}\dot{\boldsymbol{q}}_{3i,k}$$

Where:
$^{b,k}\boldsymbol{J}_{3i,k}$: is the Jacobian matrix (3×3) of kinematic chain i of module k;
$^{b,k}\boldsymbol{v}_{Pi,k}$: is the linear velocity vector (3×1) of point $P_{i,k}$ whose components are: $[v_{xi}, v_{yi}, v_{zi}]_k^T$

The inverse of $^{b,k}\boldsymbol{J}_{3i,k}$ leads to calculate $^{3i,k}\boldsymbol{J}^{-1}_{b,k}$. The inverse kinematic model of a kinematic chain i can be written as follows:

(11.3.2)
$$\dot{\boldsymbol{q}}_{a,k} = {}^{b,k}\boldsymbol{J}^{-1}_k\,{}^{b,k}\boldsymbol{V}_k = {}^{b,k}\boldsymbol{J}^{-1}_k \begin{bmatrix} \boldsymbol{v}_k \\ \omega_k \end{bmatrix}_{b,k}$$

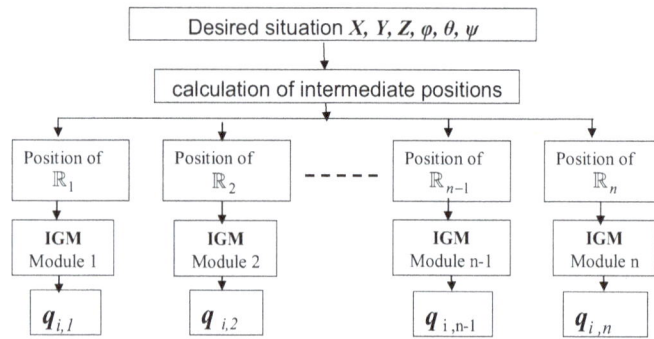

FIGURE 11.2.2. Algorithm for calculating the inverse geometrical model of the hybrid hyper-redundant robot

$^{b,k}\boldsymbol{J}_k^{-1}$: is the inverse of the Jacobian matrix of module k ;

$^{b,k}\boldsymbol{V}_k$: Cartesian velocity of the origin of the platform k relative to the frame $\mathbb{R}_{bk}$, whose components are $\boldsymbol{v}_k$ the linear velocity and $\boldsymbol{\omega}_k$ the angular velocity.

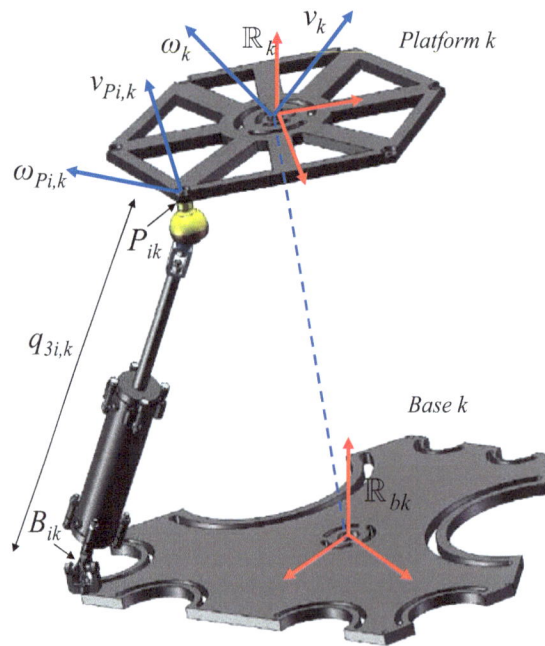

FIGURE 11.3.1. The linear velocity $\boldsymbol{v}_k$ and angular velocity $\boldsymbol{\omega}_k$ of the k platform

11.3.2. Calculation of the inverse Jacobian matrix. The calculation of the inverse Jacobian matrix is based on the determination of the velocity joint $\dot{\boldsymbol{q}}_{3i,k}$ by the projection of the velocity of point $\boldsymbol{P}_{i,k}$ on the axis $\boldsymbol{Z}_{3i,k}$ (Figure 11.3.1).

$$(11.3.3) \qquad\qquad \dot{\boldsymbol{q}}_{3i,k} = {}^{b,k}\boldsymbol{a}_{3i,k}\,{}^{b,k}\boldsymbol{v}_{Pi,k}$$

$^{b,k}\boldsymbol{a}_{3i,k}$: is the unit vector along the axis $\boldsymbol{Z}_{3i,k}$

$\boldsymbol{v}_{Pi,k}$ linear velocity of point $\boldsymbol{P}_{i,k}$ calculated from $^{b,k}\boldsymbol{v}_k$ and $^{b,k}\boldsymbol{\omega}_k$ by the relation:

(11.3.4)
$$^{b,k}v_{Pi,k} = {}^{b,k}v_k + \left({}^{b,k}\omega_k \times {}^{b,k}L_{i,k}\right)$$

With $^{b,k}L_{i,k}$ vector representing the components of vector $0_kP_{i,k}$ expressed in the frame $\mathbb{R}_{bk}$.

(11.3.5)
$$^{b,k}L_{i,k} = [^{b,k}L_{xi,k} \quad {}^{b,k}L_{yi,k} \quad {}^{b,k}L_{zi,k}]$$

Therefore

(11.3.6)
$$\dot{q}_{3i,k} = {}^{b,k}a_{3i,k}^T \, {}^{b,k}v_k + \left({}^{b,k}\hat{L}_{i,k} \, {}^{b,k}a_{3i,k}\right)^T \, {}^{b,k}\omega_k$$

With $^{b,k}\hat{L}_{i,k}$ denotes the skew-symmetric matrix (3×3) associated with the vector $^{b,k}L_{i,k}$.
Finally, the K line in inverse of the Jacobian matrix of the kth module is written as:

(11.3.7)
$$^{b,k}J_{k[K,:]}^{-1} = [^{b,k}a_{3i,k}^T \quad \left({}^{b,k}\hat{L}_{i,k} \, {}^{b,k}a_{3i,k}\right)^T]$$

11.3.3. Calculation of velocities of modules platforms. Knowing the components of linear and angular velocities of the platform $k-1$, we can determine those of platform k by the following recursive equations:
for $k = 1,2,..., n$:

(11.3.8)
$$\Omega_k = \omega_{k-1} + a_{2k}V_k$$

(11.3.9)
$$W_k = W_{k-1} + \omega_{k-1} \times P_k + a_{1k}V_k$$

With:
W_k: linear velocity of the kth platform,
Ω_k: angular velocity of the kth platform,
P_k: vector between the origin of the frame $\mathbb{R}_{k-1}$ and that of the frame $\mathbb{R}_k$,
$a_k(6 \times N_k)$: the matrix which expresses the velocity components corresponding to the six degrees of freedom in Cartesian space according to the independent degrees of freedom of the robot, given by:

(11.3.10)
$$a_k = \begin{bmatrix} I_3 & 0_3 \\ 0_3 & \Omega_r \end{bmatrix} \text{ such that : } \Omega_r = \begin{bmatrix} C\varphi\tan\theta & S\varphi\tan\theta & 1 \\ -S\varphi & C\varphi & 0 \\ C\varphi/C\theta & S\varphi/C\theta & 0 \end{bmatrix}$$

It can be decomposed as follows:
$a_k = [a_{1k}^T \; a_{2k}^T]^T$ where: $a_{1k} = [I_3 \; 0_3]$ and $a_{2k} = [0_3 \; \Omega_r]$
The equations (11.3.8) and (11.3.9) can be grouped into the following equation:

(11.3.11)
$$\begin{bmatrix} W_k \\ \Omega_k \end{bmatrix} = \begin{bmatrix} I_{d3} & -\widehat{P}_k \\ 0_3 & I_{d3} \end{bmatrix} \begin{bmatrix} W_{k-1} \\ \Omega_{k-1} \end{bmatrix} + \begin{bmatrix} a_{1k}V_k \\ a_{2k}V_k \end{bmatrix}$$

These equations are projected in the frame $\mathbb{R}_k$, as follows:

(11.3.12)
$$\begin{bmatrix} {}^kW_k \\ {}^k\Omega_k \end{bmatrix} = \begin{bmatrix} {}^kA_{k-1} & 0_3 \\ 0_3 & {}^kA_{k-1} \end{bmatrix} \begin{bmatrix} I_{d3} & -{}^{k-1}\widehat{P}_k \\ 0_3 & I_{d3} \end{bmatrix} \begin{bmatrix} {}^{k-1}W_{k-1} \\ {}^{k-1}\Omega_{k-1} \end{bmatrix} + \begin{bmatrix} {}^ka_{1k}V_k \\ {}^ka_{2k}V_k \end{bmatrix}$$

We can rewrite Eq.(11.3.12) by:

(11.3.13)
$$^k\mathbb{V}_k = {}^k\mathbb{T}_{k-1} \, {}^{k-1}\mathbb{V}_{k-1} + {}^k a_k \, {}^kV_k$$

Where $^k\mathbb{T}_{k-1}$ is the (6×6) transformation matrix between the modules k and $k-1$:

(11.3.14)
$$^k\mathbb{T}_{k-1} = \begin{bmatrix} {}^kA_{k-1} & -{}^kA_{k-1}{}^{k-1}\widehat{P}_k \\ 0_{3\times3} & {}^kA_{k-1} \end{bmatrix}$$

$^kA_{k-1}$ is the rotation matrix (3×3), which defines the orientation of the frame $\mathbb{R}_{k-1}$ relative to $\mathbb{R}_k$. $^{k-1}\widehat{P}_k$ denotes the skew-symmetric matrix (3×3) associated with the vector $^{k-1}P_k$. Then, matrix $^k\mathbb{T}_{k-1}$ is calculated as follows:

(11.3.15)
$$^k\mathbb{T}_{k-1} = {}^k\mathbb{T}_{b,k} \, {}^{b,k}\mathbb{T}_{k-1}$$

$^{b,k}\mathbb{T}_{k-1}$: is a matrix defining a constant transformation between the base of the platform k and the platform k-1. We note that if it is interesting to determine the platforms accelerations we can be made the differentiation of equations (11.3.8) and (11.3.9) with respect to time.

11.3.4. Simulation Results. In this section a numerical example is solved. It's consists of solving the kinematic analysis of 18-dof hyper-redundant manipulator. The general shape of the structure is conic that is, the modules are not perfectly the same in term of dimension. Parameters for each platform are chosen as follows (SI units):

- Coordinates of points B_i of the base
  $B_{ix,0} = [288.68; 144.34; 144.34; -288.68; -144.34; 144.34]$
  $B_{iy,0} = [0; 250; 250; 0; -250; -250]$
  $B_{iz,0} = [0; 0; 0; 0; 0; 0]$
- Coordinates of points P_i of the platform n in relation to the end effectors frame
  $P_{ix,n} = [150; 75; -75; -150; -75; 75]$
  $P_{iy,n} = [0; 129.9; 129.9; 0; -129.9; -129.9]$
  $P_{iz,n} = [0; 0; 0; 0; 0; 0]$

The principle is that, knowing the general movement of the platform n generated by the operational coordinates ($X_n, Y_n, Z_n,$ $\varphi_n, \theta_n, \psi_n$), we determine the displacements and velocities of the actuators and the mobile platforms. The movement being generated in the operational space where, we use the polynomial interpolation of degree 5:

(11.3.16)
$$r(t) = 10(t/t_f)^3 - 15(t/t_f)^4 + 6(t/t_f)^5$$

Where t is time variable varies from 0 to 10 sec. In order to validate simulation results, the following settings of the trajectory is considered:

(11.3.17)
$$\begin{aligned} X_n(t) &= 300 \times \sin(t/2) & \varphi_n(t) &= 0 \\ Y_n(t) &= 300 \times \cos(t/2) & \theta_n(t) &= 0 \\ Z_n(t) &= 0 & \varphi_n(t) &= 0.15 \times t \end{aligned}$$

For this configuration, we can see the corresponding movement of the 3-modules robot (Figure 11.3.2). Figure (11.3.3) shows displacement and orientations of mobile platform, (11.3.4) and (11.3.5) represent displacements and velocities of the actuators. To demonstrate the effect of kinematic redundancy, we consider Figure (11.3.6) which represents an hyper-redundant mechanism composed of 10 mobile platforms. It is obvious that the reachability and flexibility properties are demonstrated. Is mainly used the calculation of the Jacobian matrix of module k and its inverse to derive the six-speed actuators matching with a trajectory imposed on the platform k in the kinematic modeling. It is worth noting that joint velocities are obtained either by direct derivation of the variables $q_{3i,k}$ or by using the analytical method, based on obtaining the inverse of Jacobian matrix, which is developed in section 11.3.2.

To validate the developed models, a virtual simulator of a class of an hybrid hyper-redundant robot is designed in order to study performances of mobility and reachability based on inverse geometric model developed in section 11.2, obtained results are compared to those using the Gough Stewart's platform as in Figure (11.3.7), (11.3.8) and the case of 3-modules assembled in series in Figure (11.3.9 a and b). According to the design fulfilled, we analyze the contribution of these mechanisms compared to conventional parallel robots. This simulator has been developed in $C++/OpenGL$ environment and it is possible to view the configuration robot in a multitude angles.

11.4. Conclusion and Discussion

The main objective of this work is to present an effective method of calculating the geometric and kinematic models of a spatial hyper-redundant manipulator constructed with parallel platforms. The models serve for the design of hybrid hyper-redundant mechanism, having several advantages. Firstly, because the platforms are identical, solving kinematics problem is performed for a one platform and led eventually in order to determine the speed of the actuators and the whole robot platforms. To illustrate the application of the described modeling, a kinematic of three platforms with six degrees of freedom for each of them has been studied in simulation using Matlab / Simulink. The ability to further amplify the kinematic redundancy by increasing the number of platforms lets us think about correlation with the flexible structures. In addition, simulation analysis shows that the structure responds well to different positions corresponding to the movement of extension, flexion and rotation in horizontal, vertical and transversal plans. This flexibility shows a remarkable harmony with movements of vertebral column of the new humanoid robots for example. This kind of structures is well suited for design a miniature manipulators and they may have great robustness. The high manipulability due to a large number of controlled degrees of freedom leads to the consideration of self-configurable robot and the possibility of the implementation of obstacle avoidance algorithms.

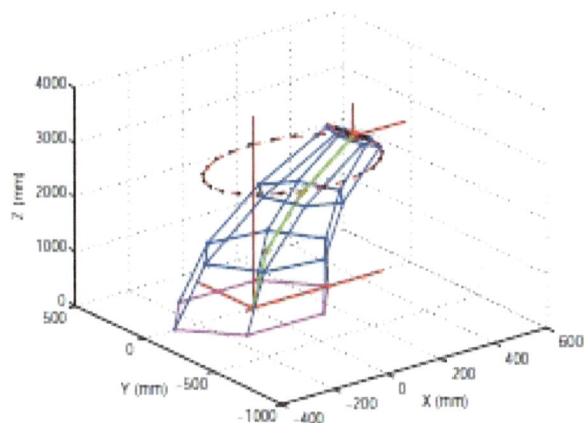

FIGURE 11.3.2. Hybrid robot with three parallel modules describing a circular path.

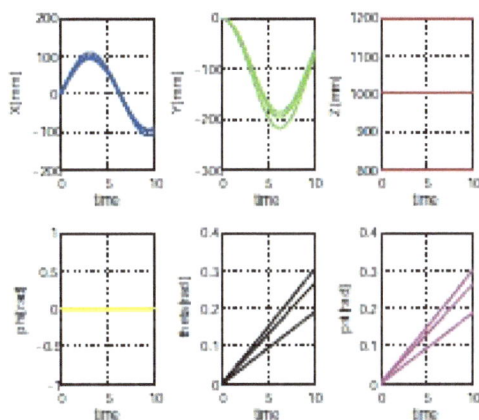

FIGURE 11.3.3. Displacements and orientations of platforms

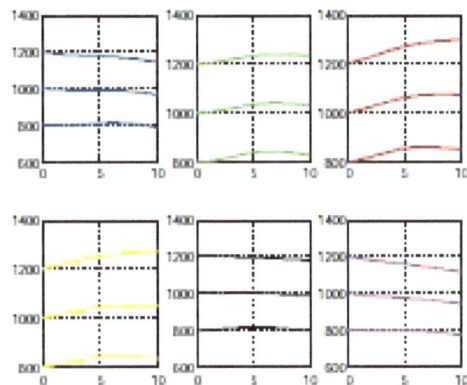

FIGURE 11.3.4. Variation of active joint variables $q_{3i,k}$ [mm]

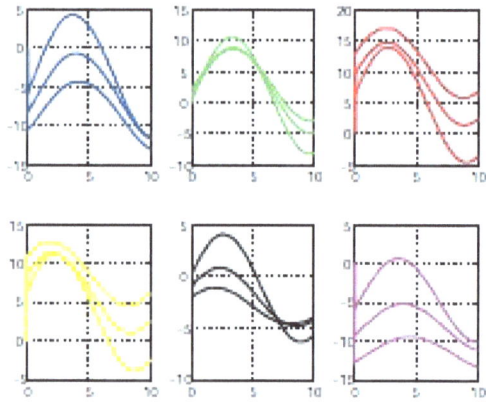

FIGURE 11.3.5. Variation of active joint velocities $\dot{q}_{3i,k}$ [mm/sec.]

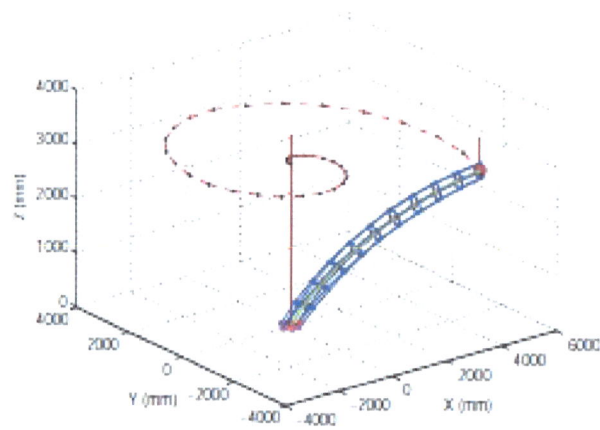

FIGURE 11.3.6. Hyper-redundant robot describing a spiral trajectory

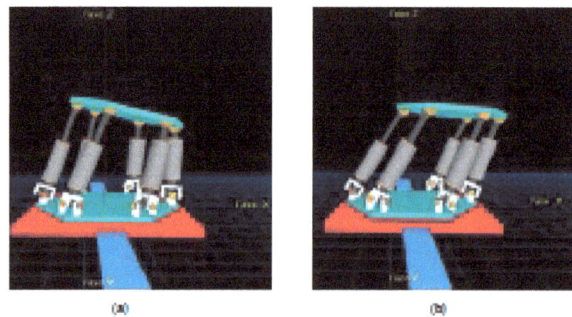

FIGURE 11.3.7. Gough Stewart platform in 3D motion

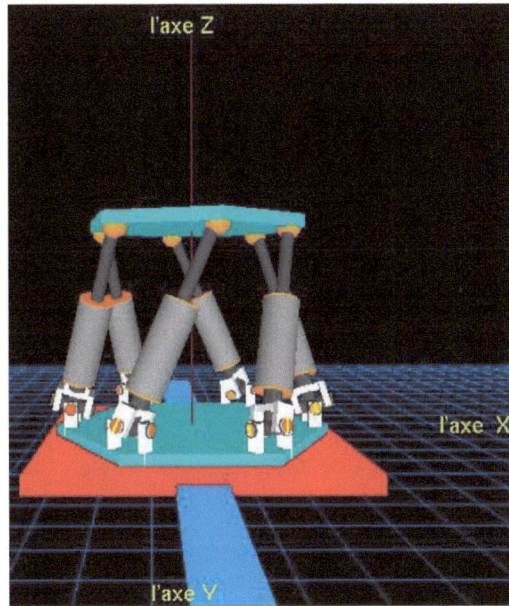

FIGURE 11.3.8. Rotation around the z axis, $\varphi = 40°$

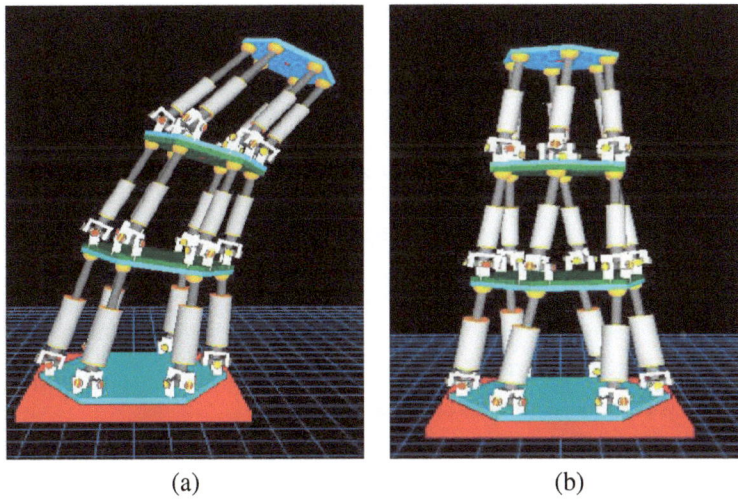

(a)   (b)

FIGURE 11.3.9. Hybrid robot in 3D motion. (a) Roll motion, (b) case of $\varphi = 40°$

Bibliography

[1] S. Charentus and and M. RENAUD, *Modelling and Control of a Modular, Redundant Robot Manipulator.* Experimental Robotics, Springer, 1990, Vol. 139, pp. 508-527.

[2] M. Ait Ahmed , Ph.D. thesis *Contribution à la modélisation géométrique et dynamique des robots parallèles.*, LAAS, Toulouse, 1993.

[3] S. Guegan , Ph.D. thesis *Contribution à la modélisation et identification dynamique des robots Parallèles.*, IRCCyN, University of Nantes, 2003.

[4] O. Ibrahim ,W. Khalil , *Inverse Dynamic Modeling of Serial-Parallel Hybrid Robots.* Proceeding: International Conference on Intelligent Robots and Systems,IEEE, October 9-15, 2006, Beijing, China. pp. 2156-2161.

[5] O. Ibrahim ,W. Khalil , *Inverse and direct dynamic models of hybrid robots.* Mechanism and Machine Theory ,Elsevier, 2009, Vol. 20, pp. 627-640 .

[6] K. Rongjie, A. Kazakidi, E. Guglielmino, D. T. Branson *Dynamic Model of a Hyper-redundant, Octopus-like Manipulator for Underwater Applications.* RSJ International Conference on Intelligent Robots and Systems IEEE, September 25-30, 2011, San Francisco, CA, USA, pp. 4054-4059

[7] G. S. Chirikjian, J. W. Burdick, *Hyper-Redundant Robot Mechanisms and Their Applications.* IROS International Workshop on Intelligent Robots and Systems, Nov. 3-5 1991, Osaka, Japan. pp. 185-190.

[8] G. S. Chirikjian, J. W. Burdick, *A Hyper-Redundant Manipulator.* Robotics and Automotion Magazine IEEE 1994, USA, pp. 22-29.

[9] J. Mintenbeck and R. Estanna, *Design, modelling and control of a hyper-redundant 3-RPS parallel mechanism.* International Conference on Robotics and Biomimetics IEEE, December 14-18, 2010, Tianjin, China

[10] G. Gallot, O. Ibrahim and W. Khalil, *Dynamic Modeling and simulation of a 3-D Hybrid structure Eel-Like Robot.* International Conference on Robotics and Automation, IEEERoma, April 10-14, 2007, Roma, Italy, pp. 1486-1491.

[11] C. Liang and M. Ceccarelli, *Design and Simulation of a Waist-Trunk System for a Humanoid Robot.* Mechanism and Machine Theory 2012, Vol. pp.50-65.

[12] T. K. Tanev, *Kinematics of a hybrid (parallel-serial) robot manipulator.* Mechanism and Machine Theory Elsevier, 2000, Vol.35, pp. 1183-1196.

[13] J. A. Gallardo , C.R. Aguilar-Najera, L. Casique Rosas, J. M. Rico-Martynez, Md. N. Islam *Kinematics and dynamics of 2(3-RPS) manipulators by means of screw theory and the principle of virtual work.* Mechanism and Machine Theory, 2008, Vol.43, pp. 1281-1294.

[14] J. A. Gallardo, C.R. Aguilar-Najera, L. Casique-Rosas, L. Pérez-Gonzalez, J.M. Rico-Martinez, *Solving the kinematics and dynamics of a modular spatial hyper-redundant manipulator by means of screw theory.* Multibody Syst. Dyn., 2008, Vol. 20, pp. 307-325.

[15] W. Khalil et E. Dombre, *modélisation, identification et commande des robots.*, 2nd ed., HERMES Sciences Publications Paris, 1999.

[16] W. Khalil et O. Ibrahim, *General Solution for the Dynamic Modeling of Parallel Robots.* Journal of Intelligent and Robotic Systems, 2007, Vol.49, pp.19-37.

[17] R.S. Hartenberg and J. Denavit, *A kinematic notation for lower pair mechanisms based on matrices.* Journal of Applied Mechanics, 1955, Vol.77, pp. 215-221

[18] A. A. Ramadan, T. Takubo, Y. Mae,K. Oohara and T. Arai, *Developmental Process of a Chopstick-Like Hybrid-Structure Two-Fingered Micromanipulator Hand for 3-D Manipulation of Microscopic Objects.* Transactions on Industrial Electronics IEEE,2009, Vol.56, N°4, pp. 1121-1135.

[19] L. Roos, F. Guenter, A. Guignard and A. G. Billard, *Design of a Biomimetic Spine for the Humanoid Robot Robota.* RAS-EMBS International Conference on Biomedical Robotics and Biomechatronics IEEE, February 20-22, 2006, Pisa, Italy.

[20] KeJun Ning and Florentin Worgotter, *A Novel Concept for Building a Hyper-Redundant Chain Robot.* Transactions on Robotics, IEEE 2009, Vol.25, N°6, pp.1237-1248.

[21] Z. Mingyang , G. Tong, C. Ge , L. Qunming ,T. Dalong, *Development of A Redundant Robot Manipulator Based on Three DOF Parallel Platforms.* International Conference on Robotics and Automation, IEEE 1995, pp. 221-226.

[22] I. Mizuuchi, T. Yoshikai, Y. Nakanishi, Y. Sodeyama, T. Yamamoto, A. Miyadera, T. Niemela, M. Hayashi, J. Urata and M. Inaba, *Development of Muscle-Driven Flexible-Spine Humanoids.* RAS International Conference on Humanoid Robots, IEEE 2005, pp. 339-344.

[23] S. Yahya, Haider A. F. Mohamed, M. Moghavvemi, and S. S. Yang, *A Geometrical Inverse Kinematics Method for Hyper-Redundant Manipulators.* Conf. on Control Automation Robotics and Vision, IEEE December 17-20, 2008 Hanoi, Vietnam, pp.1954-1958.

[24] Yi Lu, Bo Hu, JianPing Yu, *Analysis of kinematics statics and workspace of a 2(SP+SPR+ SPU) serial-parallel manipulator.* Multibody Syst Dyn, Springer 2009, Vol.21,pp. 361-374.

[25] L. Romdhane, *Design and analysis of a hybrid serial-parallel manipulator.* Mechanism and Machine Theory, Elsever 1999, Vol.34, ,pp. 1037-1055.

[26] B. A. Jones, and I. D. Walker, *Kinematics for Multisection Continuum Robots.* Transaction on Robotics,IEEE 2006, Vol. 22, N°2 pp. 43-57

[27] D. B. Camarillo, C. F. Milne, C. R. Carlson, M. R. Zinn and J. K. Salisbury, *Mechanics Modeling of Tendon-Driven Continuum Manipulators.* Transaction on Robotics,IEEE 2008 Vol.24, N°6 pp. 1226-1273.

[28] D. B. Camarillo, C. F. Milne, C. R. Carlson, M. R. Zinn and J. K. Salisbury, *Behavior System Design and Implementation in Spined Musle-Tendon Humanoid 'Kenta'.* Journal of Robotics and Mechatronics,2003 Vol. 15, N°2 pp. 143-152.

[29] N. Iwatsuki, N. Nishizaka, K. Morikawa and K. Kondoh, *Motion Control of Hyper-redundant Manipulator Built by Serally Connecting Many Parallel Mechanism Units with a Few DOF.* Int. Journal of Automotion Technology,2010, Vol. 4, N°4.

[30] D. Chablat , *Kinematic analysis of the vertebra of a eel-like robot,* ASME Design Engineering Technical Conferences,August 3-6, 2008, New York, USA

CHAPTER 12

Optical Measurement for Robotic Grinding and Polishing of Turbine Vanes

Danwei Wang[1],
EXQUISITUS, Centre for E-City, Division of Control and Instrumentation,
School of Electrical and Electronic Engineering, Nanyang Technological University,
Singapore 639798.
edwwang@ntu.edu.sg

Xiaoqi Chen
Department of Mechanical Engineering, College of Engineering,
University of Canterbury,
Christchurch 8140, New Zealand.
xiaoqi.chen@canterbury.ac.nz

ABSTRACT. Robot grinding/polishing of an unknown surface requires its 3D measurement . An existing method is based on linear variable differential transformer (LVDT) that has limitations such as contact with object and low speed. In this work, a non-contact laser sensor is studied to provide faster and more accurate measurement. Measuring an unknown 3D surface of a turbine vane is challenging because some positions of the turbine vane surface are specular after being polished or grinded. In these situations, little or no measurement data are received and the reconstruction of the 3D surface profile is difficult. Here, we develop a measurement algorithm to process the laser sensor measurement data, and to reconstruct 3D profiles of these surfaces. In particular, the corrupted data due to the specular reflection are processed to obtain a nominal 3D surface profile. Three interpolation algorithms are investigated for restoring corrupted data due to specular reflection. It is found that the linear compensations of control points must be carried out before interpolation. It is necessary to obtain approximately real data of specular reflection edge by linear compensation according to true surface shape. The results of 2D profiles processed by three interpolation algorithms are displayed and compared in the same coordinates. The findings indicate that the piecewise cubic Hermite interpolation of turbine vane surface is the most promising method to restore corrupted data. The results indicate that the measurement system can meet the repairing requirement of turbine vanes .

Keywords : 3D surface profile; interpolation algorithm; high speed scanning; non-contact laser sensor, turbine vane, 3D measurement, specular reflection, piecewise cubic Hermite interpolation.

12.1. Introduction

Although industrial robots are gaining widespread applications in manufacturing process automation, robotic precision processing of unknown 3D surface is still a formidable task, especially when it comes to jet engine turbine vanes repair in the aerospace industry. The repair process includes welding of the damaged areas of vanes and grinding and polishing down of their dimensions. The most difficult and critical phases in the overall repair process are the grinding and the polishing phase, due to two main reasons. First, the grinding and polishing process must be precisely adapted to the complex shape of each type of turbine of vanes not to affect the aerodynamic performance of vanes after repair. Second, the shape of used turbine vanes can vary from their original design dimension because of their exposure in the jet engine environment. Therefore, it is important to measure turbine vanes before and after repair.

To mimic the human capabilities, robotic automatic grinding and polishing systems can be used for overhauling turbine vanes in a constrained robotic environment. The structure diagram of the grinding and polishing system is shown as Figure 12.1.1. It includes five feedback loops and two control loops. The five feedback loops supply signal on control and detection respectively, as a part of an automatic grinding and polishing system, LVDT (Linear variable differential transformer) based on touch probe is used to measure each turbine vane's profile, of which, the data are processed to automatically reconstruct the 3D profile and supplied

---

[1]Corresponding author

218 *Mechatronic & Innovative Applications*   *Wang et al.*

data to the system to generate the robot grinding and polishing path. During turbine vanes grinding and polishing, the LVDT sensor detects the height of excessive braze material at current grinding and polishing point. A host PC will then decide whether the desired processing precision has been reached and determines the respective required contact force between the workpiece and the tool in accordance with its grinding and polishing process database. Both a force/moment sensor at the robot end-effectors and a force gauge at the grinding and polishing machine collaboratively provide force feed-back information telling the actual value of contact force exerted, whilst a displacement contact probe is used to measure the position of the grinding and polishing wheel. With desired and actual data of contact force and grinding and polishing wheel position, the host PC will control a servo motor to drive forward or backward a cam mechanism which will in turn maintain the desired contact force and compliance at current grinding and polishing point for an optimum grinding and polishing process. In this grinding and polishing system, the LVDT sensor acts as the function of "human sense" to measure the profile of the workpiece to be processed.

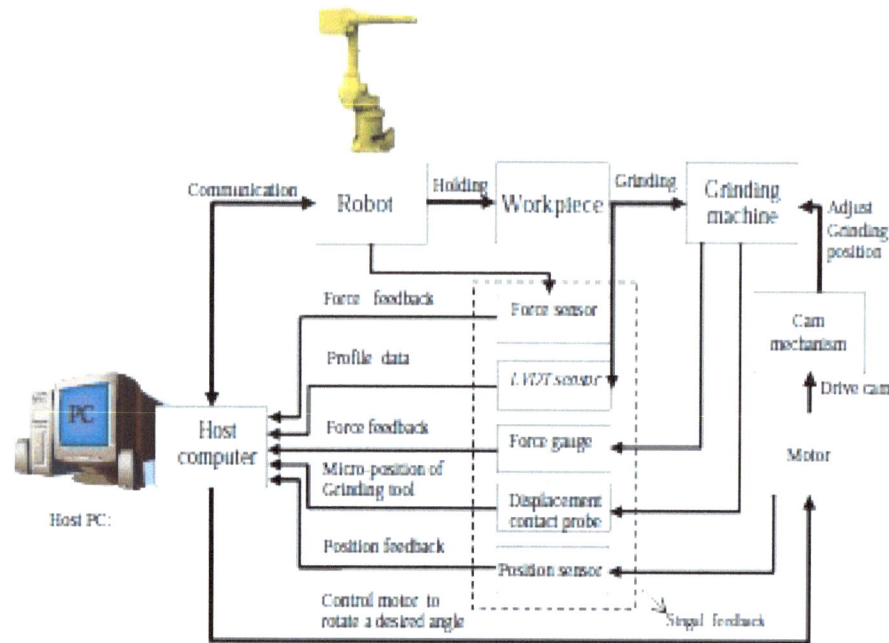

FIGURE 12.1.1. Grinding and polishing system structure diagram

12.2. Related Research Efforts

To date, much progress has been made in the measurement of turbine vanes and related field. Chen [6] and Huang [12] measured the 2D profile of turbine vanes by LVDT sensor. Where, the LVDT sensor is one of contact measurement probe. It produces analogue signals upon touching the object. The output signal of the LVDT was digitized by A/D card. The touch probe in its protection case is reliable and has a resolution of $1um$ [6, 12]. The profile geometric data are a compound of the robot coordinates and the sensor reading. The combined repeatability of the probe and the robot at the measuring location is around $10\mu m$ [12]. After measurement data were collected, a template was used to optimize the profile-fitting module with a minimum sum of errors. The template profile is established based on the design data using cubic Spline interpolation. To achieve a complete 3D airfoil profile, three 2D sectional profiles must be computed individually and then combined through interpolation. The method needs to have a template of sectional profile created using cubic Spline interpolations [6, 5, 12, 10]. Chen, Huang and Gong's approaches only measured a few special discrete points. The 2D profile is generated through profile fitting. The result shows a good achievement in 2D and 3D profile. Several other researchers [3, 8] measured 2D and 3D profile with a contact probe. However, it produces sparse data sets and requires contact with the object surface to perform measurement [7]. When the number of sample points is insignificant, the results cannot match the design profile exactly. It is known that the lower the simple points, the more difficult to obtain true fitting curve. The shortcomings for this method also include a low measurement speed due to repetitive motion into and out of the surface, a need for compensation of the contact probe radius, and deformations of the measuring point by the contact pressure which damages the surface texture. Moreover, it takes more time for the computer to calculate the fitting curve and the interpolation.

Our motivation is to find new techniques that are faster and more accurate to acquire the measurement of 3D surfaces. Laser scan sensor has emerged as a promising alternative. Its ability to accurately determine the dimension of an object without contact allows measurement of such parts with high accuracy. The measured parts can be made of any material, from ranging from to rubber [21]. Therefore, the industries are increasingly turning to non-contact optical measurement and regard it as an alternative for online measurement. In this chapter, we intend to use such a sensor to measure a turbine vane surface.

Non-contact optical measurement systems can be approximately divided into four basic categories [17]. They are described as following:

- Single point scans

  Single point systems combine a laser or LED probe with a line detector to gather highly accurate 3D measurements [7]. This technology is precise, but has limited on-line applications, as only one measurement can be made at a time, and either the measurement head or object to be measured must be moved. Single point systems are best for situations where only a few, precise measurements are needed, such as a few points on a designated.

- Single line scans

  Single line scans move a line of laser light across the object, measuring by means of PSD (Position Sensing Detector) or CCD (Charge Coupled Device). Several hundred points of measurement can be collected simultaneously along the line. It is able to obtain 3D information by combining the 2D images of line profiles. However, if measurements are needed beyond the boundary of the line, the sensor or the part must be moved.

- Area scanners

  Area scanners image a selected area. One such technique is Moiré Interferometer, which projects a sinusoidal pattern of light on surfaces, then looks at pattern distortions with variations of z height. This technique only works well on smooth single color surfaces and in controlled light settings, as it is sensitive to ambient light [23].

- Passive vision

  The passive optical technique is to recover the scene shape and reflectance characteristics from images. The objective is to acquire images of a scene observed from different viewpoints and possibly different illuminations, and from these images to compute the scene shape and reflectance at every surface point. As such, duplicate 3D models using techniques like stereo lithography can be fabricated.

The above four categories provide various degrees of speed, accuracy, and reliability. Most systems are based on triangulation, laser triangulation is based on the principle that when the distance between the laser source and the sensor (usually PSD or CCD) and the angles from either the laser or the sensor to the measuring point are known, the location of the point can be calculated from a triangulation algorithm; then the 3-D shape can be constructed by scanning the probe and assembling the measured points.

There are two ways of establishing a vision model of the above non-contact optical measurement. One is based on simple triangulation methodology, and another is based on perspective. However, almost all of the modeling methodologies originate from the pinhole camera model [20, 22, 25], which is only an approximation of a real imaging system [22]. Strictly speaking, the pinhole camera model may be accurate enough for pattern- or object-recognition purposes, but it is not adequate for dimensional measurement with high precision [20].

An increasingly popular alternative to "pin-hole" techniques is optical laser triangulation. Laser triangulation is a mature and reliable non-contact technique used in a variety of metrology application such as gauging, profiling and 3D surface mapping. Among the various vision inspection techniques, structured light 3D vision measurement has gained the widest acceptance [15, 13, 18, 19, 26]. At the present time the most common 3-D vision systems employ scanning laser line triangulation because this method currently has the greatest benefit in fast, precise, and inexpensive high resolution systems. Laser line scan sensor can provide a two dimensional measurement of height and width of objects, which makes it suitable for geometric profile recognition and distance measurement. This laser line sensor, scans parts from all orientations, then rotates the data back into a common coordinate system. It is able to digitize unknown 3D surfaces and parts of all sizes, especially those with complex geometry. This technology dramatically reduces scanning time by collecting data significantly faster than conventional non-contact measuring technologies. Laser line sensor captures scan data at 1000 points per cycle, $30ms$ per cycle (such as LJ-080 and LJ-3000) [1] and contains advanced automation features. The digitized scan data can then be processed further with optional meshing, surface modeling, color error map, easy scanning of unknown 3D surfaces and complex shapes.

12.3. Problem Statement And The Proposed Approach

The laser line sensor has special the following features: 1) Precision accuracy and reliability; and 2) High-speed scanning capability. Hence the laser line sensor will be chosen for the measurement of the turbine vane surface in this chapter. However, one main disadvantage of the non-contact optical sensor is related to the problems of precise measurement of specular reflection surfaces. Several researchers have proposed certain methods to address this problem. Kwan [16] examined the turbine vane surface containing burrs, holes or overspray, with Fuzzified Lowness Filter. Fuzzy rules are used in choosing the f value during execution. If the scanned curvature of the surface is large; f is chosen to be small. If the curvature is small, f is chosen to be large, where f is a number between 0 and 1. However, such filter cannot restore the signal of specular reflection area. This is because the signal of specular reflection area is random, composed of different frequencies. The maximum frequency of the signal randomly changes with the width and height of the specular reflection gap. In other words, we cannot get the frequency for such a filter according to its curvature. Another process approach for corrupted data is painting [**4, 9**]. Fan [**9**] reported that painting the surface could always ensure the laser probe performed accurately. However, the painting approach is not applicable for online measurement profile. Fonathan Wu [**24**] introduced the Algorithm of center location that used to process specular reflection to a structured light range scanner. However, he used CCD camera as a sensor. Meanwhile, it is not clear whether it can adapt to all structured light range scanner.

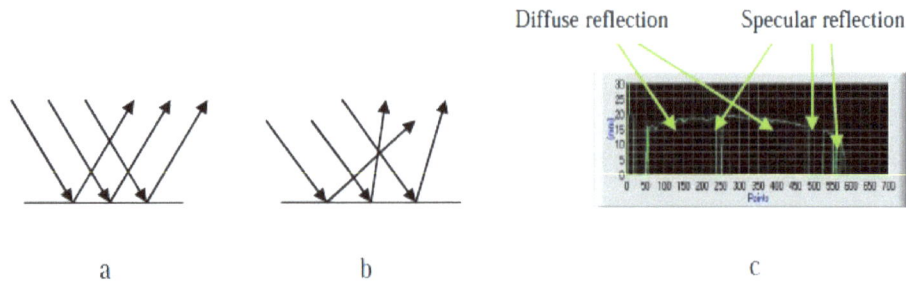

a   b   c

FIGURE 12.3.1. Reflection type. a. Specular reflection; b. Diffuse reflection; c. Scan for turbine vane.

Figure (12.3.1) shows an experiment where the grinding and the polishing of a turbine vane indicates two types of laser light. When the turbine vane surface has specular reflection, the sensor cannot receive reflection light. Thus, 2D Profile curve has corrupted the data. Usually, the reflection of turbine vane surface is not pure and simple specular reflection or diffuse reflection (shown in Figure 12.3.1c). The reflection of the turbine vane surface is diffuse reflection in most cases. Only when the turbine vane is placed at a special angle, its reflection is specular reflection . Therefore, with the help of a computer software, it is possible to restore the corrupted data using an interpolation method for the specular reflection gap.

Although some works have been done on corrupt data for specular reflection, little success is achieved on online measurement for the turbine vane of the specular reflection surface. Our current work aims to restore the corrupt data of the turbine vane using interpolation algorithm s. Before interpolation, the linear compensation is used to repair the edge data of the specular reflection gap.

12.4. Online Profiling Approaches

The turbine vanes to be repaired have severe distortions and twists after operations in a high-temperature and high-pressure engine environment. Due to severe part distortions and part-to-part variations of the turbine vanes for repairing, the design (nominal) profile cannot be used directly in a robotic surface finishing process. It is critical and necessary to have a profile sampling and distortion compensation system in this specific application to deal with part-to-part variations. Therefore, it is very important that the turbine vanes are measured after processing.

There are two approaches for profile measurement: off-line profile measurement and on-line profile measurement. The former utilizes an external instrument to measure the profile on a separate fixture. Then, the workpiece is unfixed, and then transferred to the robotic grinding and polishing system. In the meantime, the measurement data are transmitted to the host controller for further processing. Accurate measurement can be obtained for off-line profile measurement. The measurements can be carried out in parallel with robotic grinding and polishing operation, hence shortening the cycle time. However, the off-line profile measurement approach suffers some drawbacks: First, a separate measurement station incurs extra costs. As a minimum configuration, there should be an XYZ table carrying the probe (a commercial Coordinate Measurement Machine (CMM) for example), and one-axis rotary table to rotate the workpiece. Secondly, changing fixture from the measurement station (held by a rotary table) to the grinding and polishing system (held by the robot end-effecter) introduces datum errors that erode the seemingly accurate measurements obtained. On the other hand, the on-line profile measurement method utilizes the robot itself as the measurement instrument together with a range or displacement sensor. Although the robot accuracy ($0.1mm$) is much worse than a CMM (about $5microns$), the same end-effecter for both profile measurement and the grinding and polishing operation ensures a common datum, hence minimizing fixture errors. Furthermore, the sensor head can be placed in a gap where the robot has a better repeatability.

Profile measurement sensors fall into two broad categories: contact and non-contact sensors. For contact sensors, a contact probe is used for measurement as in the case for a CMM or a machining center. Such a measurement system is of slow response but of high cost, and the physical contact between the probe and the workpiece ensures the reliability of the measurement. As an alternative approach, an optical sensor can be used to measure 2D or 3D profile. The range sensing principle is based on triangulation. In order to obtain reliable readings from the laser sensor, the laser beam is kept in the normal direction to the surface. By measuring and processing data, a very good repeatability can be obtained. Since there is no contact between the sensor and workpiece, the measurement process is continuous. However, the turbine vane has rough surface with specular spots, and the laser sensor does not always produce reliable reading. Therefore, an investigation of laser sensing capability for turbine vane is needed before a laser sensor is deployed for profile measurement.

12.4.1. Laser Scanning System. The industry needs and uses a variety of laser sensors in all kinds of applications and for different measurement tasks. The height or the x, y, z Dimensional profiling of parts has to be measured accurately to improve quality inspection off-line or online. There is the absolute position or the position feedback information for closed loop control that is critical for the system. General applications are found in production and in test labs as well. There are physical dimensions to be measured, which can be obtained indirectly or derived from a sensor reading; even it does not look like a true displacement application in the first place, such as thickness, vibration, velocity, acceleration, torque, force, pressure, warp and so on.

It is well known that contact measurement techniques such as mechanical gauges, calipers can measure displacement as well as the contact sensor performs such measurements. Sometimes, contact measurement is not only desired but also not possible to contact the specimen and apply any force to the target. The surface is so sensitive that it cannot be touched and the required speed for online measurement does not allow contact measurement. Compared with the contact sensor, the laser sensor performs measurements in a non-contact mode, speeds up the measurement, gives a greater density of data and has a longer life. In this project, the laser sensor used is the laser scanner LJ-3000/LJ-080 from Keyence Company as shown in Figure 12.4.1. It comprises a laser controller LJ-3000 and a sensor head LJ-080. It is based on the principles of Position Sensitive Detection (PSD). The PSD gives an output in terms of the center of gravity of the total light quantity distribution on the active gap. PSDs are purely analog devices and rely on a current generated by a photodiode divided in one or two resistive layers. This simple design gives the advantages of stability and reliability.

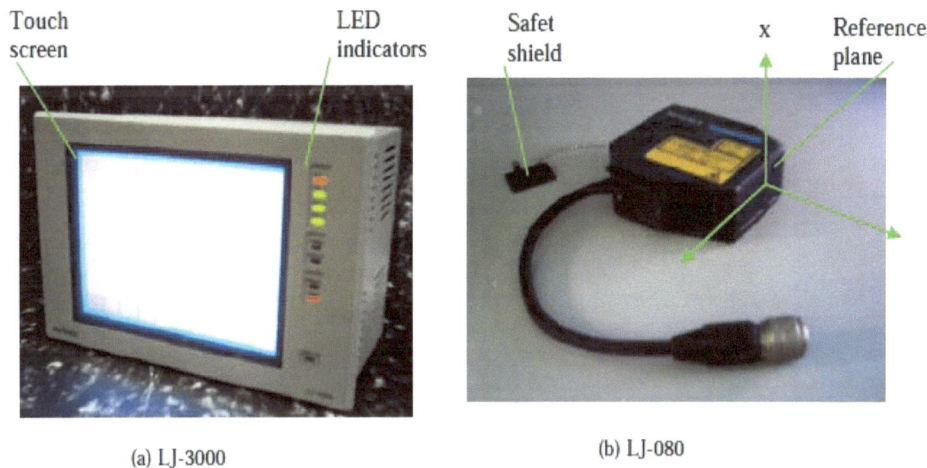

(a) LJ-3000   (b) LJ-080

FIGURE 12.4.1. Laser scanner controller LJ-3000 and sensor head LJ-080

LJ-080 sensors, the diagram of principle shown in Figure 12.4.2 determine the position by measuring light reflected off a target. A transmitter (laser diode) sends out a visible ($685nm$) light spot onto the target. The light is reflected off the object and focuses via an optical lens system onto a light sensitive detector built into the sensor head. If the target moves and when the point of light falling on the object moves closer to or farther from the reference point, the projected spot position on the detector changes.

LJ-3000 controller unit of the laser determines the position of the focused spot on the receiver and after linearization and signal conditioning a signal (analogue or digital) is provided from the sensor output that is proportional to the target position. The controller has analogue and digital signal outputs. The digital signal is an output from the RS232, which the maximum speed of communication is 9600 b/s. To speed up measurement, the analogue signal is used in this chapter. The time sequence relative of the output analogue signal for the laser controller LJ-3000 is as follows:

As shown in Figure 12.4.3, on the X-axis direction, the analogue voltage output ranges from $-5V$ to $+5V$. The measurement duration time is 20 ms, the number of sampling points is 1000 points during 20 ms. The relationship between the X-axis output voltage signal and sample points is linear. Therefore, if the sample position is known, X-axis position value can be worked out by

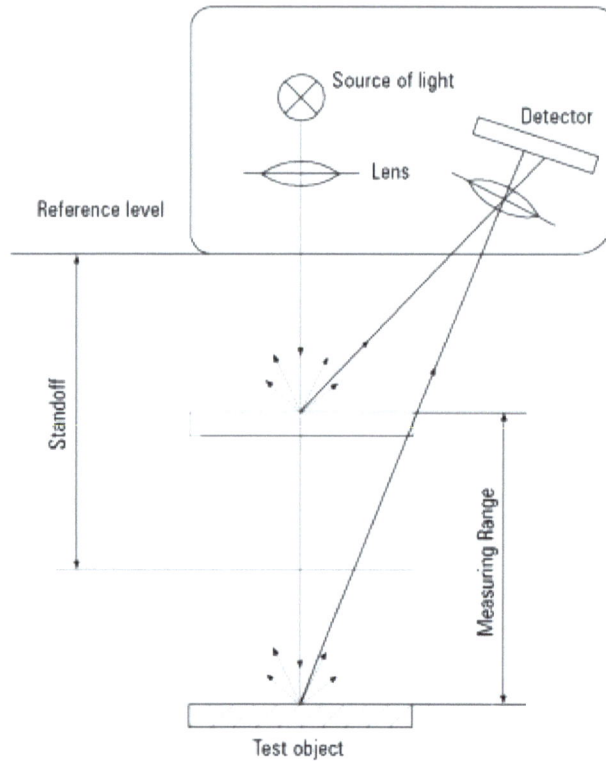

FIGURE 12.4.2. The principle of L-080 triangulation

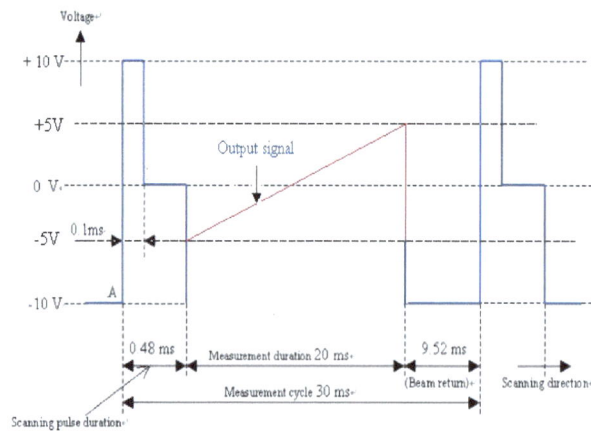

FIGURE 12.4.3. X-axis timing sequence diagram of LJ-3000

calculation, and not measurement. Moreover, X and Z axis directions have the same time sequence interval. Actually, if sample point position and its Z value are known, it is possible to get X-axis position only by computing. That is to say, there is no need to measure the X-axis output. Its output value can be obtained from the sample point's position.

The Z-axis direction signal output is shown in Figure 12.4.4. The analogue voltage output ranges from $-10V$ to $+10V$. The measurement duration time and number of sample points are the same as the X-axis direction.

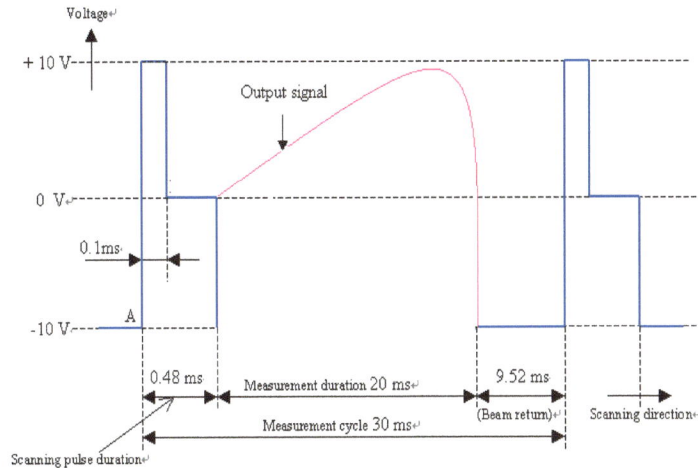

FIGURE 12.4.4. Z-axis timing sequence diagram of LJ-3000

The measurement system is shown in Figure 12.4.5. Firstly, the analogue signal of the sensor output is transferred into LJ-3000 controller, which pre-processes the signal. Secondly, the analogue signal input computer via a data acquisition (DAQ) card PCI-6024E from National Instruments (NI) and then the sampled digital signal is processed and analyzed in the program called LScan programming, which was developed in this project.

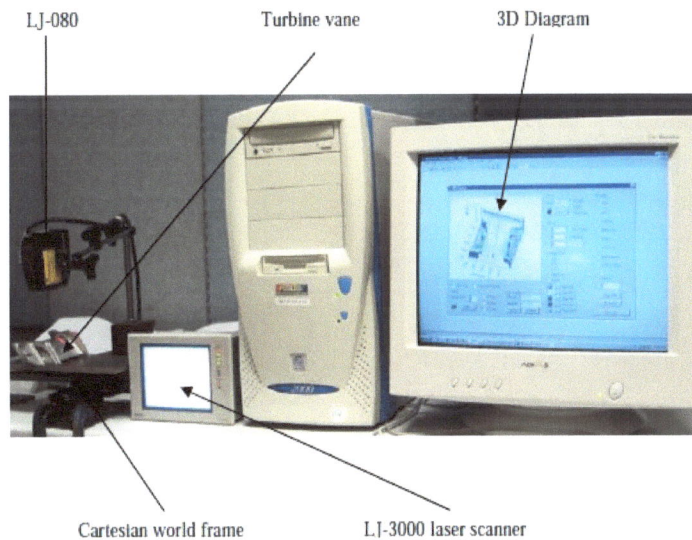

FIGURE 12.4.5. Laser sensing experiment station

During the experiments, two turbine vanes standard gauges are available and used to test the scanning capability of the laser scanner (shown in Figure 12.4.6).

Experiments showed that the laser scanner LJ−3000/LJ−080 has a good output signal for objects whose surfaces are not highly specular, such as the turbine vane T00, and standard gauges. A 2D profile can be constructed from the obtained measurement signals which reflect the surface profiles very well, as illustrated in Figure 12.4.7 and Figure 12.4.8. These figures show the sectional profile of the turbine vane when the laser is scanning the concave side of the turbine vane T00 (still unpolished). Figure 12.4.7 shows the sectional profile of the turbine vane when the laser is scanning the convex side of turbine vane T00 (still unpolished). Both of them are typical results when the laser is scanning the highly diffuse surface of the object.

Figure 12.4.9 shows the sectional profile of the turbine vane when the laser is scanning the convex side of the turbine vane T01 (already polished) and Figure 12.4.10 shows the respective experimental condition in this case. If the surface is specular

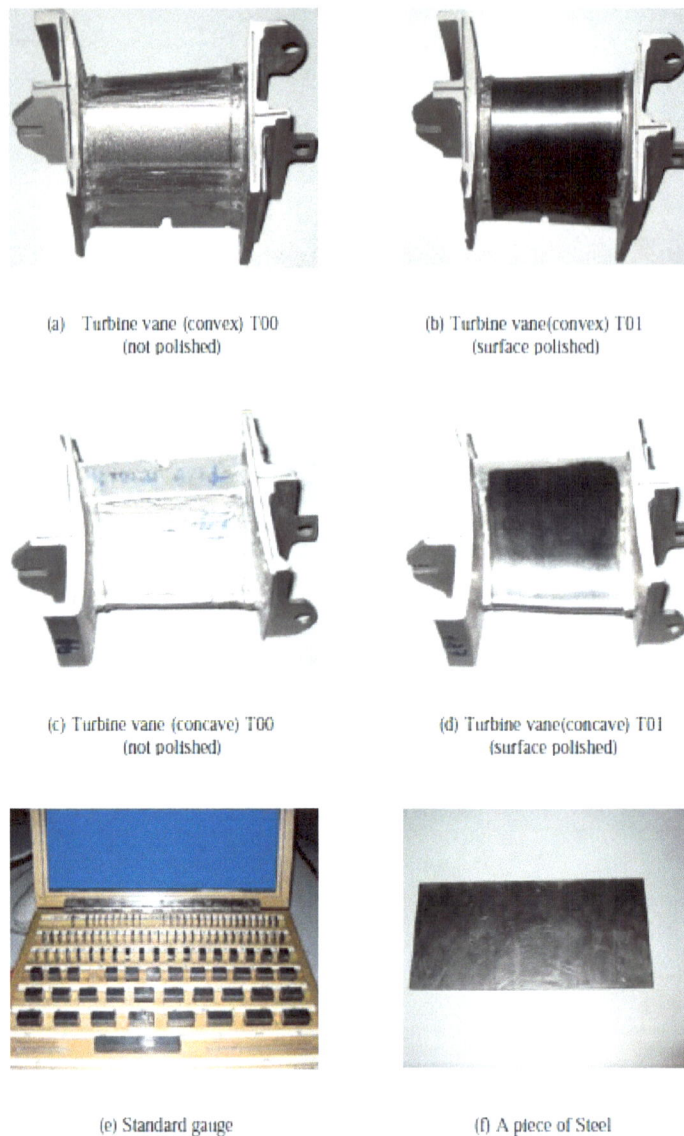

(a) Turbine vane (convex) T00
(not polished)

(b) Turbine vane(convex) T01
(surface polished)

(c) Turbine vane (concave) T00
(not polished)

(d) Turbine vane(concave) T01
(surface polished)

(e) Standard gauge

(f) A piece of Steel

FIGURE 12.4.6. Parts used in laser sensing experiments

reflection, the scanning result is worse due to the specular reflection property of the surface, as a case of turbine vane T01 (surface polished). From the experiments, it is found that if the laser beam incident direction is controlled properly, such that the laser incident direction is near the normal direction at the scanning point of the object's surface, then the scanning quality can be much improved. Although after properly controlling the incident direction, sometimes the signal profile for specular reflection surface is still distorted as shown in Figure 12.4.9, the information of the profile remains incomplete. It can be extracted and restored if certain appropriate profile recovery techniques are used. For example, the interpolation approach to the specular reflection gaps, which will be discussed later.

12.4.2. Approaches to Profile Reconstruction. If a turbine vane has a highly diffuse surface, 2D and 3D profiles can be obtained from an empirical investigation of LJ−3000/LJ−080. This study focuses on how to deal with a signal from a specular reflection of the turbine vane. It is helpful here to introduce linear compensation for the signal and three interpolation approaches [2],[11] are used on profiles in this project.

12.4.2.1. *Linear Compensation.* After conducting numerous experiments, we found that specular reflection gaps signals have distinct characteristics:

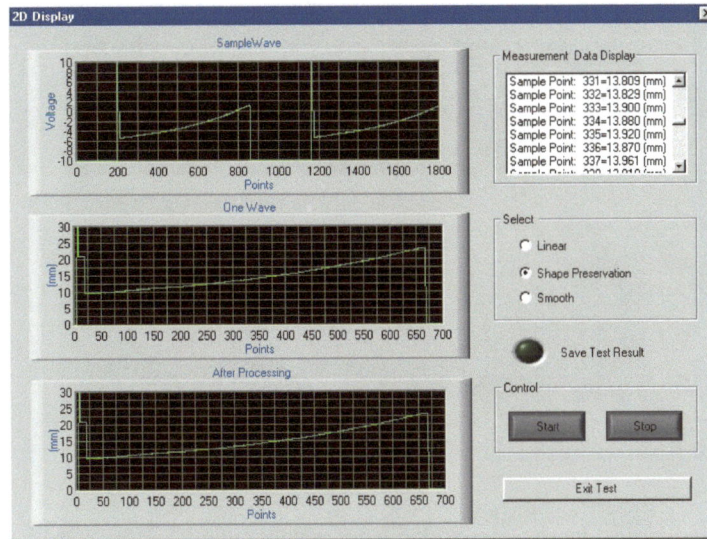

FIGURE 12.4.7. Laser scanning section of turbine vane T00 (concave side)

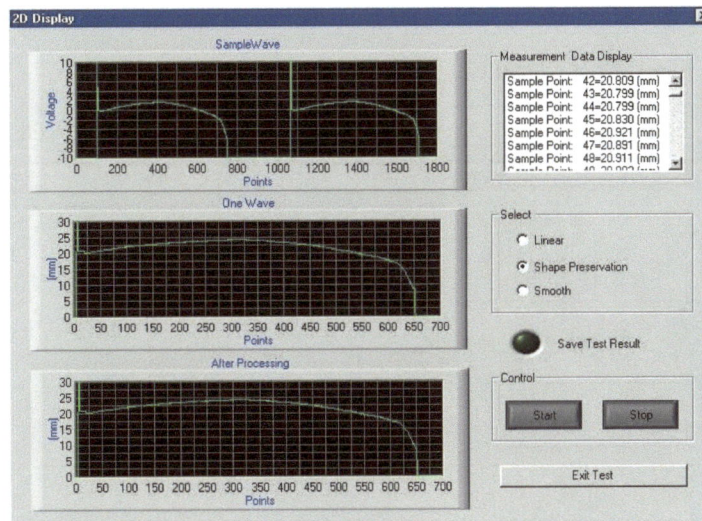

FIGURE 12.4.8. Laser scanning section of turbine vane T00 (convex side)

As shown in Figure 12.4.11, it is observed that the specular reflection position has no output signal if the turbine vane surface is a specular reflection. That is to say, the data points have been corrupted. It is found that the edge signals of the specular reflection gaps can be divided into four types by experimentation. We will use a linear compensation method to repair the edge data of the specular reflection gaps before interpolation.

- Where m, n are the index of the sample points, $f(m)$ and $f(n)$ are values with respect to the index, $0 < m < n$; $1 < n - m$. We say that the m is transition points as shown in Figure 12.4.12. We set one threshold value which is obtained experimentally, If $f(m-1) - f(m) > threshold$, then $f(m) = 2f(m-1) - f(m-2)$ formula is used to repair the mth data point.
- Here m, n are the index of the sample points, $f(m)$, $f(n)$ are values with respect to index, $0 < m < n$; $1 < n-m$. It is obvious that the n is a transition points as shown in Figure 12.4.13. If $f(n+1) - f(n) > threshold$, where the threshold is obtained from experiment. In this case, the nth point data can be obtained using $f(n) = 2f(n+2) - f(n+1)$.

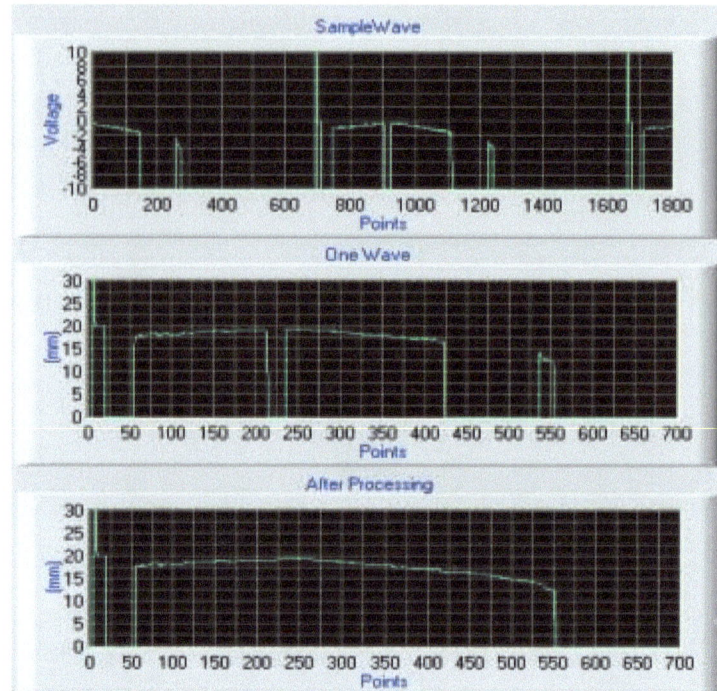

FIGURE 12.4.9. Laser scanning specular section of turbine vane T01 (Convex side, polished region)

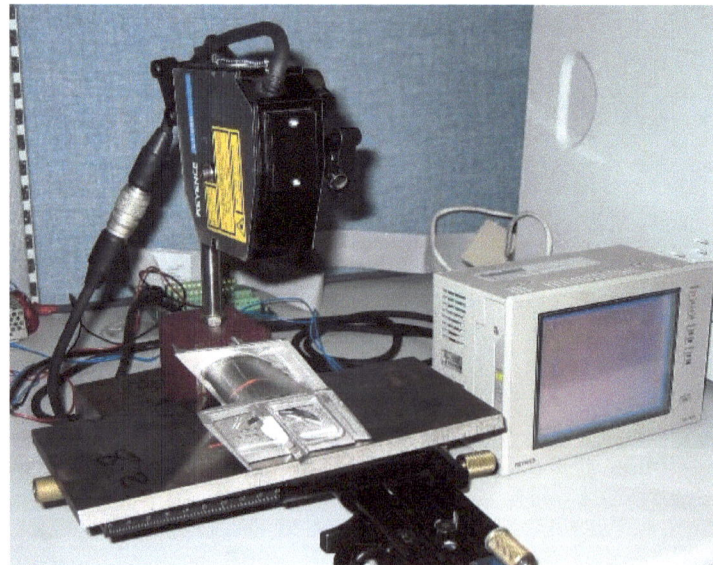

FIGURE 12.4.10. Laser scanning specular section of turbine vane T01

- Here m, n are the index of the sample points, $f(m), f(n$ are values with respect to index, $0 < m < n$; $1 < n - m$. This is undoubtedly the case where the m and n are transition points as shown in Figure 12.4.14. If $f(m-1) - f(m) > threshold$ and $f(n+1) - (n) > threshold$, where the threshold is obtained from experiment. Actually this case combines the above two. Therefor:

(12.4.1) $$f(m) = 2f(m-1) - f(m-2) \qquad f(n) = 2f(n+2) - f(n+1)$$

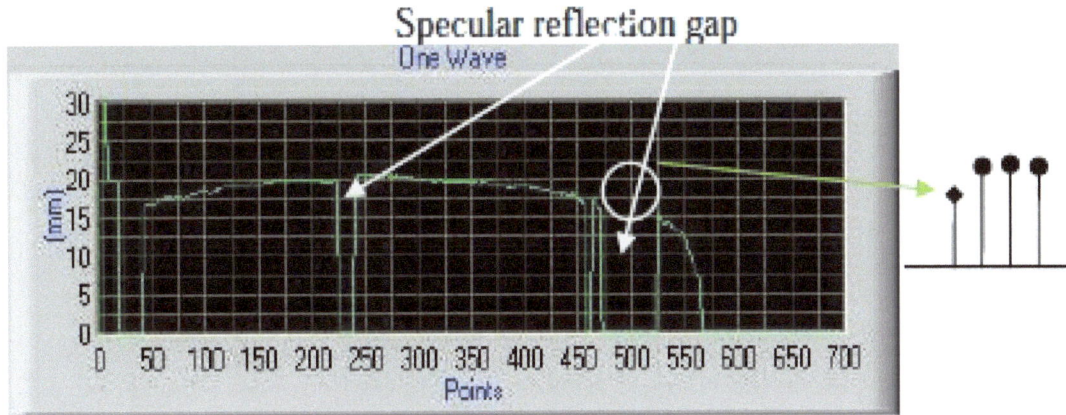

FIGURE 12.4.11. Laser scanning specular section of turbine vane T01

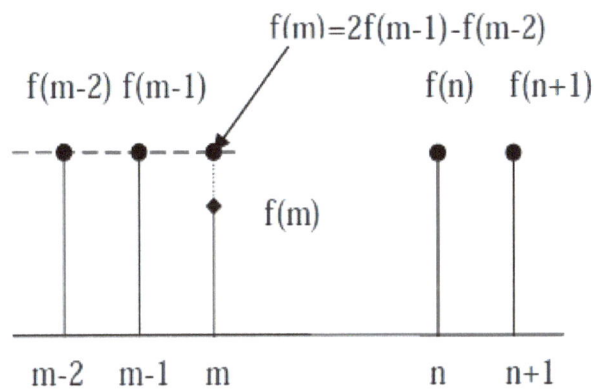

FIGURE 12.4.12. Linear compensation (case 1)

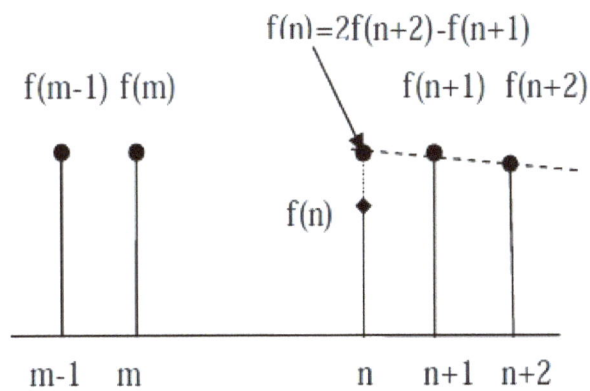

FIGURE 12.4.13. Linear compensation (case 2)

- Here m, n are the index of the sample points, $f(m)$, $f(n)$ are values with respect to index, $0 < m < k < n$; $1 < n-m$. We can say that k is an isolated point, as shown in Figure 12.4.15. If both $f(m) - f(k)$ and $f(n) - f(k)$ are larger than the threshold, which is obtained experimentally, then $f(k)$ is equal to zero.

$$f(m)=2f(m-1)-f(m-2)$$

f(m-2) f(m-1)   f(n) f(n+1)

f(m)

m-2 m-1 m   n   n+1

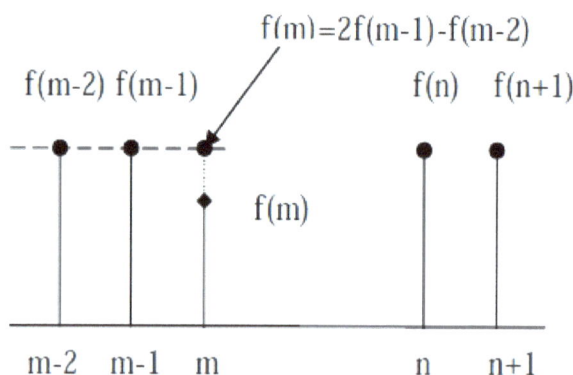

FIGURE 12.4.12. Linear compensation (case 1)

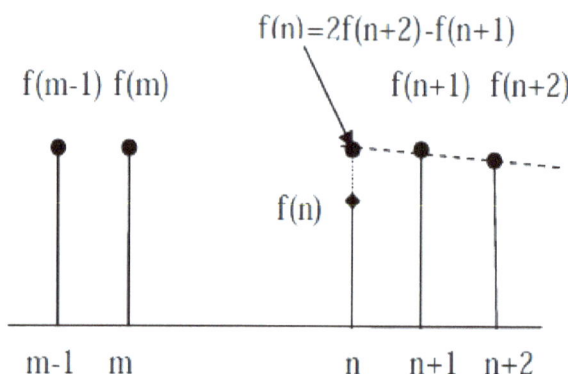

$$f(n)=2f(n+2)-f(n+1)$$

f(m-1) f(m)   f(n+1) f(n+2)

f(n)

m-1   m   n   n+1 n+2

FIGURE 12.4.15. Linear compensation (case 4)

If the turbine vane has specular reflection, the above four linear compensation methods will be employed to repair the edge of the specular reflection gap, and then the interpolation algorithm is used to restore the specular reflection gap. It is obvious that the linear compensation is actually reasonable. This is because the turbine vanes surfaces are continuous slippery surfaces. When the distance of two adjacent measurement points is very small, it will not have jumping change.

12.4.3. Interpolation Approaches. When the measurement of the turbine vane has corrupt data, the interpolation technique is first considered and used to restore the data. There are many types of interpolation algorithm s. Which type is best suited? The following actual measurement data are simulated by MATLAB6.0 with three interpolation methods respectively (shown in Figure 12.4.16). $X = [0 \quad 1 \quad 2 \quad 3 \quad 4 \quad 5 \quad 6 \quad 7 \quad 8 \quad 9 \quad 10 \quad 15 \quad 16 \quad 17 \quad 18 \quad 19 \quad 20]$; $Y = [19.241$ 19.230 19.223 19.231 19.216 19.233 19.241 19.230 19.227 19.241 19.231 19.213 19.234 19.221 19.210 19.243 19.225]$; where vector X denotes sample points, vector Y represents height values. As shown in Figure 12.4.16, it is found that linear and piecewise Hermite interpolation is better than the Spline interpolation. However, a lot of references reported that Spline interpolation is one of the best. Because it has continuous first and second derivatives. In theory, it is smoother than the other two interpolations. So it does, the second derivative is easily affected by control points at the edge of the interpolation gap. Nevertheless, the piecewise Hermite interpolation has a special process for the first derivative at control points. So it has the function of 'shape preserve' [2, 11]. It is much possible that it is utilized to restore corrupted data.

In this work, the control points at the edge of the specular reflection gap, randomly change. If it does not process the edge of the specular reflection gap in advance, all interpolation methods cannot obtain normal surface from the specular reflection surface. So the linear compensation method is introduced to repair the control points at the edge of the specular reflection gap, before interpolation.

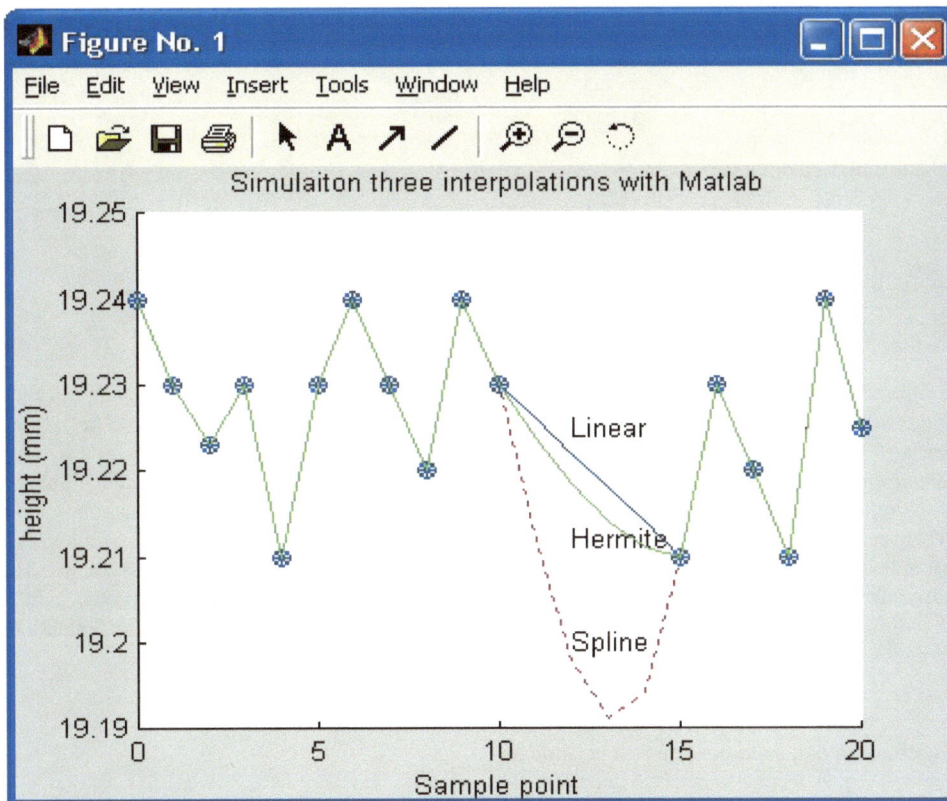

FIGURE 12.4.16. Simulation three interpolations with Matlab

12.4.3.1. *Piecewise linear Interpolation Approach.* To generate lines connecting from points, the graphic routines use piecewise linear interpolation. As shown in Figure 12.4.17.

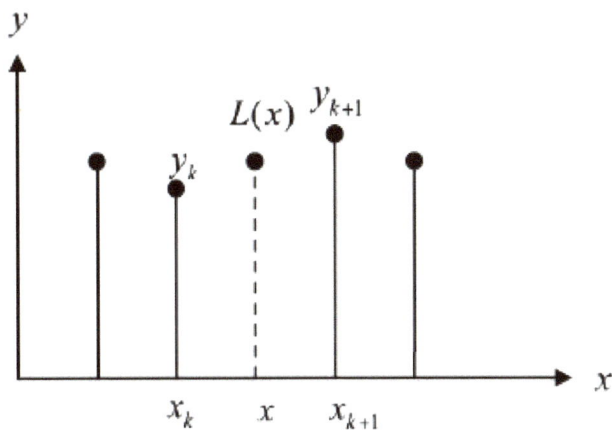

FIGURE 12.4.17. Piecewise linear interpolation

The formula for piecewise linear interpolation sets the stage for more sophisticated algorithms.

$$L(x) = y_k + s^* \partial_{k+1}$$

Here three quantities are involved. In the above equation, k is interval index, it must be determined so that

$$x_k \leq x < x_{k+1}$$

And s is the local variable

$$s = x - x_k$$

And ∂_k is the first divided difference

$$\partial_k = \frac{y_{k+1} - y_k}{x_{k+1} - x_k}$$

With these quantities in hand,

$$L(x) = y_k + (x - x_k)\frac{y_{k+1} - y_k}{x_{k+1} - x_k}$$

is clearly a linear function that passes through (x_k, y_k) and (x_{k+1}, y_{k+1}). The points (x_k) are sometimes called breakpoints or breaks. The piecewise linear interpolation $L(x)$ is a continuous function of x, but its first derivative $(L'(x))$, is not continuous. The derivative has a constant value, ∂_k, on each subinterval and jumps at the breakpoints.

12.4.3.2. *Spline Interpolation Approach.* Another interpolating function is a cubic Spline. The term "Spline" refers to an instrument used in drafting. It is a thin, flexible wooden or plastic tool that is passed through given data points and that defines a smooth curve in between. The physical Spline minimizes potential energy, subject to the interpolation constraints. The corresponding mathematical Spline must have a continuous second derivative, and satisfy the same interpolation constraints. The breakpoints of a Spline are also referred to as its knots. The world of Splines extends far beyond the basic one-dimensional, cubic, interpolation Spline we are describing here. There are multidimensional, high-order, variable knot, and approximating Splines. In this project, the Spline interpolation used is one-dimensional. The formula of Spline interpolation is:

$$p(x) = y_k + sd_k + s^2 c_k + s^3 b_k$$

Where the coefficients of the quadratic and cubic terms are

$$c_k = \frac{3\partial_k - 2d_k - d_{k+1}}{h}$$

$$b_k = \frac{d_k - 2\partial_k + d_{k+1}}{h^2}$$

Where h, ∂_k, d_k are the length of the subinterval, the first divided difference, and the slope of the interpolation point, respectively. Now, it is necessary to explain how to obtain the slope. The first derivative $p'(x)$ of our function is defined by different formulas on either side of a knot x_k. Both formulas yield the same value d_k at the knots, so $p'(x)$ is continuous. On the k_{th} subinterval, the second derivative is a linear function of $s = x - x_k$; it can be obtained from the formula of Spline interpolation and cubic terms above.

$$p''(x) = \frac{(6h - 12s)\partial_k + (6s - 2h)d_k + (6s - 4h)d_k}{h^2}$$

When $x = x_k$, $s = 0$ and

$$p''(x_k+) = \frac{6\partial_k - 2d_{k+1} - 4d_k}{h_k}$$

Where the plus sign in x_k+ indicates that this is a one-sided derivative. When $x = x_{k+1}$, $s = h_k$ and

$$p''(x_k-) = \frac{-6\partial_k + 4d_{k+1} + 2d_k}{h_k}$$

On the $(k-1)th$ interval, $p''(x)$ is given by a similar formula involving ∂_{k-1}, d_k, and d_{k-1}. At the knot x_k

$$p''(x_k-) = \frac{-6\partial_k + 4d_{k+1} + 2d_{k-1}}{h_{k-1}}$$

$p''(x)$ is required to be continuous at $x = x_k$ leading to

$$h_k d_{k-1} + 2(h_{k-1} + h_k)d_k + h_{k-1}d_{k+1} = 3h_k \partial(k-1) + 3h_{k-1}\partial_k$$

When the knots are equally spaced, the h_k does not depend on k, it becomes

$$k_{k-1} + 4d_k + 4dk + 1 = 3\partial_{k-1} + 3\partial_k$$

Similar to other interpolation, the slopes d_k of a Spline are closely related to the differences ∂_k. In the Spline case, they are a kind of running average of the ∂_k's. The preceding approach can be applied at each inside knot x_k, $k = 2, \ldots n-1$ to create $n-2$ equations involving the n unknowns d_k. As with cubic interpolating polynomials, a different approach must be used near the ends of the interval. One effective strategy is known as "not-a-knot". The idea is to use a single cubic on the first two subintervals, $x_1 \le x \le x_2$, and on the last two subintervals, $x_{n-2} \le x \le x_n$. In effect, x_2 and x_{n-1} are not knots. When the knots are equally spaced with all $h_k = 1$, this leads to

$$d_1 + 2d_2 = \frac{5}{2}\partial_1 + \frac{1}{2}\partial_2$$

and

$$2d_{n-1} + d_n = \frac{1}{2}\partial_{n-2} + \frac{5}{2}\partial_{n-1}$$

With the two end conditions included, we have n linear equations in n unknowns:

$$Ad = r$$

$$d = \begin{pmatrix} d_1 \\ d_2 \\ M \\ d_n \end{pmatrix}$$

The coefficient matrix A is tri-diagonal.

$$A = \begin{pmatrix} h_2 & h_2 + h_1 & & & & & \\ h_2 & 2(h_2 + h_1) & h_1 & & & & \\ & h_3 & 2(h_3 + h_2) & h_2 & & & \\ & & 0 & 0 & 0 & & \\ & & & & h_{n-1} & 2(h_{n-1} + h_{n-2}) & h_{n-2} \\ & & & & & (h_{n-1} + h_{n-2}) & h_{n-2} \end{pmatrix}$$

The right-hand side is

$$r = 3 \begin{pmatrix} r_1 \\ h_2\partial_1 + h_1\partial_2 \\ h_3\partial_2 + h_2\partial_3 \\ M \\ h_{n-1}\partial_{n-2} + h_{n-2}\partial_{n-1} \\ r_n \end{pmatrix}$$

The two values r_1 and r_n correspond to the terminal points. When the knots are equally spaced with all $h_k = 1$, the coefficient matrix is quite simple.

$$A = \begin{pmatrix} 1 & 2 & & & & \\ 1 & 4 & 1 & & & \\ & 1 & 4 & 1 & & \\ & & 0 & 0 & 0 & \\ & & & 1 & 4 & 1 \\ & & & & 2 & 1 \end{pmatrix}$$

The right-hand side is

$$r = 3 \begin{pmatrix} \frac{5}{6}\partial_1 + \frac{1}{6}\partial_2 \\ \partial_1 + \partial_2 \\ \partial_2 + \partial_3 \\ M \\ \partial_{n-2} + \partial_{n-1} \\ \frac{1}{6}\partial_{n-2} + \frac{5}{6}\partial_{n-1} \end{pmatrix}$$

In this project function, the linear equations defining the slopes are solved with the function introduced in NI (national instrument) library. In fact, the slopes are calculated by $d = A/r$.

12.4.3.3. *Piecewise Hermite Interpolation Approach.* Most effective interpolation techniques are based on piecewise cubic polynomials. Let h_k denote the length of the kth subinterval,

$$h_k = x_{k+1} - x_k$$

Then the first divided difference, $\pm\partial_k$, is

$$\partial_k = \frac{y_{k+1} - y_k}{h_k}$$

let d_k denote the slope of the interpolate at x_k

$$d_k = p'(x_k)$$

For the piecewise linear interpolate, $d_k = \partial_{k-1}$ or ∂_k, but this is not necessarily true for a higher order interpolation. Considering the following function on the interval $x_k \leq x \leq x_{k+1}$, expressed in terms of local variables $s = x - x_k$, equation (next) and $h = h_k$

$$p(x) = \frac{3hs^2 - 2s^3}{h^3}y_{k+1} + \frac{h^3 - 3hs^2 + 2s^3}{h^3}y_k + \frac{s^2(s-h)}{h^2}d_{k+1} + \frac{s(s-h)^2}{h^2}d_k$$

This is a cubic polynomial in s, and hence in x, that satisfies four interpolation conditions, two on function values and two on the possibly unknown derivative values.

$$p(x_k) = y_k; \quad p(x_{k+1}) = y_{k+1}$$

$$p'(x_k) = d_k; \quad p'(x_{k+1}) = d_{k+1}$$

Functions that satisfy interpolation conditions on derivatives are known as piecewise cubic Hermite interpolate, because of the higher order contact at the interpolation sites. If we happen to know both function values and first derivative values at a set of data points, then piecewise cubic Hermite interpolation [2, 11] can reproduce that data.

For example, if we have known both function values and first derivative values at 2 and 4 point as shown in Figure 12.4.18, we want to obtain 3 point function values.

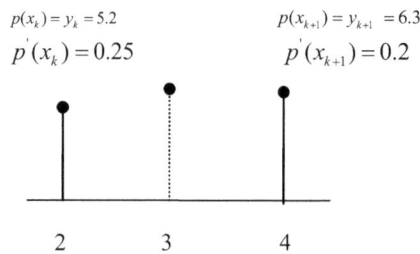

FIGURE 12.4.18. Example for piecewise Hermite interpolation

$$Q \quad 2 < 3 < 4$$
$$s = x - x_k = 3 - 2 = 1, h_k = x_{k+1} - x_k = 2,$$
$$p(x) = \frac{3hs^2 - 2s^3}{h^3}y_{k+1} + \frac{h^3 - 3hs^2 + 2s^3}{h^3}y_k + \frac{s^2(s-h)}{h^2}d_{k+1} + \frac{s(s-h)^2}{h^2}d_k$$
$$= \frac{3 \times 2 \times 1^2}{2^3} \times 6.3 + \frac{2^3 - 3 \times 2 \times 1^2}{2^3} \times 5.2 + \frac{1^2(1-2)}{2^2}0.25 + \frac{1(1-2)^2}{2^2}0.25$$

B 5.76.

Therefore, the function value of point 3 is 5.76. But if the derivative values are unknown, we need to define the slopes d_k somehow. Of the many possible ways to do this, we will describe one, piecewise cubic Hermite interpolating polynomial. Spline interpolation is also one of cubic interpolating polynomials, but with different slopes. Our particular piecewise cubic Hermite interpolating polynomial is a shape-preserving, "visually pleasing" interpolation. This is why we use it to repair corrupt data. The key idea is to determine the slopes d_k so that the function values do not overshoot the data values, at least locally. If ∂_k and ∂_{k+1} have opposite signs, or if either of them is zero, then x_k is a discrete local minimum or maximum, so we set

$$d_k = 0$$

If ∂_k and ∂_{k-1} have the same signs and the two intervals have the same length, then d_k is taken to be the harmonic mean of the two discrete slopes.

$$\frac{1}{d_k} = \frac{1}{2}(\frac{1}{\partial_{k-1}} + \frac{1}{\partial_k})$$

In other words, at the breakpoint, the reciprocal slope of the piecewise cubic Hermite interpolates is the average of the reciprocal slopes of the piecewise linear interpolation on either side. If ∂_k and ∂_{k+1} have the same signs, but the two intervals have different lengths, then d_k is a weighted harmonic mean, with weights determined by the lengths of the two intervals.

$$\frac{w_1 + w_2}{d_k} = (\frac{w_1}{\partial_{k-1}} + \frac{w_2}{\partial_k})$$

Where $w_1 = 2h_k + h_{k-1}$, $w_2 = h_k + 2h_{k-1}$. This defines the piecewise cubic Hermite interpolating polynomial slopes at interior breakpoints, but the slopes d_1 and d_n at either end of the data interval are determined by a slightly different, one-sided, analysis. The details are shown in the software. In mathematics, the trade off between smoothness and a somewhat subjective property might be called local monotonic or shape preservation. The piecewise linear interpolation is at one extreme. It has hardly any smoothness. It is continuous, but there are jumps in its first derivative. On the other hand, it preserves the local monotonic of the data. It never overshoots the data and it is increasing, decreasing, or constant on the same intervals with the data. The Spline is smoother than the piecewise cubic Hermite interpolating polynomial in the theory. But we will see that it is not appropriate for this project. The reason is that the Spline has two continuous derivatives, and is easily interfered by control points. While piecewise cubic Hermite interpolating polynomial has only one. A discontinuous second derivative implies discontinuous curvature. The human eye can detect large jumps in curvature of graphs and in mechanical parts made by numerically controlled machine tools. On the other hand, piecewise cubic Hermite interpolating polynomial is guaranteed to preserve shape, but the Spline might not.

12.5. Results And Discussions

Testing is performed to make sure that the interpolation approach is chosen properly. Meanwhile, it is to check whether or not results meet measurement requirement. Most of measurements are performed on 2D profiles. Because 3D profiles have 70×700 pixels points for an image, it is not necessary that each point on 3D profiles is measured. But the coordinate point of 3D profile can be measured with cursor on 3D profile surface. The effects can be illustrated with 3D profiles. Since there are 700 sample points per frame on 2D profile, it is too difficult to calculate all points and to analyses them. In this chapter, this study focuses on the specular reflection gap. Measurement results and discussions are presented. Then the effects of convex and concave surfaces on measurement results are discussed by changing different turbine vane positions.

12.5.1. Accuracy. Before the accuracy is measured, it is necessary to explain two definitions.

- Accuracy:
    Accuracy is the closeness of agreement between the average of a large number of experimental measurements of a characteristic and the master value of that characteristic.
- Bias (12.5.1):
    (1) Bias is the difference between the average value of all the measurements and the master value.
    (2) Bias is the numerical value used to measure accuracy.

It is important to study the measurement accuracy of the system developed in this project. It can verify whether our measurement system is accurate or not. Since the standard turbine vane with known profile height is not available, the standard gauge is employed to test the measurement accuracy [1]. Here the standard gauge with $11.500mm$ in height, was used as standard (shown in Figure 12.5.2). The standard gauge was placed on the platform shown in Figure 12.5.2. The measurement was repeated 10 times. The average value z_1 is shown in Table 1; z_3 is calculated from the formula $z_3 = z_1 - z_2$, where, z_3 is the measured value for the standard gauge. The standard gauge was marked $11.5mm$, here $z_4 = 11.5mm$, the measuring accuracy z_5 is obtained by calculating z_3 minus $z4$. It is significant that the range of accuracy values is from $0.004\ mm$ to $0.028\ mm$.

Actually, the accuracy of one point is not appropriate for us to measure 2D and 3D profiles due to the fact that line scan is used in this project. As stated earlier, 700 sample points are obtained for every frame. Thus, the lowest accuracy of $0.028mm$ is regarded as the accuracy for laser scan systems. That is to say, the accuracy of the measurement system is less than $30\mu m$, which meets the system requirement.

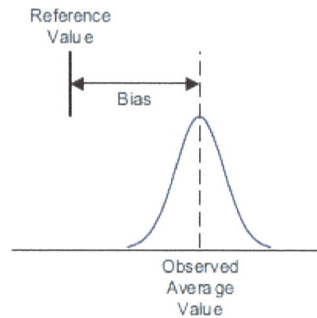

FIGURE 12.5.1. 21 Bias

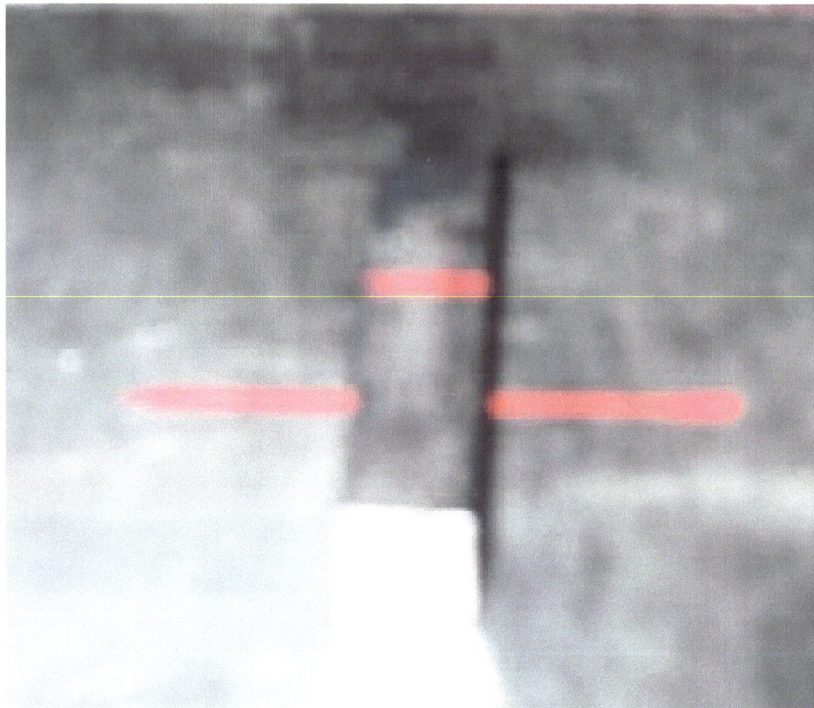

FIGURE 12.5.2. Standard gauge measurements

12.5.2. Repeatability. The repeatability and reproducibility of the laser sensor studies determine how much of our observed process variation is due to measurement system variation. The overall variation is broken down into three categories (Figure 12.5.3): part-to-part, repeatability, and reproducibility. The reproducibility component can be further broken down into its operator and operator by part (or components).

Repeatability is the variation due to random error introduced by the differences of the repeated measurements of the same part, by the same operator. On the other hand, the reproducibility is the uncertainty among operators for measuring the same part. Here, we are interested in repeatability experiments of this sensor. Reproducibility experiments have not been done in this study. We will use following method which is introduced by Johnson [14] to study the repeatability of this measurement system.

$$C_p = \frac{USL - LSL}{6S}$$

Where USL refers to upper specification limit, LSL is lower specification limit, S is standard deviation, C_p is the capability index of repeatability. If $C_p < 1$, the repeatability is poor; if $C_p = 1$, the repeatability is considered as marginal; If $C_p \geq 1.33$, the repeatability is considered as good. Here, the setup of the repeatability experiment is the same as the accuracy experiment (shown

TABLE 1. Data of measuring accuracy (mm)

Sample points	z_1	z_2	z_3	z_4	z_5
245	20.707	9.179	11.528	11.500	0.028
246	20.702	9.175	11.527	11.500	0.027
247	20.699	9.173	11.526	11.500	0.026
248	20.698	9.172	11.526	11.500	0.026
249	20.701	9.178	11.523	11.500	0.023
250	20.704	9.180	11.524	11.500	0.024
251	20.707	9.193	11.514	11.500	0.014
252	20.699	9.195	11.504	11.500	0.004
253	20.701	9.192	11.509	11.500	0.009
254	20.698	9.189	11.509	11.500	0.009
255	20.703	9.188	11.515	11.500	0.015
256	20.709	9.193	11.516	11.500	0.016
257	20.703	9.188	11.515	11.500	0.015
258	20.710	9.192	11.518	11.500	0.018
259	20.699	9.194	11.505	11.500	0.005
260	20.693	9.189	11.504	11.500	0.004

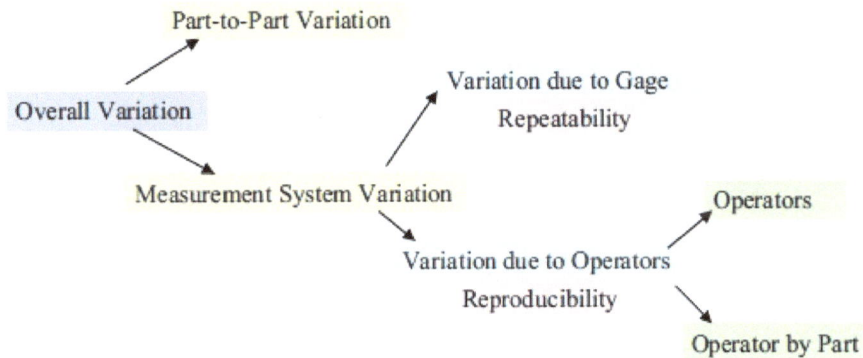

FIGURE 12.5.3. Measurement system variation diagram

in Figure 12.5.2). Therefore, we can use the same experiment data to analyze the repeatability of the system. The standard gauge with 11.500mm in height was used as standard. We set $USL = 11.5 + 0.05(mm)$, $LSL = 11.5 - 0.05(mm)$, and then the C_p in Table 2 can be obtained from the above formula. When $1 < C_p < 1.33$, repeatability is considered as marginal, when $C_p > 1.33$, repeatability is considered as good. Therefore, the repeatability of the laser measurement system is considered as marginal from Table 1. To improve the repeatability, it is necessary that the laser measurement system is calibrated.

12.5.3. Comparison experiment.

12.5.3.1. *Comparison of three interpolation result.* The Comparison experiment is tested for the specular reflection gap after interpolation; each interpolation was tested ten times. The mean, standard deviation, maximum, and minimum for three kinds of interpolations, are summarized in Table 3, Table 4 and Table 5, respectively. The 2D profiles of linear interpolation are shown in (Figure 12.5.3). The advantage of this approach is that it can restore the corrupt data of the flat surface despite the specular reflection gap. It is noted that the specular reflection gap should be very small as shown in (Figure 12.5.4).

However, if the specular reflection surface is convex or concave, linear interpolation will create incorrect results. The profiles are shown in Figure 12.5.5. It is found that the profile is linear for the specular reflection gap instead of the real convex.

Table 4 shows that the results of the Spline interpolation method. When the specular reflection gap is small, its 2D profiles are shown in Figure 12.5.6; it is found that the profile is inadequate. However, it is better than that of the wide specular reflection gap, which will be discussed later. For the sake of smoothness in the interpolated curve, the Spline interpolate approach should make the curve not only continuous and smooth, but also continuous at the first and the second derivative. In fact, Spline interpolation is also one kind of cubic interpolation polynomials. However, piecewise cubic Hermite interpolation is continuous at the first derivative

TABLE 2. Data of measuring repeatability (mm)

Sample points	z_1	z_2	z_3	Std.dev	Range	C_p
245	20.707	9.179	11.528	0.013	0.102	1.282
246	20.702	9.175	11.527	0.015	0.098	1.111
247	20.699	9.173	11.526	0.014	0.103	1.190
248	20.698	9.172	11.526	0.016	0.099	1.042
249	20.701	9.178	11.523	0.012	0.100	1.389
250	20.704	9.180	11.524	0.014	0.102	1.190
251	20.707	9.193	11.514	0.013	0.103	1.282
252	20.699	9.195	11.504	0.012	0.095	1.389
253	20.701	9.192	11.509	0.011	0.103	1.515
254	20.698	9.189	11.509	0.015	0.101	1.111
255	20.703	9.188	11.515	0.012	0.103	1.389
256	20.709	9.193	11.516	0.016	0.104	1.042
257	20.703	9.188	11.515	0.013	0.102	1.282
258	20.710	9.192	11.518	0.014	0.105	1.190
259	20.699	9.194	11.505	0.011	0.097	1.515
260	20.693	9.189	11.504	0.013	0.106	1.282

TABLE 3. Linear interpolation results (mm)

Sample points	Average	Std.dev	Max	Min
195	19.280	0.034	19.355	19.238
196	19.284	0.046	19.359	19.189
197	19.294	0.055	19.395	19.197
198	19.296	0.053	19.394	19.205
199	19.299	0.052	19.394	19.213
200	19.302	0.051	19.393	19.221
201	19.305	0.050	19.393	19.229
202	19.307	0.049	19.392	19.237
203	19.310	0.049	19.392	19.245
204	19.312	0.048	19.391	19.250
205	19.315	0.049	19.391	19.249
206	19.318	0.049	19.390	19.249
207	19.321	0.049	19.393	19.248
208	19.323	0.050	19.396	19.247
209	19.326	0.051	19.399	19.246
210	19.329	0.052	19.402	19.245
211	19.331	0.054	19.405	19.242
212	19.334	0.055	19.408	19.238
213	19.337	0.057	19.411	19.234
214	19.340	0.059	19.415	19.230
215	19.342	0.061	19.418	19.226
216	19.345	0.063	19.421	19.223
217	19.348	0.066	19.424	19.219

instead of the second derivative, which will be discussed later. The boundary of control point on the specular reflection gap is very important for the Spline kind of interpolation. Changing one of the control points will affect the overall appearance of the curve in segments far removed from the point changed. When the specular reflection gap is too wide to restore, the useful data cannot be obtained (shown in Figure 12.5.7). It is found that the oscillatory wave was displayed on the specular reflection gap. It is obvious

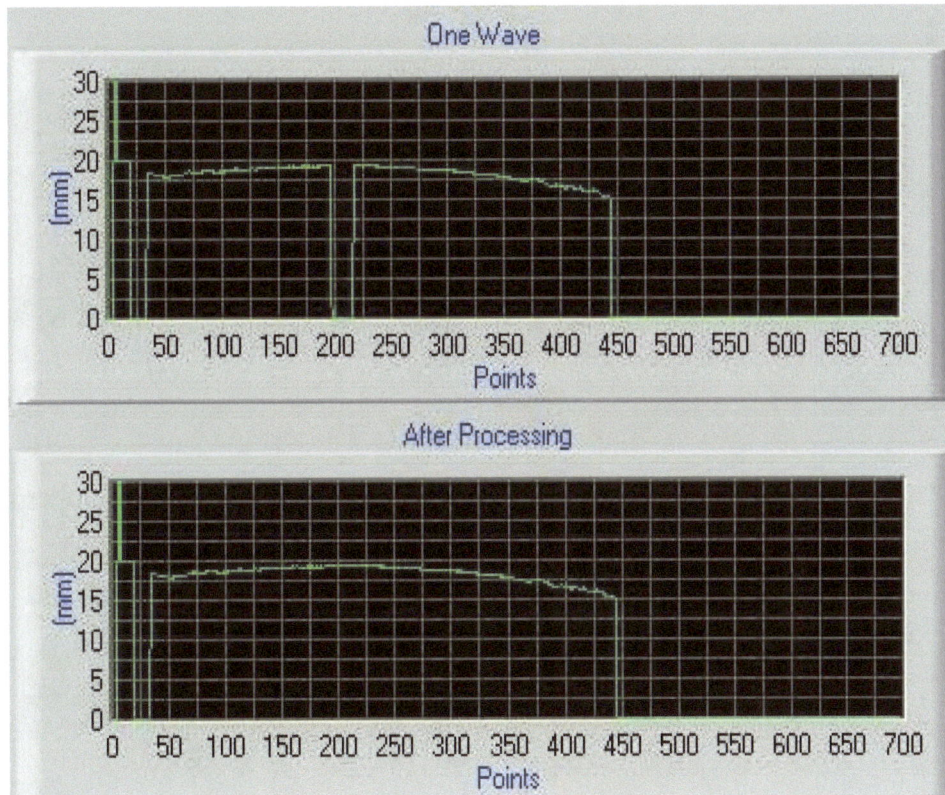

FIGURE 12.5.4. Linear interpolation for specular reflection gap

that Spline interpolation is also not an appropriate method for this project. Table 5 shows the piecewise cubic Hermite interpolation results. The 2D profiles are shown in Figure 12.5.8.

As stated earlier, the piecewise cubic Hermite interpolation approach is continuous at the first derivative. A discontinuous second derivative implies discontinuous curvature. The human eye can detect large jumps in the curvature. To overcome this problem, we use the particular function stated in section 4.4.1, which limits the slope so that the function values do not overshoot the data values, at least locally. Thus, the real shape is preserved. It is similar to the "visually pleasing" interpolation introduced in chapter 4. When the specular reflection gap is wide, it is found that the piecewise cubic Hermite interpolation yields good interpolation results (data not shown). Figure 12.5.9 shows that 2D profile is smooth. It is precisely the result required.

To evaluate the effect of three interpolations intuitively, the experiment data of Table 3, Table 4 and Table 5 were used to plot in the same coordinates. The displayed curves only are the gaps that have been restored with the interpolation method in the Figure 12.5.10. The x axis denotes the sample point; the y axis represents the average height of the turbine vane. It is found that the result of Spline interpolation was clearly different from that of the other two interpolations (shown in Figure 12.5.10). Although the result of the linear interpolation is similar to that of piecewise cubic Hermite interpolation, the linear interpolation is only suitable for restoring flat surfaces rather than convex and concave surfaces. The piecewise cubic Hermite interpolation was regarded as the best method due to the fact that it can restore both flat surfaces and non-flat surfaces.

12.5.3.2. *Comparison Experiment Between piecewise cubic Hermite interpolation and Paint.* It was reported that the painting surface could always ensure the laser sensor performed accurately on the specular reflection surface [9]. The purpose of the comparison experiment was to compare the effect of different methods on the relationship between the height of turbine vane and the sample points. It is noted that the specular reflection gap ranged from sample point 400 to 450 in this experiment. Firstly, the turbine vane surface was tested 10 times using piecewise cubic Hermite interpolation method. After that, the average value was calculated. The curve below is shown in (Figure 12.5.11). Secondly, the turbine vane is fixed, and is painted white. It was also tested 10 times, after the average value was obtained, the curve above can be plotted by the same method (shown in Figure 12.5.11). Interestingly, it is observed that the trend line of these two curves shown in Figure 12.5.9 is similar. This strongly indicates that piecewise cubic Hermite interpolation method can be used to restore corrupt data. However, it is found that the two curves don't match each other exactly. The possible reasons for the difference between interpolation and paint of curve are listed as following:

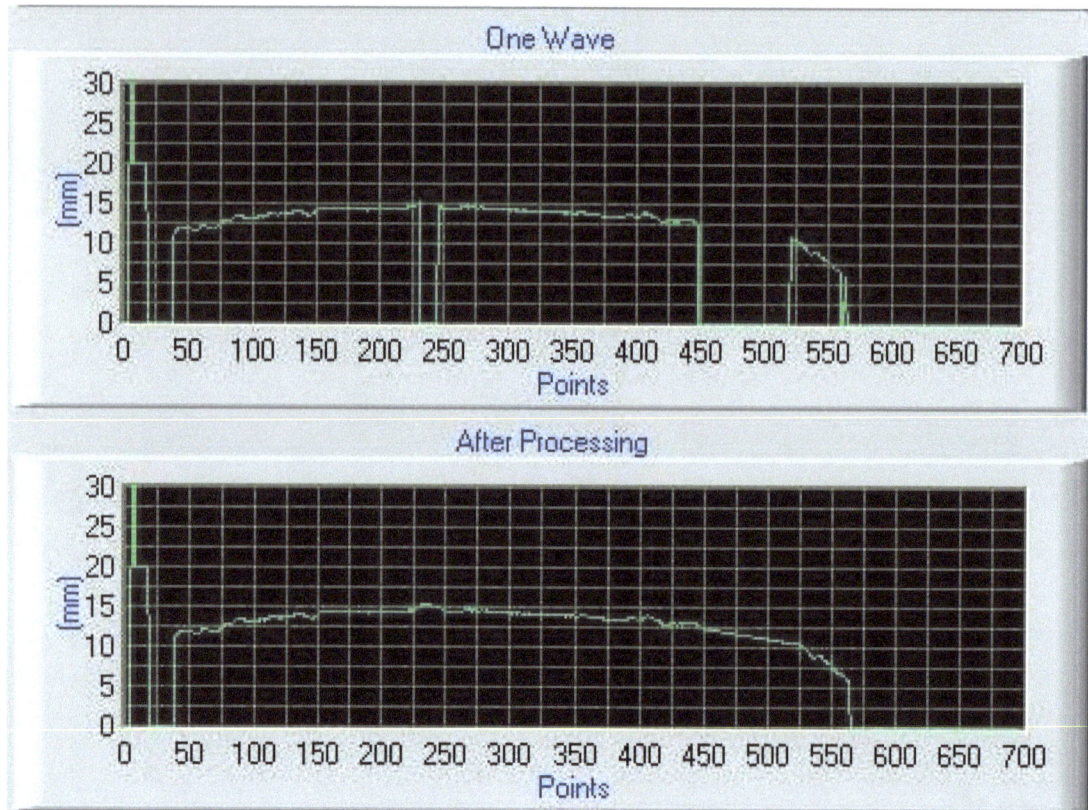

FIGURE 12.5.5. 2D profiles for specular reflection gap before and after linear interpolation

- After painting, the paint has some extent of thickness. So the value of paint curve is higher than that of the interpolation curve.
- The height difference around sample point 400 is greater than that around sample point 450. This is because the paint flow along with the curve surface, after painting. Moreover, the flowing speed and the thickness of the paint are different on the surfaces of turbine vanes due to different curvatures. The more the paint, the bigger the difference.
- The paint curve is not very smooth, because the manual paint is not equably distributed on the turbine vane surface.

12.5.4. 3D measurement and Application.

12.5.4.1. *3D measurement* . As stated before, with the function Plot3DParametericSurface (xData, yData, zData) built in Measurement Studio for Visual c+ + 6, 3D surface profile was successfully reconstructed by analyzing 2D surface profile and using an external coordinate platform. Now, the surface height of 3D profile is shown in Figure 12.5.12. The cursor can move arbitrarily on a 3D profile, and then 3D coordinate is obtained. It is found that the coordinate is $(493, 45, 16.3281)$ shown in Figure 12.5.12. Interestingly, it is observed that the corresponding point of 2D profile is sample point 493, whose 2D coordinate is $(493, 16.3281)$. It is observed that the measuring value on 3D profile matches that of 2D profile. So we can move the cursor with the coordinate plane to compute the 3D profile data throughput. It is found that the data throughput have highly improved when we employ interpolation.

12.5.4.2. *Application.* 2D profiles are used to reconstruct 3D profiles. Seventy frames are captured and processed continuously by rotating the Y axis knob on the coordinate platform. Each frame has 700 sample points, so the total number of points have 49000 (70700) pixel points. Of course, the number of frames can be arbitrary determined according to required accuracy and computer processing speed. Some methods to get a good scan are listed as follows:

- Ensure that the coordinate platform is flatly placed on the table.
- Ensure that the starting scan line is parallel with the turbine vane edge.
- Ensure that the sensor is properly placed over the turbine vane, i.e. the sensor is placed with its axis approximately normal to the turbine vane surface.
- Ensure that the scanning distance of Y axis is 1mm for each frame scan.

TABLE 4. 4 Spline interpolation results (mm)

Sample points	Average	Std.dev	Max	Min
195	19.269	0.034	19.316	19.199
196	19.260	0.077	19.355	19.127
197	19.243	0.121	19.446	19.022
198	19.216	0.149	19.494	18.940
199	19.183	0.166	19.505	18.879
200	19.143	0.173	19.483	18.838
201	19.100	0.178	19.434	18.815
202	19.055	0.186	19.363	18.700
203	19.010	0.201	19.276	18.545
204	18.966	0.225	19.176	18.392
205	18.925	0.256	19.125	18.247
206	18.890	0.291	19.146	18.119
207	18.863	0.326	19.170	18.011
208	18.843	0.358	19.196	17.932
209	18.836	0.383	19.224	17.887
210	18.840	0.398	19.253	17.884
211	18.859	0.400	19.281	17.928
212	18.895	0.387	19.308	18.025
213	18.948	0.355	19.332	18.183
214	19.022	0.302	19.354	18.407
215	19.117	0.228	19.372	18.705
216	19.237	0.137	19.385	18.960
217	19.358	0.073	19.434	19.180

The scan direction is illustrated in Figure 12.5.13. The red color line represents the laser scan line, which is parallel with the right edge. turbine vanes move from left to right during the reconstruction 3D profile. However, the sensor is fixed.

Figure 12.5.14 shows a group of 3D convex images before interpolations. The convex picure observed from the Z axis direction is illustrated in Figure 12.5.14-(e). It is found that one specular reflection gap appeared on the middle gap and the serration-shaped edge appeared on the left and right edge gap. This is due to the fact that the reflected light cannot be received by PSD sensor, which is caused by the turbine specular reflection surface deformation. It is observed that similar phenomena occurred in Figure 12.5.14-(f). If a surface that does not lose any data is defined as a good scan surface, then the pixel point's number of the good scan surface from Figure 12.5.14-(a), (b), (c), (d) can be calculated according to the following formula:

$$70(302 - 187 + 496 - 380) = 16170$$

That is to say, if we do not employ interpolation, we can obtain a good surface for the specular reflection turbine vane with 16170 pixel points. To obtain an entire 3D surface profile , another 5460 pixel points $(70 (380 - 302))$ are needed. However, it would take a lot of time to acquire and process data. Moreover, it reduces good surface data throughout, without interpolation. However, if we employ interpolation for processing the specular reflection gap of the turbine vane, it is found that the interpolation improves the data throughout a lot. Convex images observed from the Z axis direction are shown in Figure 12.5.15-(b) (d). Interestingly, it is observed that the specular reflection gap in the middle section has been restored. The specular reflection gap disappears, but the serration-shaped edges on the left and right remain. From Figure 12.5.15-(a), (c), the number of data points on the good scan surface were calculated as following:

(12.5.1) $$70(501 - 187) = 21980$$

It is apparent that we can obtain much more pixel points after interpolation than before interpolation from the calculated results. It is helpful to improve processing speed for the 3D surface profile reconstruction.

The above 3D reconstruction, mainly focuses on the convex surface of the turbine vane. Now, the reconstruction of the concave surface of the turbine vane under the same condition is discussed. It is found that the specular reflection gap located at the edge of the concave surface as shown in Figure 12.5.16. The corrupt data points can be calculated as follows:

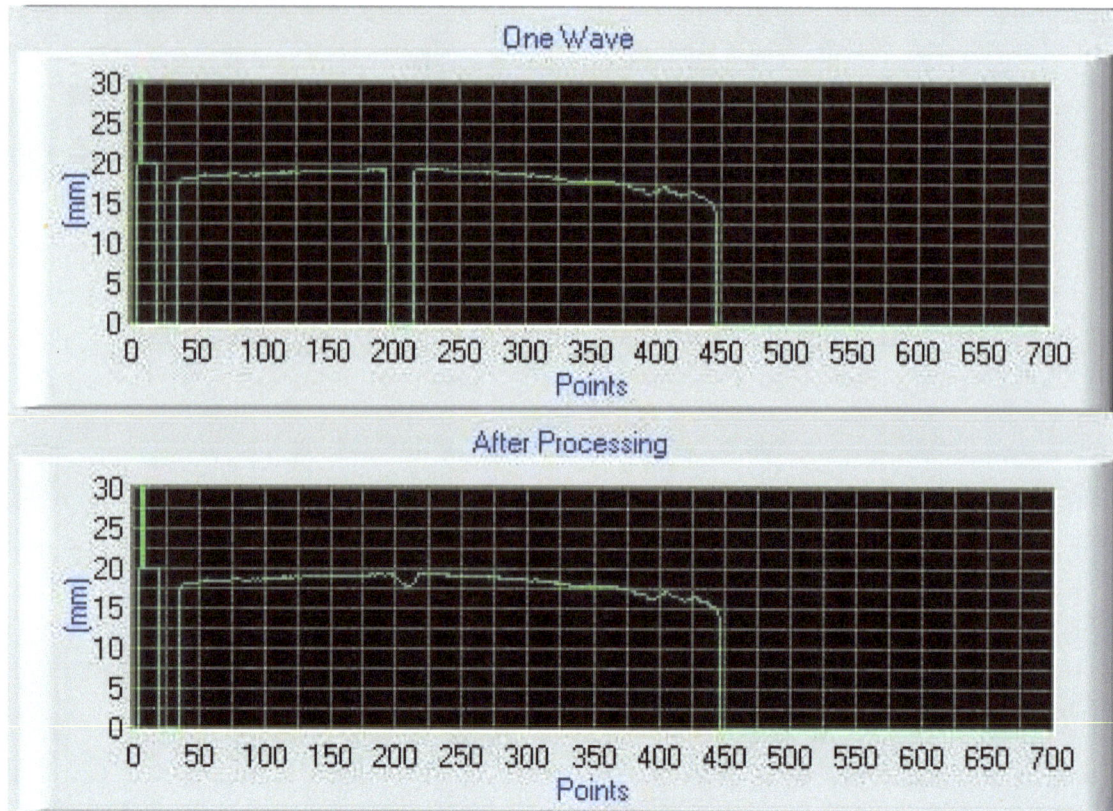

FIGURE 12.5.6. Spline interpolations for specular reflection gap

(12.5.2) $70(158 - 52) = 7420.$

Here we suppose that all conditions are the same before and after interpolation. It is calculated that 7420 more data points existed after interpolation than before interpolation. The 3D surface concave picures after interpolation are illustrated in Figure 12.5.17-(a) (b) (c) (d). The result of the above analyses shows that not only the convex surface but also the concave surface specular reflection gap can be restored by piecewise cubic Hermite interpolation. Meanwhile, it is significant that the interpolation can highly improve the output of pixel points. It is important to reconstruct the 3D entire prototype quickly.

12.6. Conclusions

Through online processing of laser acquisition data, non-contact online profiles were obtained. The results of experiments show that the laser sensor measurement system in this project can be used to measure the surface of the turbine vane. Their main advantages are:

- Precision accuracy and reliability
- High-speed scanning capability

It is observed that the laser sensor has also some limitations; for example, it is difficult to measure the specular reflection surface of the turbine vane. This is because specular reflection causes data corruption and results in distorted 2D profiles. Therefore, the interpolation of the specular reflection gap of the turbine vane is necessary. However, direct interpolation of the specular reflection gap does not yield a correct surface profile. Because the edge data of the specular reflection gap cannot represent a true surface, that is to say, the control points interpolation is not correct. To overcome this problem, linear compensation must be applied before interpolation. Comparison of three interpolation methods shows that the piecewise Hermite interpolation is the best method. By using piecewise cubic Hermite interpolation and linear compensation for the specular reflection surface, an entire 2D profile can be obtained, although the surface of the turbine vanes has specular reflection After obtaining a nominal 2D profile, it is necessary to reconstruct a 3D profile using the algorithm of 2D profile reconstruct 3D profile. It is found that the measurement data of a 3D

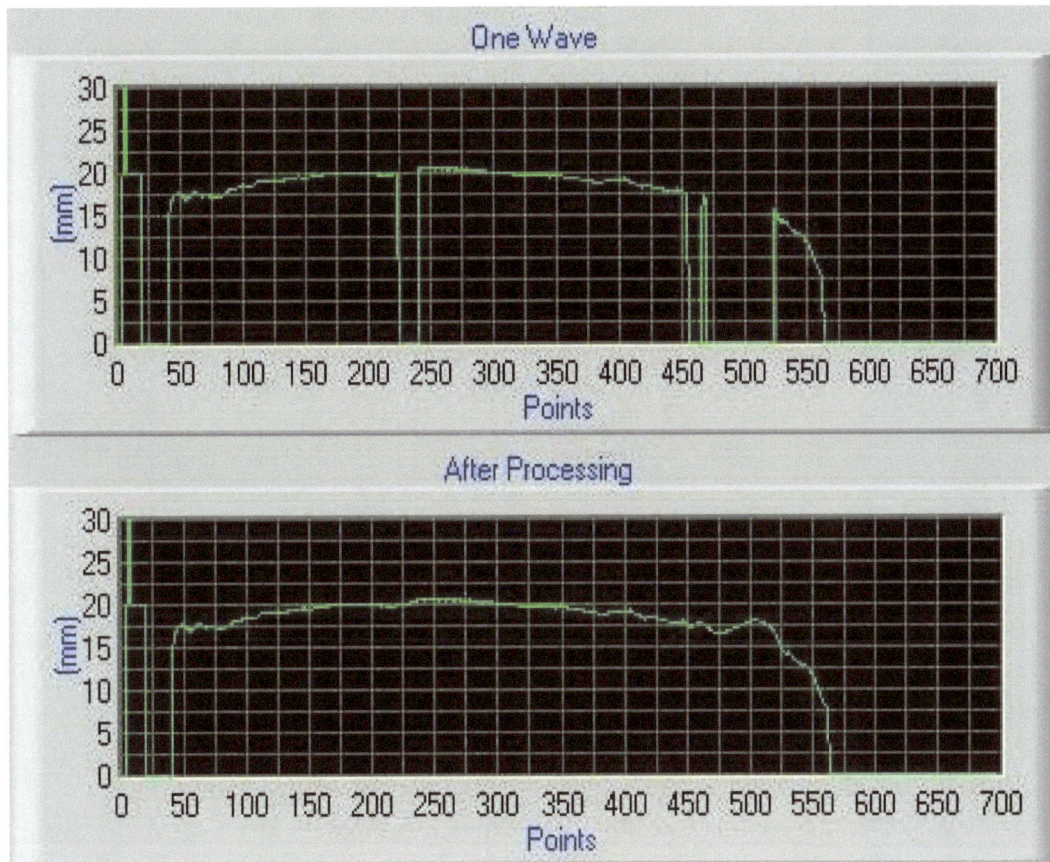

FIGURE 12.5.7. 2D profiles for specular reflection gap before and after Spline interpolations

profile matches that of the 2D profile. At the same time, it confirms that the 3D profile can be used for robot path planning and feedback control purpose. The measurement results and comparisons show that the system is reliable, the measurement accuracy is better than $30\mu m$, that the range of repeatability is $\pm 0.050mm$. The results indicate that the measurement system can meet the repairing requirements of turbine vane .

TABLE 5. Piecewise cubic Hermite interpolation results (mm)

Sample points	Average	Std.dev	Max	Min
195	19.299	0.040	19.365	19.229
196	19.324	0.044	19.365	19.238
197	19.333	0.053	19.414	19.247
198	19.333	0.044	19.402	19.259
199	19.337	0.044	19.416	19.269
200	19.341	0.044	19.425	19.280
201	19.347	0.038	19.431	19.293
202	19.349	0.039	19.435	19.301
203	19.349	0.040	19.435	19.307
204	19.352	0.039	19.434	19.315
205	19.354	0.037	19.430	19.313
206	19.355	0.036	19.425	19.310
207	19.357	0.035	19.418	19.306
208	19.358	0.035	19.411	19.302
209	19.360	0.036	19.412	19.299
210	19.361	0.038	19.413	19.295
211	19.362	0.040	19.414	19.293
212	19.363	0.043	19.414	19.290
213	19.364	0.045	19.414	19.289
214	19.366	0.048	19.417	19.288
215	19.367	0.050	19.420	19.287
216	19.378	0.043	19.424	19.297
217	19.383	0.056	19.482	19.297

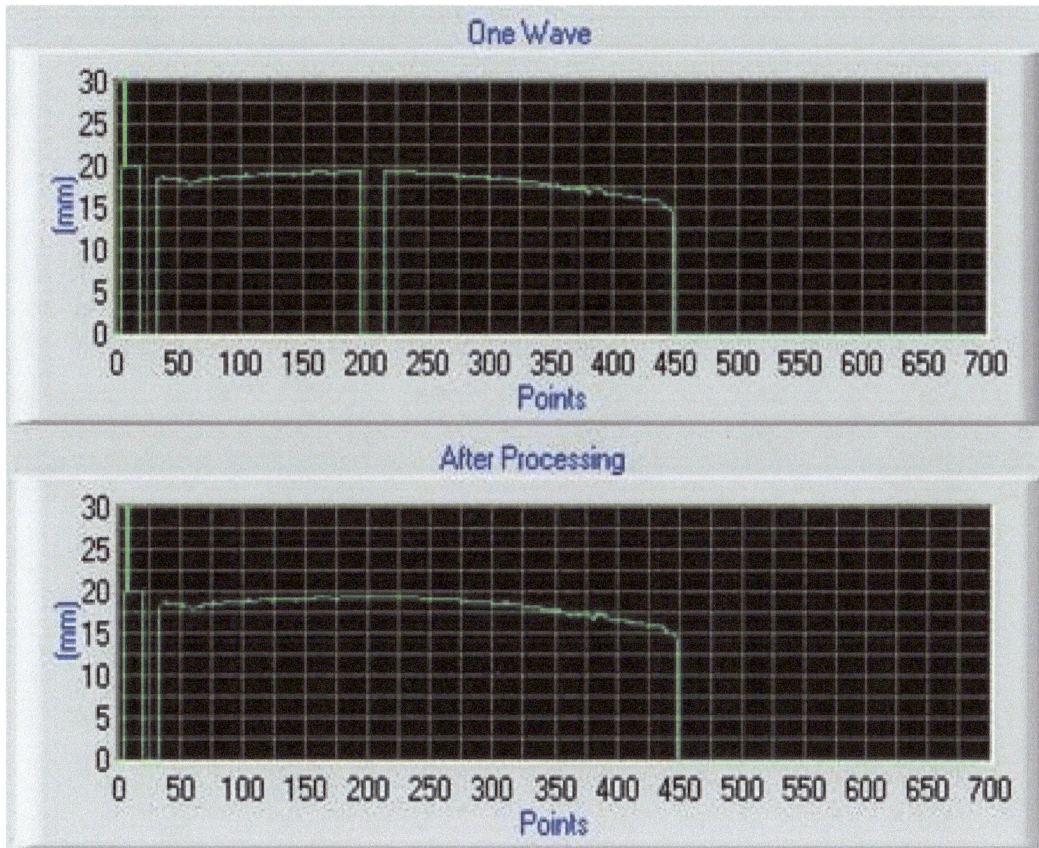

FIGURE 12.5.8. Piecewise cubic Hermite interpolation for specular reflection gap

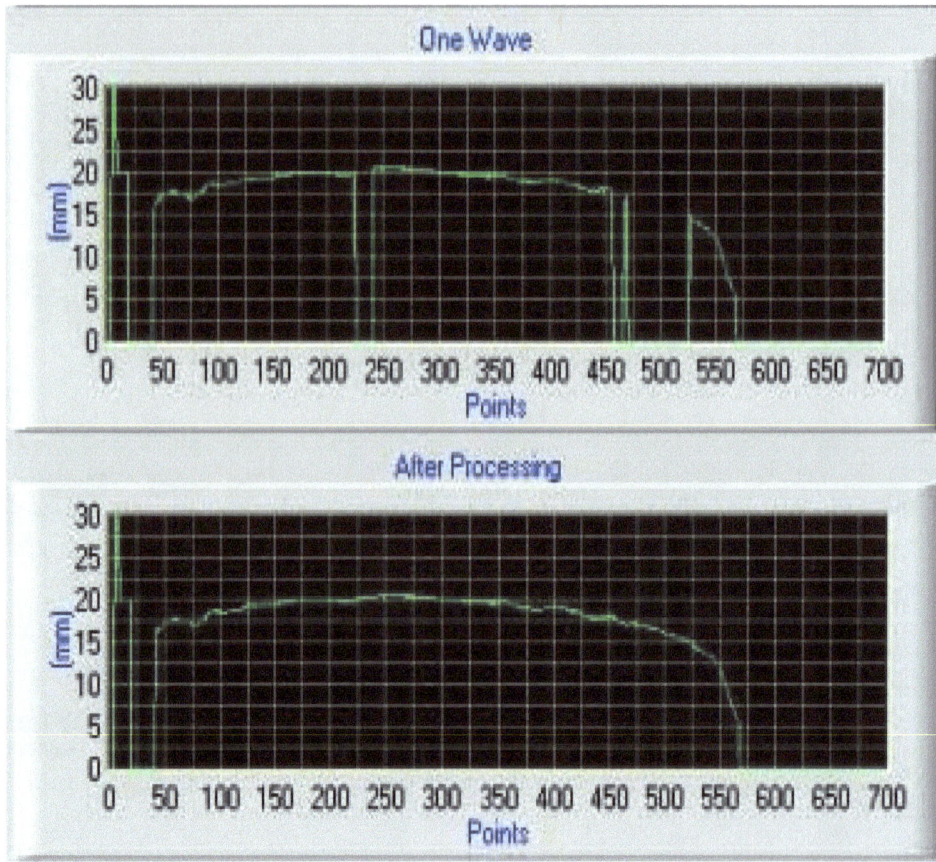

FIGURE 12.5.9. 2D profiles for specular reflection gap before and after piecewise cubic Hermite interpolation

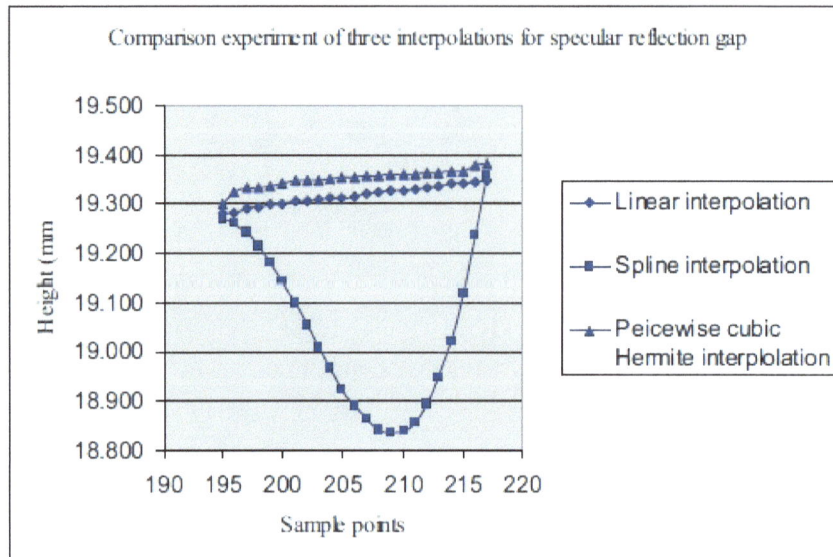

FIGURE 12.5.10. Comparison experiment for different methods

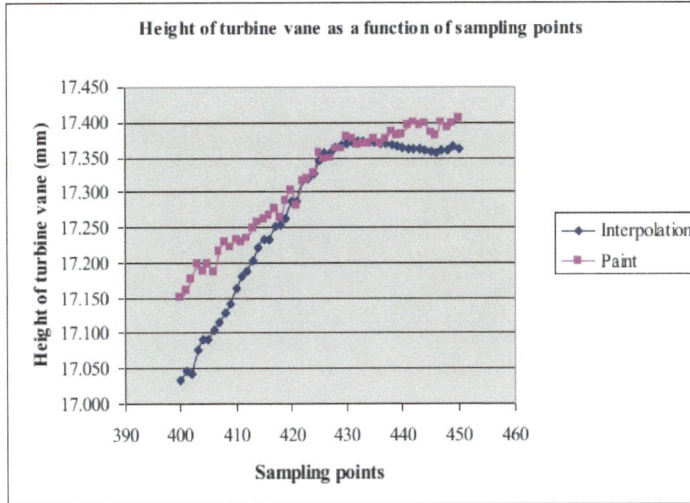

FIGURE 12.5.11. Comparison between interpolation and painting

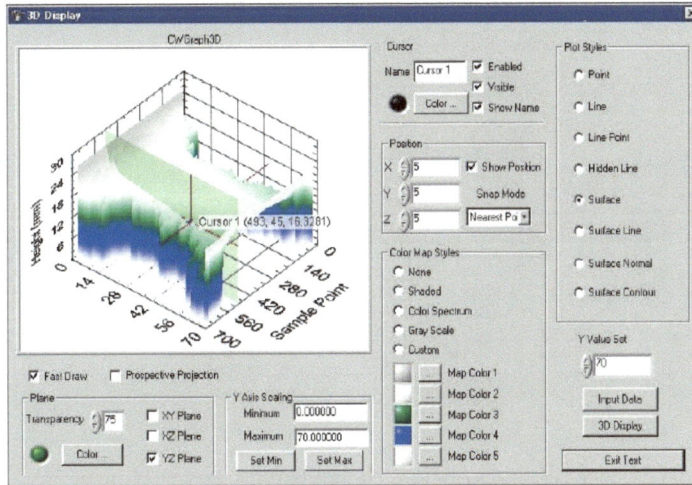

FIGURE 12.5.12. Measurement with 3D profile

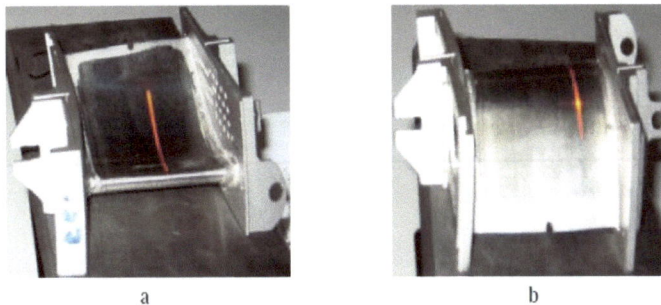

FIGURE 12.5.13. Scan directions. a. Scan direction on concave surface; b. Scan direction on convex surface

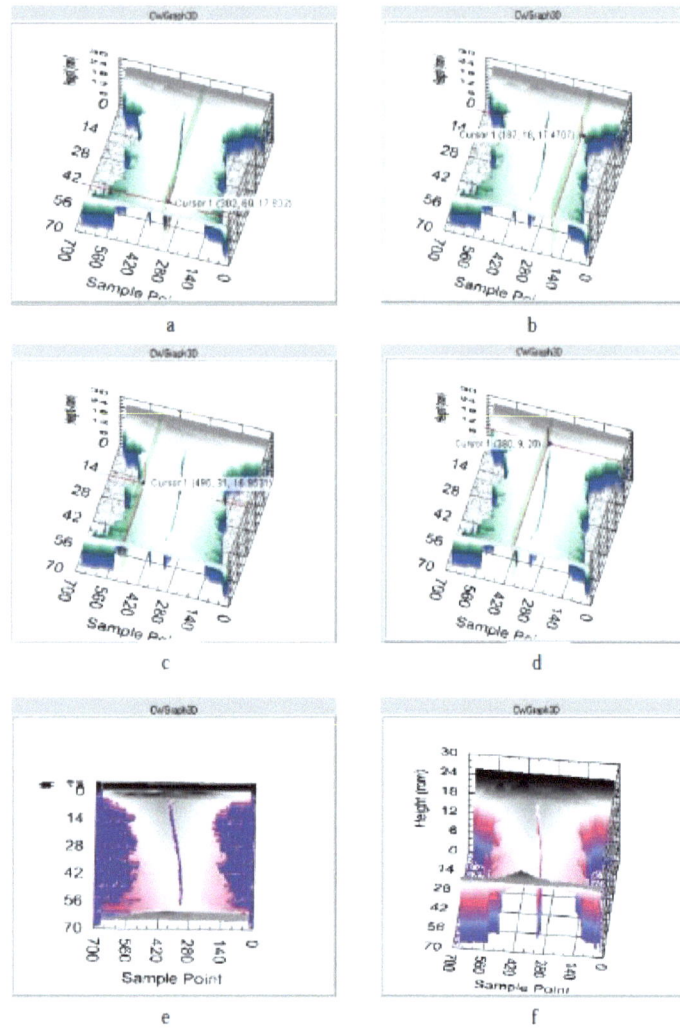

FIGURE 12.5.14. 3D convex images before interpolation. a. Cursor measurement on $(302, 60, 17.832)$; b. Cursor measurement on $(187, 18, 17.470)$; c. Cursor measurement on $(496, 31, 16.953)$; d. Cursor measurement on $(380, 9, 20)$; e. Convex image from the Z axis direction observation; f. 3D convex image after rotating coordinates.

a   b

c   d

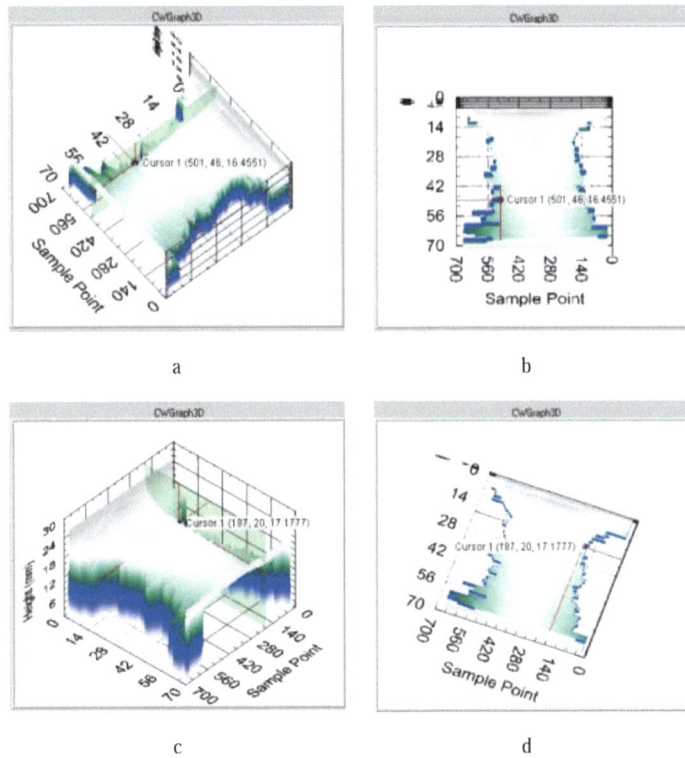

FIGURE 12.5.15. 3D convex images after interpolation.   a.   Cursor measurement on $(501, 46, 16.455)$; b. Cursor measurement on $(501, 46, 16.455)$ from the Z axis direction observation; c. Cursor measurement on $(187, 20, 17.177)$; d. Cursor measurement on $(187, 20, 17.177)$ from the Z axis direction observation.

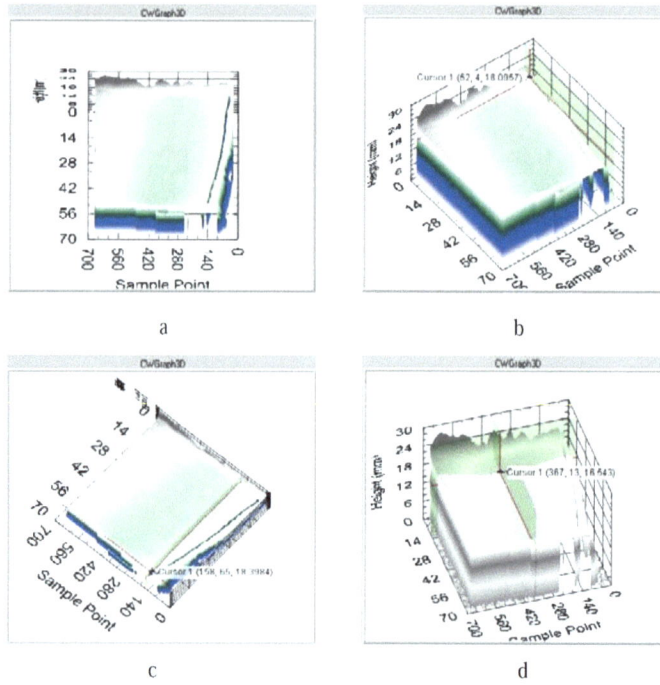

FIGURE 12.5.16.   3D concave images before interpolation. a. specular reflection gap on right edge; b. Cursor measurement on right edge of specular reflection gap; c. Cursor measurement on left edge of specular reflection gap; d. Show XZ and YZ planes image.

a

b

c

d

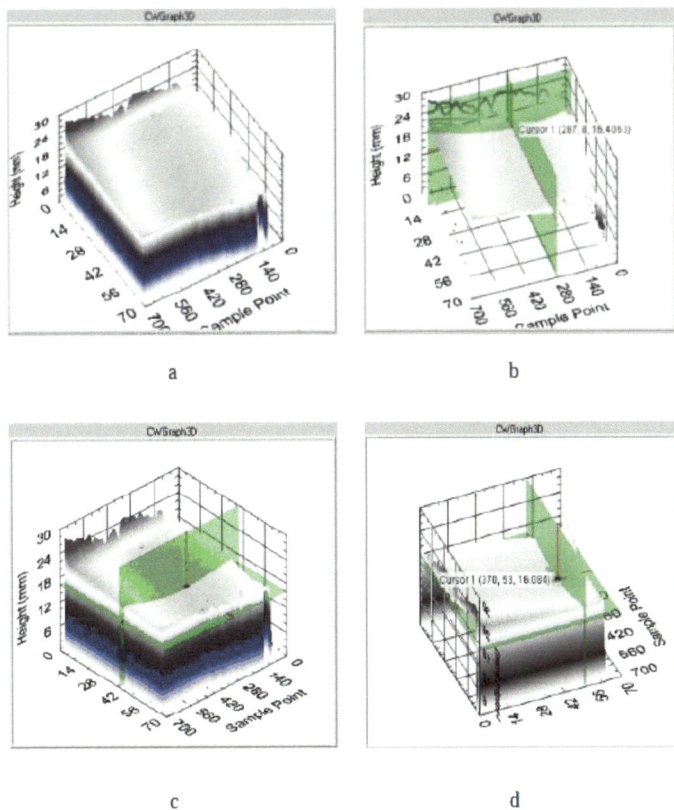

FIGURE 12.5.17. 3D concave images after interpolation. a. Show 3D concave image after interpolation; b. Show 3D image with XZ and YZ planes image; c. Show 3D image with XY and XZ planes; d. 3D concave image after rotate coordinate.

Bibliography

[1] *LJ-3000/LJ-080 User Instruction Manual.*

[2] V. B. Anand. *Computer Graphics and Geometric Modelling for Engineers.* John Wiley and Sons, incorporation, 1993.

[3] J. G. Bollinger. Two dimensional tracing and measurement using touch trigger probes. *Annals CIRP*, 31:415–419, 1982.

[4] C. Che and J. Ni. A ball-target-based extrinsic calibration technique for high-accuracy 3D metrology using off-the-shelf laser-stripe sensors. *Journal of International Societies for Precession Engineering and Nanotechnology*, 24:210–219, 1997.

[5] X. Q. Chen, Z. M .Gong, H. Huang, and L. Zhou. An automated 3D polishing robotic system for repairing turbine airfoils. In *Proceedings of 3rd International conference on Industrial Automation*, June 1999.

[6] X.Q. Chen and R. Devanathan. *Advanced automation techniques in adaptive material processing.* World Scientific Publishing Co. Pte. Ltd., 2002.

[7] J. Clark. 3D scanning systems for rapid prototyping. *Assembly Automation*, 17(3):206–210, 1997.

[8] K. A. Donnell. Effects of finite stylus width in surface contact profilometry. *Applied Optics*, 32:4922–4928, 1993.

[9] K. C. Fan. A non-contact automatic measurement for free-form surface profiles. *Computer Integrated Manufacturing Systems*, 10(4):277–285, 1997.

[10] Z. Gong, X. Q. Chen, and H. Huang. Optimal profile generation in distorted surface finishing. In *International Conference on Robotics and Automation*, pages 1557–1562, San Francisco (USA), April 2000.

[11] http://www.mathworks.com.

[12] H. Huang, Z.M. Gong, X.Q. Chen, and L. Zhou. Smart robotic system for 3D profile turbine vane airfoil repair. *The international Journal of Advanced Manufacturing Technology*, 21:275–283, 2003.

[13] R. A. Jarvis. A perspective on rang 2nding techniques for computer vision. *Vision Computer*, 14:659–666, 1996.

[14] R. A. Johnson. *Probability and Statistics for Engineers.* Prentice Hall International Incorporation, 2000.

[15] B. F. Jones and P. Plassmann. An instrument to measure the dimensions of skin wounds. *IEEE Transactions on Biomedical Engineering*, 42:464–470, 1995.

[16] Kwok K., Loucks C., and Driessen B. Automatic tool path generation for finish machining. In *IEEE International Conference on Robotics and Automation*, volume 2, pages 1229–1234, 1997.

[17] W. P. Kennedy. No contact measurement: Can laser triangulation help you? *Quality*, 37:36–37, 1998.

[18] P. Kim, S. Rhee, and C. H. Lee. Automatic teaching of welding robot for free- formed seam using laser vision sensor. *Optics and Laser in Engineer*, 31:173–182, 1999.

[19] E. Mouaddib, J. Batle, and J. Salvi. Recent progress in structured light in order to solve the correspondence problem in stereo vision. In *Proceedings of the 1997 IEEE International Conference on Robotics and Automation*, 1997.

[20] J. A. Munoz-Rodriguez, R. Rodriguez-Vera, and M. Servin. Direct object shape detection based on skeleton extraction of a light line. *Optical Engineering*, 39:2463–2471, 2000.

[21] J. Peoples and L. B. Weinstein. The use of noncontact laser gauging systems for online measurement. *Sensor Review*, 19:260–264, 1999.

[22] V. E. Theodoracatos and D. E. Calkins. A 3D vision system model for automatic object surface sensing. *International Journal Computer Vision*, 11:75–99, 1993.

[23] W. Wolfson and S. J.Gordon. Three-dimensional vision technology offers real-time inspection capability. *Sensor Review*, 17(4):299–303, 1997.

[24] Q.M. Fonathan Wu, Min fan Richy Lee, and Clarence W. de Silva. An imaging system with structured lighting for on-line generic sensing of three-dimensional objects. *Sensor Review*, 22(1):46–50, 2002.

[25] G. Zhang and H. Wang. Method of establishing general mathematical model of tructured light 3D vision. In *SPIE Proceedings*, pages 662–666, 1996.

[26] Y. Zhang and R. Kovacevic. A real-time sensing of sag geometry during gta welding. *Manufacturing Science Engineering*, 119:151–160, 1997.

Index